ON PHYSICS AND PHILOSOPHY

❧❧

ON PHYSICS AND PHILOSOPHY

⸙

Bernard d'Espagnat

PRINCETON UNIVERSITY PRESS

PRINCETON AND OXFORD

COPYRIGHT © 2006 BY PRINCETON UNIVERSITY PRESS
PUBLISHED BY PRINCETON UNIVERSITY PRESS, 41 WILLIAM STREET,
PRINCETON, NEW JERSEY 08540
IN THE UNITED KINGDOM: PRINCETON UNIVERSITY PRESS,
3 MARKET PLACE, WOODSTOCK, OXFORDSHIRE OX20 1SY

ORIGINALLY PUBLISHED IN FRENCH AS *TRAITÉ DE PHYSIQUE ET DE
PHILOSOPHIE BY LIBRAIRIE ARTHÈME FAYARD, PARIS 2002*

ISBN-13: 978-0-691-11964-9
ISBN-10: 0-691-11964-3

LIBRARY OF CONGRESS CONTROL NUMBER: 2006926978

BRITISH LIBRARY CATALOGING-IN-PUBLICATION DATA IS AVAILABLE.

THIS BOOK HAS BEEN COMPOSED IN SABON

PRINTED ON ACID-FREE PAPER. ∞

PUP.PRINCETON.EDU

PRINTED IN THE UNITED STATES OF AMERICA

7   9   10   8

CONTENTS

§§

PREFACE TO THE ENGLISH EDITION

TRYING TO UNDERSTAND what contemporary physics is truly about unavoidably raises philosophical problems. Such, anyhow, is the thesis that underlies the whole content of this book, the French edition of which came out just a few years ago. Since then, at least to my knowledge, no basically significant relevant material has appeared. For preparing this English edition only quite minor changes had therefore to be made. Practically all of them just came from the fact that I took the opportunity here to be more explicit in certain accounts and explanations, thus rendering this book clearer.

B. d'Espagnat

ON PHYSICS AND PHILOSOPHY

FOREWORD

☙

GREAT PHILOSOPHICAL riddles lie at the core of present-day physics and most people, by now, are aware of their existence even if but a few have a precise idea of their nature. The present book should, hopefully, make the issue clearer. But its ambition is, in fact, greater, for its aim is not merely to make these problems explicit. Nor is it just to try and relieve other physicists of conceptual perplexities liable to slow down the further elaborating of their discipline. Its more far-reaching purpose is to make it clear that, nowadays, *any* tentative philosophical approach to a world-view should take information coming from contemporary physics into account quite seriously. The book should make it manifest that, while, admittedly, philosophers have to go on pondering on the deep nature of knowledge—for such is their primordial role—still they may not ground their systems just on their own cogitation. They have to consider what emerges from other sources. For example, it is sometimes claimed that any beginner in philosophy must, in a first approximation, choose to draw his or her inspiration from either Plato or Aristotle: an opinion that may sound acceptable. But the choice in question is often said to be fully arbitrary, to depend exclusively on the mental bent of the concerned individual. This is not really correct. Some elements of present-day scientific knowledge cast serious shadows on such and such Platonic intuitions, but they bring even more clearly to light the anthropocentric—hence relative—nature of the very notions that Aristotle and his followers got us into the habit of considering to be fully primary and basic. This of course should not remain without effect on the relative degrees of acceptance of these two thinkers' positions.

To be sure, it might be objected that intrusion of science in the field of philosophy is presumptuous and inappropriate. And in support of this, some would pertinently point out that past scientific discoveries gave rise to simplistic extrapolations that are nowadays proved wrong. Admittedly, this is quite true. The risk that such a mishap should happen again is minimal, however, when (as is implicitly the case in the foregoing example) the information science yields serves to *limit* possible options, rather than put forward the allegedly correct one. And we shall find that in the philosophical realm such is indeed its major role. In fact, it is not inappropriate to consider that while Nature—in the broadest possible sense—

refuses to explicitly tell us what she is, she sometimes condescends, when we press her tenaciously enough, to let us know a little about what she is not. Admittedly, it must be granted that, in the cases in which physics falsifies such and such views or shows them to be too naive, the arguments leading to these judgments are, more often than not, susceptible of being circumvented. But, as will appear in the book, this can only be at the price of shifting to theories that, for a host of good reasons, are considered by most physicists to be very likely untrue.

Unfortunately, the split that presently exists between scientists and philosophers means that many of the latter tend to simply ignore recent scientific advances, which makes them miss the point that, by blocking some traditional perspectives, such findings may well open promising new ones. For example, some philosophers do still make unrestricted use of classical notions of quite a general nature, such as locality, distinguishability, etc., taken to be obvious ever since Galileo's and Newton's times. Most of them do so without realizing that the domains of validity of such notions are known, nowadays, to be severely limited. Ironically, some even condemn—as "unduly exploiting scientific data"—any use made of the findings of such great twentieth-century physicists as Louis de Broglie, Dirac, Feynman, and Heisenberg. They thereby overlook the fact that the said findings are the very ones that invalidate unrestricted use of the classical notions in question and thereby open bright new fields for philosophical thinking. And more generally it is somewhat shocking to see conjectures that, daring as they are, still are in the wake of present-day knowledge being set on an equal footing with rash speculations resting on naive conceptions.

True, between science and philosophy a bridge theoretically exists. It is epistemology. Epistemology is a relatively young discipline. As a collective undertaking it has hardly been in existence for more than one century. It is meant to describe the ways in which the various sciences build up and test their theories. It tries and evaluates their logical and cognitive value, as well as the reliability of the validation processes they use. It questions the nature of the knowledge thus acquired, with the aim of determining in what sense it may be said to be "true." These are important activities, which did yield quite significant results in many domains. On the other hand, the activities in question are of a very general nature. And this generality may bring about a misconception in that, in view of it, some epistemologists might consider that their enterprise, directed as it is at providing a suitable framework for research, should be carried over quite independently of the factual results of the latter. Now, it is true that sticking to such a conception may at first sight seem to be quite a reasonable standpoint. For indeed it enables the research workers who take it up to keep aloof from a certain instability that characterizes science proper and is an unavoidable consequence of its very development. Besides, it may be added that the human mind, when confronted with multifarious and often rather technical data, has a natural tendency to consider that their very technicality restricts their bearing to details. And this opinion may well lead some to the reassuring view that just taking a rapid superficial glance at the technical data in question should be enough. . . .

Maybe this is, even now, the attitude of mind that a few epistemologists entertain toward present-day physics. However, if this is the case I claim that these searchers are in the wrong. To me it seems quite clear that whoever expresses himself or herself, even on matters as general as those epistemology deals with, tacitly refers to a world view, be it only to draw from it the concepts his or her statements will bind together. Now, during the twentieth century, science—mainly though one of its branches, physics—brought about such momentous changes in the possible representations of the world, it so drastically ruled out notions hitherto considered obvious, that reasoning on epistemological matters without taking such developments into account has become inconsistent. I hope this point will become crystal clear to any reader of this book. In fact, we live in an age in which, due to the findings in question, the picture of the world entertained by the general public has entered a process of extensive evolution—not to say "revolution." But such great moves take much time (one or two centuries in general). And it is quite normal that they should arouse reticence (it is, alas, also normal that they should give rise to fancy extrapolations: against the latter we should therefore keep on our guard).

As could well be guessed, the content of this book will eventually make us go so far as to pose questions of a distinctly metaphysical nature; and we shall even engage in some reflection on them, essentially aiming at a better understanding of their true scope and meaning. However, no really extensive, systematic analysis of them is to be taken up. The reason for this reticence is simple. While metaphysics in a broad sense—including the question of its very legitimacy—is, in the opinion of most people after all, an extremely important topic, it so happens, as we shall see, that, concerning it, present-day physics can, so to speak, but utter some warning signals. What it can tell us is essentially negative: "No, Being is not this," "No, it is not that either!" "No, it is in no way what you imagined it to be!" (An elder scientist once told me that, during his life, he had had to switch from a traditionally religious World picture, with angels, fall, Satan, etc., to one in terms of DNA, quarks, and so on. I retorted that he still had quite a long road ahead, with some turns that he would find even sharper.) Such important indications, some of which follow from quite recent discoveries, should certainly encourage reflection, so that, in their light, renewed careful explorations of the whole metaphysical domain seem appropriate. But certainly investigations of this kind will become most daring and adventurous as soon as they enter, as they will have to, into specifics and details. None of them is to be carried through here.

Still, in the present book the metaphysical domain has willy-nilly to be approached in one respect. For indeed one piece of information that contemporary physics clearly yields, as we shall see, is the absolute necessity of carefully distinguishing between two concepts of reality. One of them is *ontological reality*, that is, the notion referred to when "what exists independently of our existence" is thought of or alluded to. The other one, *empirical reality*, is the set of phenomena, that is, the totality of what human experience, seconded by science, yields access to. In common language the word "reality" designates indiscriminately these two notions and, still at present, many consider that distinguishing between them is in no way imperative. According to some, the notion of possible experience is the only one that can possibly impart a meaning to the word "reality"; so that, they claim, the empirical reality notion is the only relevant one. Alternatively, according to others, the purpose of science can only be to lift the veil of appearances and know with ever increasing accuracy how reality truly is—is "by itself"—so that, for them, only mind-independent reality may appropriately be called "reality." (While the individuals composing the first set are mostly philosophers, those constituting the second one are predominantly scientists.) However, it will, I believe, be clearly shown in this book that, with regard to the content of contemporary physics, both views are pertinent to some degree. Consequently, on whatever side our personal bias lies, we should at least carefully keep in mind that both conceptions exist.

One of the main arguments that should convince scientists of the need to distinguish between empirical and mind-independent reality lies in the nonseparability of whatever sensible notion we may form of the latter. Nonseparability is a surprising piece of knowledge that could only recently be shown to rest on strict, scientific data of an experimental nature. When the scientific community has completely assimilated it, it will realize at the same time the importance of the fact—equally rigorously proved—that nonseparability does not provide us with new means of operating at a distance. Consequently, it constitutes a feature of any sensible representation of mind-independent reality that, unquestionable and significant as it is, still does not fully and genuinely extend to the empirical reality domain. In other words, in sharp contrast with mind-independent reality, which, to repeat, can hardly be thought of as constituted of distinct parts, most of the phenomena that compose empirical reality exhibit no features that could be called nonseparable. Clearly, this confirms the necessity of at least distinguishing between the two *notions* of mind-independent and empirical reality.

Partly as a consequence of this, the material developed in this book will naturally lead us to the view that the ultimate "ground of things" escapes analytical description. Such an outcome may well appear most unpalatable to—at least—two quite different families of minds: the Hegelians and the scientists. For Hegel what was ineffable was nothing but muddled thinking; it had no meaning until words were found. However, Hegel himself abundantly produced words and sentences, without (judging by the variety of his followers' opinions) his message having being thereby rendered truly unambiguous. Finding words is certainly not a sufficient condition for valid thinking, and, after all, it is not absolutely sure that everything meaningful is, ipso facto, analyzable.

Possible reservations coming from my scientist colleagues would certainly worry me more. Quite a number of the latter seem to believe science has a genuine aptitude at—eventually—truly reaching the "core," or "ground," of things (or at least important features thereof), and in view of what is to be explained in this book it is impossible for me to grant them that much. However, at the same time I also definitively brush aside the view according to which the significance of our discipline is merely practical; that pure science is nothing but a technology focalized on the long term. Quite on the contrary, I consider it most plausible that the multifarious regularities and symmetries science reveals in all domains do correspond—albeit in a highly hidden manner—to some form of the absolute. Moreover, I consider, as will be explained in the text, that the proper domain of scientific knowledge, empirical reality, is far from being a mere mirage. And finally—a trivial but essential point—I perceive science as being an irreplaceable school of exactness and consequently a kind of "last bulwark," sheltering advanced societies from the puerile spirit considered by wise thinkers to be presently invading all mediated parts of cultural activity.

Quantum mechanics comes in support of the above-stated view since, as will be verified in the body of the book, within it the set of observational predictive rules is—by far—what is established on the most firm grounds. There are therefore quite serious reasons to suspect that it is only in appearance that physics, as taught and made use of in universities and research centers, describes mind-independent reality. In fact, what it describes is first of all communicable human experience. Curiously enough, this fact is very rarely mentioned in textbooks dealing with quantum mechanics. In private talks, most of the involved physicists willingly grant that its truth is hard to deny, but, as a rule, they show some reticence at saying so much in public.

On the other hand, an idea that, among them, is almost unanimously accepted (and for good reasons as will become apparent) is that the predictive quantum mechanical rules are of unrestricted validity. What best illustrates this is the difference between the attitudes of mind of the physicists and the general public (media included) in situations in which these rules generate verifiable experimental predictions that turn out to be at variance with some deeply ingrained views. In the opinion of the public and the media, what would be an "astonishing discovery" would be that these ingrained views, which seem to go without saying, should be found at variance with experimental outcomes. Whereas, on the contrary, for the physicists' community, because of its strong belief that the rules in question are universal, the astonishing fact, the "tremendous discovery," would be just the reverse. It would be that such experiments should show these ingrained views to be correct, since the universality of quantum mechanics would thereby be disproved. (It remains that when the said community sees the quantum rules being confirmed, as was hitherto always the case, it should have the intellectual courage to mentally break with the "ingrained views" thus falsified.)

Several books of mine have already appeared dealing with the impact quantum physics has or should have on the theory of knowledge. The present one does not merely recapitulate their content, it goes beyond it in at least five domains.

The first one has to do with nonseparability; and, more precisely, with the philosophical bearing of the experimental proofs of nonlocality. In particular, I think I show here more explicitly than had previously been done (at least, more explicitly than I did before) the implicit role of counterfactuality in the characterization of what is usually called "philosophical realism," and also, therefore, its role concerning the conceptual significance of the violation of the Bell inequalities.

The second one is about justifying the view—taken for granted, as just noted, by a considerable majority of physicists—that the (predictive) quantum rules are universal. Essentially, I ground my argument concerning this point on the recently appeared decoherence theory and on the experimental findings confirming it. But concomitantly I also stress the fundamentally nonontological character of the universality in question.

The third domain has to do with quantum measurement theory and the immensely puzzling riddle of the nature of consciousness. On the conceptually very difficult problems of the Schrödinger cat and Wigner's friend I think that the developments to be found in this book should shed some additional light. And I have a glimmer of hope that some new perspectives on what inner consciousness actually is should materialize through this a priori unexpected channel.

My fourth purpose is to make use of the whole material just mentioned in order to revisit the concepts of cause, explanation, and law, that is, to critically reexamine under this light the main philosophical theories concerning them.

Finally, the fifth domain concerns more specifically my former work. I had the good luck that the latter underwent the critical analyses of highly competent philosophers of radically opposite tendencies: one materialist and two neo-Kantians. Their remarks and objections efficiently enabled me to sharpen my views. The reader will find here my detailed answers to their criticisms as well as my reactions to their own—quite inspiring—conceptions.

As is apparent, the book is meant as a venture in reflection, grounded on an analysis meant to be extensive enough to take into account all significant ideas and facts, and aiming, if possible, at substantially improving our understanding of the field. When I wrote it, no concern about offering the general public an easy access to the latter could prevail over this ambitious goal. But on the other hand, while thus striving to reach up to the conceptual core of the problems, I observed once more that the core in question can be appropriately analyzed only by making use of common, everyday language, free from formulas and expressions merely accessible to those "in the know." Now, this common language is the one of people who are neither scientists nor deep philosophers. And I considered it appropriate not to neglect an opportunity arising thereby: I mean the possibility of providing the said public—or at least, its enlightened part—with an access to the set of problems considered here. Accordingly, I but rarely introduced technical words, and I saw to it that they should in no cases create understanding difficulties for nonexperts. In the same spirit, I took care to introduce the various topics by focusing on the simple and accessible facts that lie at their roots, and I limited their study to their conceptually significant aspects. In particular, I avoided unfolding the multifarious quantitative or just descriptive elements in them that, fascinating as they are for the specialists, hardly contribute, when all is said and done, to a genuine understanding of the subject under study. Consequently, even though the analysis of such and such specific points may, in it, appear difficult, the book should, on the whole be quite accessible to any educated person. (Incidentally, let me mention my "counter-models." They were a few contemporary texts whose style is so tortuous that when we finally succeed in discovering some content they possibly have we would not on any account jeopardize our elation by—additionally!—inquiring whether the said content is justified or not. As for me, I may at least certify that the data and views presented in this book have been stated with a great *yearning* for maximal simplicity and clarity).

The book is made up of two parts. The first one is mainly informative. It describes the relevant facts and the conceptual problems they raise, and, concerning the various contemplated solutions to the latter, it reports— and merely reports—what is useful for gaining an unbiased understanding of their real significance. The second part is more philosophically oriented, and hence more personal. Grounded on the data just mentioned, a critical analysis is made of the views developed by several philosophical schools, accompanied with an account of the conception that, all things considered, I think is the most coherent.

A practical remark may here be useful. While, as a material object, a book is intrinsically linear, or otherwise said monodimensional (the words follow one another), such themes as those investigated here are essentially multidimensional (the study carried on in chapter z is grounded on points analyzed in chapter y but also, independently, on notions introduced in chapters $s,t,u, \ldots ,x$, etc.). Consequently, for clarity's sake many cross references are here in order. Concerning the latter, and at the risk of making the text look a little clumsy, I did not just write, "we saw that." In most cases I stated in what section the information was given. I considered that this procedure would help any reader anxious to really capture the actual concatenation of the arguments. Clearly, however, such indications are just hints. A mention such as "see section x" definitely does *not* imply that, in order to understand the argument, it is, in any case, imperative to carefully reread section x.

Let it be added that, linked together as they are, the various themes the book deals with are, of course, distinct. Quite normally, the description of any one of them begins with its most easily apprehensible features and continues with those that are less so. On first reading it seems therefore advisable to skip some sections and several remarks exclusively bearing on precision issues. A few hints along this line are made here and there. They may be extrapolated to other passages. Some philosophically oriented readers may even find it convenient to begin by simply skipping most of the first part. In fact I could even imagine that, on first reading, they would pass the whole of it, with the sole exceptions of chapter 1, chapter 5, and section 10-4. The rationale within this procedure would be that when, in following it, they come upon a statement that surprises them they have a choice. Either they just accept it and go on reading or, making use of the cross references alluded to above, they go back to the section in the first part in which considerations are developed that justify, or at least back up, the statement at hand.

As the reader may well suspect, notwithstanding its title, this book is not meant to deal, at one and the same time, with the whole of physics and the whole of philosophy! Indeed it focuses on the problems raised by the notion of reality and is just an attempt at understanding a little better the information stemming, on this issue, from contemporary physics. Is it an essay? Is it a treatise? I find that it is hard to tell. On the one hand several ways presented themselves of trying to decipher the information in question, so that I had to clear a way for myself through a forest of evidence and suggestions. In this respect the book is principally of the nature of an essay. But on the other hand these bits of evidence and suggestions had to be impartially spotted, without any significant one being left aside, and it was then necessary to objectively compare them with one another, this being the only way to appreciate the pertinence of each of them. My feeling is that the forthcoming pages primarily result from the latter type of activity, which I strove to exert, at every stage, with the highly critical attitude of mind the subject obviously calls for. So that if some of their readers, taking their (resultant) somewhat severe aspect into account, tell me that they view them as constituting a sort of treatise I shall not be taken aback. However that may be, my dearest hope is that this book should help some of the said readers to carry somewhat further still their inner reflection on the truly basic existential issues.

PART 1

PHYSICAL FACTS AND RELATED

CONCEPTUAL PROBLEMS

CHAPTER 1

BROAD OVERVIEW

§☙

1-1 A General Picture

ADMITTEDLY THE new world-view that soared in the early seventeenth century originated from Copernicus' discovery but it is not to be questioned that the change from Aristotelian to Galilean physics played an essential part in its development. Aristotle had kept close to raw data obtained through the senses. He saw that all moving bodies not subjected to any force finally stop, and he raised this observed fact to the level of a basic principle of knowledge. He observed that living beings have all sorts of different shapes, qualitatively differing from one another, and he therefore took the notion of a wide variety of fundamental forms as a guideline for his philosophy. Hence, for him, there were a great abundance and variety of concepts, all of them lying, so to speak, on the same level. Hence also he gave considerable care to detailed qualitative descriptions, counterbalanced by a marked weakness concerning anything physically quantitative. With Galileo, Descartes, etc., on the contrary, the idea that came to the forefront was that of a hierarchy of concepts. Within their approach there are basic and nonbasic ones, and the latter must be accounted for in terms of the former, so that, in the end, the description of the physical world should be entirely expressed in terms of just a few basic notions linked together by quantitative laws. And we know, of course, that within classical physics as well as in all other sciences this is the conception that finally prevailed.

The difference just sketched between the Aristotelian and the Cartesian-Galilean approaches is quite well known. But what often remains unnoticed is that, notwithstanding this difference, the approaches in question had an important feature in common. In fact, they shared the view that the basic concepts (the nonderived ones) are either obvious ones or, at least, idealizations of obvious ones; that they are familiar notions—"clear and distinct ideas" as Descartes said—whose unquestionable validity is fully guaranteed by commonsense (i.e., by God, according to the same). It is often—and rightly—stressed that Galileo, Descartes, and Newton brought mathematics into physics. But an often overlooked point is that they made use of mathematics primarily for imparting quantitative content to developments exclusively bearing on objects designated by means of familiar concepts. Descartes was bent on describing the whole of the physical world by "figures and motions" and referred to "the pipes and springs that cause the effects of the natural bodies." Newton spoke of material points, that is, (basically) idealized grains or specks. Even Pascal, in his fable of the mite, clearly took it for granted that the domain of validity of the familiar concepts extends to the whole range of conceivable scales, from the infinitely great to the infinitesimally small.

Were they right? Yes of course, in a sense. Pioneers they were and, as such, their most urgent task was to explore the ins and outs of such a natural idea. Moreover, the idea in question proved spectacularly fruitful. Still today, there are many fields of study in which it is possible to describe data and processes by the sole means of basically familiar concepts and in which this is obviously the best way to carry on fruitful research. Consider, for example, molecular biology. Molecular biologists have to do with large molecules whose behavior—for well-known reasons following from the very rules of quantum physics—practically obeys the laws of classical physics. Consequently it is possible and natural to think of them as having rigid atomic structures, fixed shapes (including hooks), and so on; in short, to reason about them as if they were component parts of a clockwork or a machine. This mode of thought opened the way to quite a host of predictions, many of which proved brilliantly successful. No wonder that many biologists are tempted to raise it to the level of an absolute; to view it as yielding *the* proper canvas for a description of "the Real itself," including thought.

All this naturally leads to a mechanistic world-view styled naïve—to be sure—within philosophical circles but stamped with commonsense and taken therefore by the majority of well-informed people to be by far the most sensible one. Think, for a moment, of an attorney, a senior executive, an engineer, or even a scientist working in some highly specialized field other than theoretical physics. Such people—most of our contemporaries indeed—have to do, all day long, with machines of all kinds, that is, with human-made devices of which the clock is the example par excellence. No wonder that they spontaneously lean toward a generalized mechanistic world-view, in spite of all that the philosophers may write and say! It is therefore fully understandable that such a type of natural philosophy should remain, so to speak, the instinctive one in the minds of both the enlightened laymen and a majority of scientists. The idea that the World—or, at least, its physical part—is of the nature of some gigantic machinery seems infinitely plausible indeed.

And still, the opposite is true! In a move that was slow at first but progressively speeded up, physics taught us, not only that the human mind is able to operate well outside the framework of familiar concepts, but also that it absolutely must do so. Of all sciences, only physics, apparently, yields this message. But it does. And the fact that it does may well be taken to constitute one of its main contributions to the development of thought.

For brevity's sake, let me illustrate this by but one example, that of particle creation in high-energy collisions. This phenomenon is explored and investigated in laboratories where large particle accelerators are available and observed in bubble chambers in which moving particles produce tracks. Two protons are accelerated. Each one has a given motion, a given velocity, and hence a given energy. They collide and they then part from one another. At that time we observe that, even though they are still in existence and did not break up, other particles have appeared, which are "really true" particles—possessing masses, electric charges, and so on—that have been *created* in the collision, at the expense of the incident protons' total energy. This, at least, is what we see. Admittedly, the phenomenon is quite in agreement with the celebrated $E=mc^2$ law expressing mass-energy equivalence. But if we insisted on describing it by the sole means of familiar concepts we would have to say that the incident particle motion was changed into particles. Now, motion is a property of objects, so that what we would thereby refer to would be nothing else than the transformation of a property of objects into objects. Such an idea lies entirely outside the realm of our familiar concepts. Within the set of the latter there are, on the one hand, objects and, on the other hand, properties of objects; and no element of either one of these subsets ever transforms into an element of the other one. The very idea of such a transformation looks just as absurd as that of changing the height of the Eiffel tower into another Eiffel tower. Or as the view that, when two taxis collide, they may both emerge undamaged, accompanied by five or six other taxis arising from the initial kinetic energy of the former. All this makes it crystal clear that contemporary physics forces upon us the use of basic concepts lying outside the realm of the familiar ones, with the sole help of which Descartes (and the other "founding fathers" of modern science) originally claimed that physics would describe the World.

Are we then faced with an enigma exceeding the powers of understanding? Quite on the contrary, theoretical physicists not only know how to deal with this creation phenomenon but also had *predicted* it, on the basis of their equations.[1] Which shows that, when applied to physics, mathematics makes it possible to really reach beyond all familiar concepts; to actually *coin* new concepts. This reminds us of Pythagoras' famous saying: "Numbers are the essence of things." A sentence to be understood, of course, as "*mathematics* are the essence—the very essence—of things."

[1] We here refer to the work of the British physicist P.A.M. Dirac.

This, to be sure, is not the proper place for entering into a detailed account of how physics manages to describe the creation phenomenon. Already at this stage let us note, however, that it offers several ways of doing so, grounded on quite different basic concepts. The existence of this diversity should not disconcert us, but it is important that we should be aware of it, since it casts quite a serious doubt on the very possibility of univocally determining, by means of physics, a list of the truly basic notions. Which, in turn, makes it unlikely that Pythagorism is the "last word" of our story. Indeed the diversity in question is a good example of a philosophically important phenomenon—called "incomplete determination of theory by experience"—that we shall frequently have to take into consideration. With respect to the case under study, it so happens that the mathematical formalism yields not one but three distinct theories, all of them grounded on the general quantum rules, yielding essentially the same observational predictions, but widely differing concerning the ideas they call forth. They are called the "theory of the Dirac sea," "Feynman graph theory," and "quantum field theory."

We shall soon get acquainted with the two first named ones (sections 2-6 and 2-7), and shall then have the opportunity of observing how widely both depart from the Cartesian ideal of a description using only familiar concepts. But, for the time being, let us focus our attention on quantum field theory.

Unquestionably more general than the first one—Dirac's—this theory is, in a way more basic than the second one—Feynman's—which primarily appears as a powerful method of calculating, grounded, as its author himself stressed, precisely on the very rules of quantum field theory. To form a broad idea of the general guiding lines of the latter let us begin by observing that the notion of creation is not a scientific one: We do not know how to capture it, and even less quantify it. It is therefore appropriate to try and reduce it to something we can master. Now we do master the notions of a system state and changes thereof. We know how to calculate transition rates from one state to another. And the brilliant idea, the breakthrough, just came from this. It consisted in considering that the existence of a particle is a state of a certain "Something," that the existence of two particles is another state of this same "Something," and so on. Of course, the absence of a particle is also a state of this "Something." Then, the creation of a particle is nothing else than a transition from one state of this "Something" to another, and therefore we may hope to be able to treat it quantitatively. It is just as simple as that! In practice—believe it—the matter is appreciably more complex. Quantum field theory textbooks are big, fat objects, full of formulas, many of which are in no way beautiful. But by plodding through the latter it proved possible to account for observed phenomena with a precision that, in some cases, extends to the seventh decimal. Which, really, is "not too bad"!

The reader will be spared the calculations. Instead, let him or her reflect on the just described basic idea of the quantum field theory. True, the problem of the "real nature" of the "Something" that it brings—at least implicitly—into play is, as we shall see, an inordinately delicate one. However, concerning it, one point at least seems rather clear. It is that the cornerstone role tacitly attributed to it somehow suggests the presence, within the core of present-day physics, of a wholeness of some sort, radically foreign to classical physics. The point is that classical science was very much in favor of what may be adequately said to be a *multitudinist* world-view. In other words, it favored a conception of Nature in which basic Reality—matter, as it was called—was constituted of a myriad simple elements—essentially localized "atoms" or "particles"—embedded in fields, and hence interacting by means of forces decreasing when distance increased. The first two of the above mentioned theoretical approaches still are more or less compatible with such a view, although, as we shall see, they considerably weaken it by attributing to the particles behaviors that the mind cannot imagine. On the other hand, the—more general— quantum field theory is radically at variance with it. Not only is it true that, in it, the particles no longer play the role of the constitutive material of the Universe. What is more, the only "entity" that, in it, might conceivably be thought to constitute basic Reality is the "Something," of which we saw that it is fundamentally the only one of its species.

The idea that, here, is seen to come to light (though dimly as yet) is, to repeat, the notion of a wholeness of some sort. Within elementary particle physics wholeness, admittedly, remains ambiguous since while, say, manifest in one formulation, it is evanescent in the other two—in spite of the fact that all three are equivalent and proceed from the same—quantum— formalism. This perhaps explains why, when quantum mechanics appeared, the notion in question was clearly apprehended by neither the epistemologists nor even the physicists, with the perhaps unique exception of Schrödinger. But theoretical as well as experimental advances gradually made people realize that it constitutes an inherent part of the very quantum formalism and has quite specific experimental consequences.

Nowadays the common name *nonseparability* serves to designate both the just mentioned mathematical features of the formalism and the corresponding observable effects. A most important point is to be noted concerning it. It is the fact that the range of validity of the notion it designates is even wider than that of the presently accepted theory. Indeed, it has been established that in one at least of its main aspects—nonlocality—it will certainly remain true, even if the quantum formalism must, one day, be replaced by some other, more general, one. As will be seen (chapter 3), this follows from the Bell theorem and the experiments—such as Aspect's—associated with it, for the results of the latter are incompatible with some consequences of the inverse hypothesis—locality, and this quite independently of any theory whatsoever.

To be sure, scientists and even physicists go on expressing themselves in terms of particles, molecules, and so on, all words calling forth the idea of individual, localized objects depending less on one another as the distances between them grow greater. In short, they go on making use of a multitudinist language. And from their angle they are right for, as we saw, this amounts to referring to a model that is, by far, the most convenient one in an enormous variety of cases. But, by now, it appears more and more clearly that it is merely a model. With due reservations a comparison could be ventured here with Ptolemy's geocentric model, which also works quite well on specific problems. In both cases, to raise the model to the level of a description of "what really is" is scientifically illegitimate.

In this respect, let it be noted that the question "reality or just model?" never comes to light in the articles physicists write. The latter wisely remain on "secure ground," which means that their theoretical constructions, elaborate as they may be at the level of equations and methods, are left by them very much "open" regarding concepts. In fact, when they work on such constructions the condition they impose on them is just that they should be highly general models, correctly accounting for what we observe in a great variety of experiments. Consequently it is without qualms that they ground them—tacitly at least—on the basic principles of "standard" quantum mechanics, without being in the least worried by the fact that, as we shall see, some of these principles impart a fundamental role to such notions as "measurement" and "preparation of system states." Now, this fact—the occurrence of a reference to human action within the very axioms of physics—is sometimes explicitly stated. Often it is kept implicit. But in any case it implies that the theories built up in this way markedly depart from a principle that was one of the main guidelines of all classical ones. I mean the rule that basic scientific statements should be expressed in a radically objectivist language, making no reference whatsoever, be it explicit or implicit, to *us* ("operators" or "measurers").

As a consequence of all this, it becomes clearer and clearer that our senses do not reveal the "real stuff," as it truly is. Indeed, let us consider an object that is more or less on the human scale, say a stone or a speck of dust. That it is not what it looks like has been known for quite a long time. Classical physics taught us already that, while we tend to take a stone to symbolize the very notion of "fullness," it is, in fact, mainly composed of vacuum (the space between the nucleus and the electrons).

But nonseparability suggests that, strictly speaking, it does not even exist as a distinct object! That its "quantum state" is "entangled" (this is the technical word) with the state of the whole Universe. How does it then come that, to us, it seems localized? Recently, a very general argument—called "decoherence theory"—was found that partly accounts for this fact. But its nature is disconcerting enough, for, as we shall see, it amounts to proving that, for all practical purposes, we are unable to measure any one of the quantities the measurement of which would show that the stone is *not* localized. It makes it clear that all such measurements are far too complex to be performed (they would necessitate inconceivable instruments, perhaps composed of more nuclei than there are in the Universe, or, alternatively, performing times longer than the life of the latter, or other, similarly unthinkable conditions). Obviously, this view is quite the opposite of the classical, commonsense one that objects truly have the shapes and positions we see, and that they have them "by themselves," quite independently of the limitations of our own aptitudes, as well as of the size of the Universe or anything else. Were some simile requested, the best one would probably consist in comparing the quantum objects to rainbows. If you are driving, you see the rainbow moving. If you stop it stops. If you start again, so does the rainbow. In other words, its properties partly depend on you. Taken literally, quantum physics, when thought of as universal, imparts to all objects such a status relative to the sentient beings that we are. It is true that some physicists strove to revert to a more classically objective standpoint but they had such serious obstacles to circumvent that, as we shall see, the outcome of their quest has finally to be considered unsatisfactory.

To sum up, the foregoing quick survey yielded glimpses at three main points. One of them is the necessity not to keep to the set of the old, familiar concepts. Another one is the necessity of going over from multitudinism to a holistic view of whatever is meant by the word Being. And the third, related to the latter, is that trying to go on using a universal objectivist language generates difficulties that, finally, make such attempts artificial. Here the expression "objectivist language" means a language that is descriptive, that is, as already stated, not merely predictive of observational outcomes. In other words, it means a language the grammatical form of which at least makes it possible to think of what it deals with—essentially contingent, space- and time-localized data—as existing quite independently of us. All this, of course, has merely been sketched and will be developed and made precise later on.

Accordingly, the chapters immediately following this one will deal with these three points, which will be examined one after the other and in detail (with emphasis, of course on the more delicate ones). Before that, however, it is suitable that some notions in rather current use be considered. The main purpose of the following section is to make the latter precise and define words or expressions for designating them, so that we can later unambiguously refer to them. Some of the definitions in question will be supplemented by commentaries aimed at describing the contexts in which they appear.

REMARK

It should be noted that within the notion that most physicists form of their science the three mentioned points do not lie at the same "obviousness level." The first one—the necessity of reaching beyond familiar concepts—is almost unanimously considered undeniable and essential. This, already, is not quite the case concerning the second point, the one relative to wholeness. In fact, while quantum field theory led to very many fruitful mathematical developments, surprisingly enough it gave rise to but few analyses explicitly bearing on its concepts. And the existing ones (such as some articles from the great physicist Erwin Schrödinger) are hardly known, even by the experts. A reason may be that, as will be shown below, a general feature such as nonlocality has no direct impact on what physicists are, as a rule, most directly interested in, namely, prediction of experimental results within their own specific field. It is therefore not surprising that, even though the Bell theorem and the corresponding experiments are by now well registered facts, nevertheless the physicists' community is still not unanimous in recognizing their full importance and bearing. Finally, with respect to the status of objects relative to us great divergences of opinions remain. True, it is rarely denied that reconciling the realist approach with the theory is difficult. But the idea remains widespread that such problems are, after all, just subtleties that the passing of time is bound to somehow remove. Being worried by them is the lot of but a minority, which, however, is numerous enough to hold international symposia and so on. The problem has many facets that lie at the core of numerous debates, and is even broader than most of the physicists engaged in such studies believe it to be. This is because—contrary to what many intuitively think—the Galilean ontology, even remodeled the Einsteinian way, is nothing like an "obvious truth" that we should have to take for granted. And in fact it is just the ambition of books such as the present one—that is, books aimed at a better philosophical understanding of present-day scientific data—to play a part in shedding light on this galaxy of questions.

1-2 Some Useful Definitions

For the above-stated purpose we shall, of course, from the next chapter on, make use of physics, that is, of essentially concrete observational data. But since we have to match such concrete material with abstract and varied human ideas, we must first allot to the latter labels making it possible to currently refer to them, without having to reiterate on every occasion the essentials of their definitions. Such is the purpose of this section, which should therefore be considered a sort of necessary parenthesis in the overall unfolding of our enterprise. It consists in alphabetically ordered definitions of words and expressions that will often occur. Following this list is a table of not so frequently used words, indicating the sections where they are defined.

Counterfactuality. See *Realism of accidents.*

Idealism (Temperate). So is to be called here Kant's conception, named by him "transcendental idealism." In it, the *thing-in-itself* notion is held to be meaningful, in spite of the fact that the said thing-in-itself is considered unknowable.

Idealism (Radical, Otherwise Known as Critical. So will be called the neo-Kantians' position, in which the *thing-in-itself* notion is rejected.

Everybody knows that idealism, in either one of its two versions, is grounded on the following remark. If objects exist of which we may acquire direct—hence sure—knowledge (it is by no means certain that there are any), these objects can only be of a mental nature: ideas, raw sense data, etc. They cannot be elements of the outside world since knowledge of the latter results from operations of the senses, and senses are likely to deceive us. Idealism claims therefore that we merely have access to representations, that is, to "phenomena," and that the only legitimate purpose of science—and of knowledge in general—is the investigation and ordering of the said representations. It is along such lines of reasoning that Kant claimed space, time, and causality are but a priori forms (of our sensibility as regards the two former and our understanding concerning the latter).

Note that phenomenalism, positivism, and pragmatism may be considered to be variants of idealism, at least in the sense that they all more or less agree with the above-stated views. Also note that Kantian idealism parts from other forms of idealism in that, according to it, raw sense data such as, say, a visual impression are neither more nor less "directly known" than external objects and are not therefore elements of "reality-in-itself" any more than the latter are. Moreover, it tends to dismiss the view that some unknown "object-in-itself" exists, corresponding to each "object for us." As Putnam put it (Putnam 1981): " On Kant's view, any judgement about external or internal objects (physical things or mental entities) says that the noumenal world *as a whole* is such that this is the description that a rational being (one with our rational nature), given the information available to a being with our sense organs (a being with our sensible nature) would construct" (my emphasis).

Language (Objectivist). The (already used) expression "objectivist language" means here a language that in no way basically refers to us. More precisely, it means a language that is essentially *descriptive*, as opposed to *predictive of observations* ; a language grounded on the assumption that either the considered objects—in a wide sense of the word, that is, particles, fields, and so on—really exist, or we can *do as if* they so existed, quite independently of us. Of course the view that the objectivist language is universally valid—at least concerning nonmental issues—is derived by us from everyday life. It is, moreover, engrained in our mind by basic scientific education. Obvious pedagogical reasons induce high-school science teachers to tell or at least to suggest to their pupils that an electrical field "exists," that atoms "exist," and so on, just as we currently say that stones and grass exist.

It is worthwhile to note that, although the Kantian and neo-Kantian philosophers denied that knowledge has any ontological meaning, the tendency to universalize the objectivist language was in no way opposed by them. Quite on the contrary, these thinkers were among those who made great—and temporarily successful—efforts at justifying the objectivist language (in the "as if" sense, of course), and its extension to the whole field of physics.

Under these conditions the conceptual difficulties raised by the advent of quantum mechanics could not but be perceived as being particularly serious. For indeed, to be expressible in the objectivist language any theory must—at the very least—make it possible to specify what, within it, is to be taken as real, or can be treated as being real. In other words, within the mathematical formulation of the theory, at least some mathematical symbols must be interpretable as describing elements, or features, of reality. In Newtonian physics this, for example, is the case concerning the symbols representing particle positions. In classical electrodynamics it is the case as regards those representing the fields. But in quantum mechanics, specifying what is real is very far from being easy. And, in fact, this was the essence of the objection raised by Einstein against quantum mechanics. Contrary to what is often said, Einstein was not craving for a return to old classical concepts. His criticism could be expressed as a question: "What, in the theory, can be considered real or treated as if it were real? Be outspoken and let me know!" Bohr's answer to this query may be considered to be a denial of the very validity of the question; a denial justified by the fact that, as we shall see, Bohr finally interpreted quantum physics and science in general as being descriptions, not of anything like a given external reality, but merely of communicable human experience: in other words, of a "reality" (if we dare venture the term) actively constructed, not only by thought but also by our operational decisions, thus obviously parting with the objectivist language.

On the other hand, few physicists willingly accept giving up a language that, in the opinion of many of them, adequately reflects the ultimate aims of physics. And, to say the truth, for quite a long time few of them even realized that the notion of a science exclusively aimed at describing collective human experience lay at the core of Bohr's approach (even though Bohr himself took great pains to make the fact clear). The worrying question "what, then, is real?" was thus left open. It is true that it long remained in latency, so to speak, precisely because of this misunderstanding. Many physicists were, somewhat naively, convinced that the Bohrian solution was compatible with the notion of a quantum formalism correctly describing reality-in-itself. They therefore did not feel incited to think over a problem that, without having personally examined it, they thought had been solved long ago by others, in a way compatible with their own conceptions of reality. It is the belated collective realization of the occurrence of this misunderstanding that explains for the most part the present-day renewal of interest in these questions.

Multitudinism and Principle of Analysis.   "To divide all the difficulties I shall examine in as many parts as possible and as needed for better solving them"—it is in this concise form that Descartes—he again!— stated a principle that, from the seventeenth century to ours, has been at the core of scientific research and may be considered one of the main sources of its success. It suggests the view that, when we have to do with a somewhat complex physical system, we should divide it by thought into simpler ones, study each one of the latter separately, take—of course— the forces connecting them duly into account, and finally make a mental synthesis of them. The fact that this procedure was very generally successful strongly suggested that, indeed, complete knowledge of the parts— and forces—ipso facto yields that of the whole, and that, therefore, the *whole* basically is a composition of *parts*. This is what will be called the *multitudinist* conception.

In section 1-1 we already had a glimpse of the fact that contemporary physics yields serious indications that multitudinism is flawed. Later we shall determine more precisely what the nature of the said flaw actually is.

Phenomenon.   In this book this word is usually taken in its etymological (and somewhat restrictive) sense: the object of some possible (human) experience.

Platonism.   As is well known, this venerable conception lies midway, so to speak, between idealism (it is sometimes called "objective idealism) and realism (it is also referred to as "realism of the essences").

Pythagorism.   See *Realism (Einsteinian).*

Realism.   As we shall see, there exist several forms of realism. But practically all the "realist" conceptions (in the philosophical sense of the word) are basically composed of two elements. The first one consists of the notion of reality-per-se—a "reality" conceived of as totally independent of our possible means of knowing it—along with the hypothesis that we do have access to the said reality, at least in the sense that "we can say something true" concerning it. At first sight this "hypothesis" looks like a mere truism (if we see an object in front of us how could it not be there, as we see it?). But the simple fact that we have dreams already convincingly shows that it is quite far from being one. In fact the hypothesis in question is one of those that, like induction and so on, are quite often intuitively assumed true (and maybe rightly so) without being scientifically provable. To try to make it plausible is, of course, quite legitimate; for example, by means of the no-miracle argument, or by referring to intersubjective agreement (both attempts will be discussed in chapter 5). But it cannot be proved correct.

The other element constitutive of (almost all) "realist" conceptions[2] (it is indeed distinct from the first one, as will appear) consists in a *representation* we build up of independent reality. A representation worked out from the phenomena, that is, from human experience.[3] Indeed, this

---

[2] "Open realism" is an exception, and so is my "veiled reality" conception (see chapter 10), which is a special case of the latter.

[3] Concerning these representations a concatenation of three almost self-evident points is worth being made explicit. One of them is that obviously, being phenomena, elementary sense data (such as having the impression that a pointer has at a certain time a certain position on a dial) may (and often do) come in as parts of such a representation. Another one is that as long as we consider a representation to be valid all its elements, including these sense data qua "impressions," are to be considered valid, or "true," by definition. And the third one is that a representation is valid only if it is self-consistent, that is, if the

representation may be (and, in science, actually *is*) constructed without a reference to the reality-per-se notion being a necessary ingredient in the process. It is a posteriori—so to speak—that we identify the elements of the said representation as elements of reality. But it must be noted that the elements of experience that we preferentially select for this identification are of a varied nature and that the realists do not all select exactly the same ones. This is what gives rise to the various kinds of realism described below.

Realism of the Accidents (Alias Objectivist Realism, Alias Galilean Ontology). The "representation" element of this realism[4] primarily stems from the importance attributed by human beings to some groups of impressions, the relative stability of which they undoubtedly noted right from the very beginning of the human species. These impressions were gathered together as ideas of objects, quantities, value possessed by these quantities, sequential rules concerning these values, and so on. It is these groups of impressions that were raised to the "dignity" of representations of elements of reality. In philosophy the word "accident" was often used to designate the contingent (and possibly changing) properties of perceived things, such as their multiplicity, their forms, their positions, and their motions (in physics the two latter are often called "dynamical properties"). This is why the name "realism of the accidents" was found here adequate for labeling this realism. But another basic feature of the representation in question is the fact that it takes into account and elevates to the status of a genuine component of reality the outcome of a very general and obviously equally primitive observation: The observation of the considerable usefulness, at least for practical purposes, of the *counterfactuality* notion. Schematically, I make a counterfactual reasoning when I say to myself: "By performing such and such an operation (for example, going and seeing) I found that such and such a quantity has such and such a value. I consider that this quantity would have this value, even if I had not performed the operation in question." Counterfactuality enables us to trust the validity of some statements referring to an observation or an action that we could perform instead of the one we actually do. Now, our impression that the things are real is largely due to this trust we feel we

---

consequences inferred from some of its elements by applying rules assumed in it to be valid are not at variance with some other elements composing it.

[4] One should of course carefully avoid mixing up realism of the accidents with the "realism" of the great medieval texts. In fact, these two conceptions are diametrically opposed. Realism of the accidents imparts the status of real entities to the individual objects that we perceive (and also of course, to their constituting parts), whereas medieval realism attributed this status essentially to the great general concepts, along the lines of Platonism.

can have. Clearly, when I am having lunch, the statement "if I were now in my office I would see my books on the bookshelf" is, for me, meaningful and true. And it is so in a totally convincing way if I assume that, during that time, the objects in my office are not subjected to any action from outside; that, in particular, no hidden forces are at work between my dining room and my office. And obviously, knowing this unconsciously backs up my faith in the *reality* of my books.[5] In contrast, within Bohr's approach, centered, as we shall see, on different (less atavistic) aspects of human experience, many counterfactual statements similar in form to this one are meaningless.

On the scientific level the emphasis put on the notions of objects, quantities, values, and form by the realism of the accidents (alias objectivist realism) must of course be made more precise. With this aim in view let the defining condition be set that, in this realism, the objective state of any physical system (corpuscle, field, macro-object, or whatnot) is specified at any time by a discrete or continuous set of known or unknown, knowable or unknowable (and possibly functionally linked) real numbers. (It is likely that, among the very many tenets put forward by various philosophical schools, some are consistent with the realism of the accidents on every point except this one[6] ; by definition we shall not give them the "realism of the accidents" label.) As for counterfactuality, it may be made precise by stating that if P is a proposition relative to some given physical system and if a *nondisturbing* measurement performed on the latter (or on a system strictly correlated with it) has shown P to be true, then P would be true even if that measurement had not been performed.

According to the realism of the accidents, space and time are real. And this holds true even though Galilean relativity renders meaningless the notion of the absolute position in space of an object. Indeed, within the realm of the said realism the instantaneous positions of objects with respect to one another are typical examples of "accidents" and are therefore paradigmatically real. Consequently, inherent in the realism of the accident is a notion of—at least relative—locality, and also a certain "principle of locality." This expression, more precisely defined in chapter 3, schematically means that the interaction between objects gets weaker when the distances separating them get larger.

---

[5] But let it be stressed that, while it is difficult to conceive of any form of (conventional) realism that would not imply counterfactuality the converse is not true. Being just an element of our representation, counterfactuality does not *imply* realism (in the ontological sense here imparted to this word) even though it does suggest it. For example, as we shall see in section 13-2-1 (note 3), the notion of *complete determination* introduced by Kant in his philosophical system, while involving counterfactuality, has, in the said system, no ontological implication.

[6] Some forms of animism may be approximate examples.

As here specified, the realism of the accidents (alias objectivist realism) seems to have been Galileo's position and it also seems to be the one most present-day scientists more or less instinctively take up. This, incidentally, remains true even though, along the lines of relativity theory, the notion of events should be considered more basic than that of objects. The *realism of the events*, which better fits relativity theory, may be held to be but a refined version of the realism of accidents.

Realism (Einsteinian) (Alias Mathematical Realism, Alias Pythagorism). This is the view that the notion of a reality independent of us (a "reality-in-itself") is indeed meaningful, that this reality is knowable, but that it cannot be reached by using familiar notions and that other ones, borrowed from mathematics, are to be used for this purpose. For example, the four-dimensional space-time and the curved space of special and general relativity, respectively, are, as a rule, considered to be real.

It should be noted that these two relativity theories are classical, that, admittedly, internal consistency of physics requires they should be quantized, but that, as long as this objective is (momentarily) set aside it remains possible to interpret them realistically in the above-specified sense. And this holds true notwithstanding the fact that it is by means of a heuristically operational procedure that the young Einstein could (in parallel with Lorentz and Poincaré) elaborate the "special" relativity theory. Sometimes, the possibility of such a realist interpretation is questioned on the grounds that, in special relativity, such an important notion as the one of simultaneity at a distance can only be defined by referring to observers. But this objection hardly stands up to the observation that the basic entities in the theory—events and space-time intervals—may, if so desired, be raised (some would say "hypostasized") to elements of reality-in-itself. Events—par excellence local beings—are, in this theory, independent of both our knowledge and the manner in which we group them together by thought. This is why Einstein, in his later years, and obviously with his whole work in mind, could legitimately write: "There is something like the "real state" of a physical system, which [this state] objectively exists independently of any observation or measurement" (Einstein 1953a).[7] (It is well known that he lamented the discrepancy between quantum mechanics and such views.) All this is of course fully compatible with the fact that both relativity theories make use of highly nonfamiliar concepts.[8]

Incidentally, let it be pointed out that Einsteinian realism is not to be confused with *mathematical realism* when this expression is understood

---

[7] See chapter 6, note 1 for a more extensive quotation of the same passage.

[8] Yet it is true that Einsteinian realism is not *necessarily* a "realism" in the strongest sense of the word. It seems that in Einstein's mind it was (although with shades of meaning). But

in the sense that pure mathematicians impart to it. For indeed the latter refer through this name to a conception some of them have of the nature of pure mathematical knowledge, and this conception has nothing to do with physics. For the same reason, the name "Pythagorism" might create confusion. Here it exclusively refers to the already quoted well-known Pythagorean dictum: "numbers are the essence of things," barring of course any mystic use of numbers. But, clearly, pure mathematicians may well give it broader acceptance, with no reference whatsoever to the theories worked out by Einstein.

Realism (Objectivist).   This realism is precisely the same as the realism of the accidents. The only shade of difference between them is that the expression "objectivist realism" should preferentially be used in cases in which it is appropriate to lay an emphasis more on the existence of the objects than on that of the properties.

Realism (Ontological).   The word "ontology" means "knowledge of Being." Ontological realism is therefore the tenet according to which we can, by means of science, gain an exact and exhaustive knowledge of ultimate reality. Nowadays, even the adherents to objectivist or Einsteinian realism grant, most of them, that such a sharp standpoint has something presumptuous in it. In other words, ontological realism, on the whole, inspires caution. But it would seem that, in the mind of not a few scientists, this reticence is a matter of form rather than substance.

Realism (Open).   This is the—indeed quite "open"—view that *there is something the existence of which does not hinge on thought.*
   Is this "something" the set of all the objects, of all the atoms, of all the events, God, the Platonic Ideas, still something else? Open realism is mute on this. In other words, it should be considered a mere starting point for some subsequent investigation aimed at reducing the spectrum of all these possibilities. It just says "something," in the widest possible sense of the word. It is "open" to the extent of being compatible with any philosophical system whatsoever, with the sole exception of radical idealism.

Realism (Physical).   This is a suitable name for the theory claiming that, in the last analysis, physics, conceived of as being the "hard core" of the other sciences, is, in principle, qualified for describing qualitatively and in detail "reality as it really is." The (nonessential) difference between physical and ontological realism(s) is that the former grounds such an

---

it may be given a nonontological interpretation, consisting in considering that it is just a language synthesizing by means of mathematical concepts the hard core of our experience.

expectation essentially on the sciences we actually know and considers it to be just a long-term, asymptotic view. Realism of the accidents and Einsteinian realism both are special instances of physical realism.[9]

Clearly, while physical realism is considerably more general than realism of the accidents, it seems that, still, both conceptions must incorporate counterfactuality as a quasi-necessary ingredient.

Realism (Near).   Near realism is the mechanistic conception that Descartes, as recalled above, put forward. In the *Principles of Philosophy* Descartes explained that he surveyed all the clear and distinct notions that may be found in our understanding concerning the material things. And he stated he could find "no other ones than those we have of figures, sizes, motions and the rules by means of which these three things may be diversified by one another, the said rules being those of geometry and mechanics." From this he inferred, as mentioned above, that the entire knowledge human beings can form of Nature must be composed of just this; that all their descriptions of the physical world must be expressed in terms of figures, sizes, and motions; and that there is therefore not the least difference between natural and manufactured objects . . . except (and this is where the strong statement alluded to above comes in) that the elements composing the latter are big whereas "the pipes and springs that cause the effects of natural bodies usually are too small to be perceived by our senses."

This mechanism regarding matter I call *near realism.* "Realism" because it is supposed to picture ultimate matter "as it really is." And "near" (to us) because the latter is described exclusively by means of "clear and distinct notions," that is, of familiar concepts (figures, sizes, motions, and combinations thereof). As we see, near realism is but a restricted version of objectivist realism, limiting the set of admissible features to the few listed ones.

Realism (Structural).   This is the conception according to which only "structures" are real (what is contingent is but phenomenal), and which claims that structures are knowable (in principle) as they really are.

Note that, admittedly, a conception could be upheld in which the second of these two conditions would be weakened, or just ignored; in which the said structures would be knowable only approximately, or not at all. However, what the expression "structural realism" generally means is the

[9] As we see, physical realism is a very general conceptual framework. In it, nothing is assumed a priori concerning the structure of physical reality, the nature of the elements constituting it, and the physical quantities attached thereto.

view according to which they are, at least theoretically, knowable in the full sense of the word.

Realism (Transcendental) and Idealism (Transcendental). As is well known, within Kantian philosophy the objects of human experience are but phenomena, that is, mere representations. They and their attributes do not possess, outside our mind, any existence by themselves. This philosophical standpoint Kant called "transcendental idealism" (*Critique of Pure Reason, Antinomies of Rational Thinking,* sixth section*).* And he set up this view in contrast with the standpoint of the transcendental realist who, he stated, considers such experienced facts to be things existing on their own, in other words, "things in themselves." It may therefore be considered that, within Kantian philosophy, the expression "transcendental realism" labels a view that is very much related to the one called objectivist realism (or realism of the accidents) above.

Reality (Mind Independent, Alias "the Real," Alias Reality-per-Se). A name has to be imparted to the "something" Open Realism refers to if we want to discuss this concept at all. Here it will be called either "the Real," or "reality-per-se," or "mind-independent reality," or simply (as already done above) "independent reality." It should be understood that these expressions are mere labels and entail the attribution of no feature whatsoever. In order to avoid possible ambiguities stemming from other meanings of the word "real" (those, in particular, that this term has in the writings of Kantians and neo-Kantians) it will here be written with a capital R whenever meant in the above explained sense.

"Real" (as an Epithet) and "Really". Just like the name "reality" the epithet "real" and the adverb "really" have two meanings, corresponding to the two concepts of *independent reality* and *empirical reality* (see below concerning the latter). But, just as ordinary thinking implicitly mixes up the two concepts, ordinary language when using the epithet or the adverb in question does not distinguish between these two meanings. It considers that any object on which we can act is real and it "instinctively" raises this feature to the level of an intrinsic quality, thought of as being fully independent of us. And, of course, exactly the same remark is in order concerning the adverb "really." Since, within analyses of the conceptual foundations of quantum mechanics this mixing up readily generates errors, it is appropriate, when reading books or articles dealing with this subject, to be on the lookout concerning the shifts in meaning that a careless use of such apparently unproblematic words may cause.

Variants.   A discourse held in realist language may have a few different senses.

First, it may be meant either as a genuinely realistic (ontological in the strongest sense) description or as one just implying an "as if" standpoint.

Second, it may implicitly assume either objectivist or Einsteinian realism. In both cases contingent data are thought of as being describable and knowable. Or it may just as well merely assume structural realism, within which this is not the case.

Other expressions, which are not of immediate use, will appear in the course of reading these pages. They are to be defined at the appropriate places but below, to facilitate reading, the sections are indicated where the main definitions are given.

Contextuality.   See section 5-2-4.

Objectivity (Strong and Weak).   See section 4-1.

Realism about Entities.   See section 6-2.

Realism of Signification.   See section 9-7.

Reality (Empirical, alias "Contextual").   See section 4-2-1 and the next to last paragraph of section 4-2-3 (schematically, it is the set of the phenomena).

Reality (Epistemological Elements of).   See section 8-1-1.

CHAPTER 2

OVERSTEPPING THE LIMITS OF THE FRAMEWORK OF FAMILIAR CONCEPTS

§§

2-1 Introduction

WE ALREADY KNOW that overstepping the limits of the framework of familiar concepts is necessary, but details on this are needed. In particular, the content of this chapter will illustrate the facts that trusting our simple ideas too much makes us take for granted views that are actually erroneous; and that our great longing for descriptive accounts sometimes ends up in our inventing nonexistent beings. But before that a few pages should be devoted to a brief review of what, historically, made such changes necessary.

2-2 From Aristotle's Ontology to Descartes' Near Realism and Galilean Ontology

The participants in the great debates of the early seventeenth century had in common a robust belief in notions directly derived from our sense data. From Aristotle's followers this, of course, was to be expected. Aristotle himself trusted such data, as we saw, to a great extent. He (meritoriously!) took care not to let himself be carried away by dazzling ideas. He made a point of basing his views on serious grounds and, just like present-day scientists, he relied, to this end, on observed facts.

But it is worth noting that Galileo, his main critic, took up a rather similar standpoint. True, in his books—which are crammed with controversies—he sometimes produced arguments based on a different approach. He stated at various places that, thanks to mathematics, we a priori have the right concepts at our disposal, which made Alexandre Koyré claim that his theory of knowledge was that of a pure Platonist. And it must be granted that, when we read Galileo's assertion "This immense book [the Universe] [. . .] is written in a mathematical language," we may be tempted to give it a Platonist, or even a "Pythagorean" significance, of the type "numbers are the essence of things." But, on the other hand, Maurice

Clavelin (1968) appropriately pointed out that, in his two main contributions to science—inertia and the relativity of motion, Galileo crucially referred to experience (of inclined planes and moving ships, respectively). Clavelin also noted Galileo's rejection of arguments built up "as if the bodies started existing when we begin to observe them" (Letter to Mgr. Dini, 1611). And he drew attention to the presence, in Galileo's writings, of sentences such as "as for me, I rather think Nature first produced the things to its own liking and then created human reason" (*Dialogue II*), which are statements of an objectivist-realist kind. Clavelin, of course, also stressed the basic distinction Galileo introduced, as is well known, between the qualities later called primary and secondary, between the properties—such as color, scent, or savor—that the object shows but whose qualitative features come from us and those, essentially form, position, and state of motion, that the object cannot be thought of as not having. In view, especially, of this distinction it finally seems clear that Galileo developed a nondogmatic rationality consisting in considering that a genuine knowledge of what "really is" does emerge from observation, provided that some of its aspects, deemed secondary and, in a way, illusory, are ignored. His statement on the role of mathematics thereby takes up a simple meaning. It is to be understood as signifying that mathematics is necessary in order to set links between simple notions that are neither pure products of thought nor mere syntheses of our impressions, that we draw from our experience, and that do refer to reality. On the whole we may therefore speak of a genuine Galilean ontology. And it is to be noted that this ontology is more or less the conception still instinctively taken for granted by the majority of present-day scientists. It is just the one described above under the name "realism of the accidents."

We may discern at least the "skeleton" of the ontology in question in the views of the other "founding father of modern science," Descartes, with just the difference that Descartes was also a philosopher, and, as such, was bent on analyzing and rationally justifying his realism. The main lines of his reasoning are known. Its starting point is the Cogito. Then comes, on this basis, the ontological argument proving the existence of an infinite Being incapable of deceiving us. Next, the corollary "everything is true that we clearly know to be true" is made explicit. And from it it follows that the notions of form, size, and motion, being clear, are true, with the additional specification that, in the field of material things, they are the only ones about which that much may be claimed.[1] The differ-

---

[1] Let it however be noted that, on this, the first Cartesian thinkers were far from unanimously following Descartes to the very end. "To be absolutely sure" Malebranche wrote "that bodies exist, we must be shown, not only that a God exists who is not a liar, but also that He assured us He created some. And I do not find this in Mr. Descartes' works" (*Re-*

ences between the reasoning of these two authors are important but their messages are nevertheless very similar. Let them be quickly restated. For the qualitative, swarming Aristotelian ontology they both substitute an approach grounded on a small number of elementary concepts viewed as primary and on mathematics considered as a kind of structuring glue. However, as we noted right from the start, their authors did not break with Aristotelism so radically as is often claimed. Like Aristotle, they were philosophically objectivist, nay, even multitudinist, realists. According to them the concepts in question referred not just to human representations but to the multifarious reality of things.

On the other hand, an appreciable difference between the two messages, due presumably to the difference in their construction, should also be noted. It lies in that, contrary to Descartes' conception, Galileo's involves the principle of relativity of motion nowadays known under the name "Galilean relativity." Descartes' great metaphysical reasoning could conceivably guarantee that as far as form, size, and motion are concerned our judgment does not deceive us. But it could hardly lead to the discovery that motions are relative. Now it so happens that Galilean relativity is of utmost importance. It should be considered the forerunner of the Einsteinian relativity theory, which lies at the root of the enormous conceptual changes we know of. But it also brought in much change by itself, especially since it also affects position. In the classical mechanics developed by Newton and others on the basis of the ideas just considered it so happens that position is no more absolute than motion is. And as early as the beginning of the eighteenth century this led a person of such wide-ranging mind as Leibniz to suggest that space itself might be ideal. All this amply shows that Galilean relativity is in no way just a trivial fact of Nature and that it is only because we are very much accustomed to it that we consider it as such.

2-3 A Small Digression on Ontology

Is it possible to proceed further along such lines? In the first chapter we noted that present-day physics questions the possibility of a universal objectivist language (and we shall come back on this at great length). Clearly, most serious doubts are thereby cast on the validity of the notion of ontology and more generally on that of a "scientific realism" of any sort. And

---

cherche de la vérité, VIe *éclaircissement*). Note also that, curiously enough, by thus limiting the list of clear and distinct ideas concerning the physical realm to a few elementary ones Descartes deprived himself of the possibility of extracting from mathematics basic concepts liable to correspond to the deep structures of Reality.

we shall have to systematically investigate the nature of such a conceptual upheaval. But this is the proper place to note that, in view of the above, some thinkers question, not actually the magnitude, but the novelty and the suddenness of this change. They claim that already at the beginning of the seventeenth century, as an implicit consequence of the Galileo-Cartesian revolution, the notion of ontology had in fact to be set aside.

There are some arguments favoring this standpoint. As we noted, Galileo as well as Descartes (and then Newton and his followers) strictly limited the number of properties they considered intrinsic (the so called "primary qualities"), with the consequence that already in their time all the other properties, those that, in fact, "constitute" the tangible world, had to be seen as originating largely in us. In addition—and just as significantly—Galileo's approach parted with the great leading idea of Antiquity and the Middle Ages according to which the sole rationally acceptable research method consisted in proceeding from the universal to the particular; in first building up fully consistent ontological conceptions and then inferring from them, by means of purely logical methods, all the shimmering aspects of Nature. One may feel inclined to consider that Galileo substituted for this conception—"focused on stating the features intrinsically possessed by natural beings" (Bitbol 1998)—an objectivity centered on law and which was less aimed at accounting for phenomena according to the structure of an underlying reality than just at "identifying their law-like ordering and defining objectivity on such a basis" (Bitbol, *loc. cit.*). Adding Galilean relativity to this, one might thus be tempted to consider, contrary to our claim above, that it was indeed already at the stage of the Galileo-Cartesian revolution that physics discarded ontology.

This view was put forward, notably by Jean Petitot (1997). According to him, "the fact that [classical] mechanics describes a spatio-temporal reality (i) dependent on a relativity principle and (ii) reducible to entities the intrinsic properties of which boil down to mass, definitely disconnects its objectivity from any ontology whatsoever." Petitot grounded his claim essentially on point (i), that is, on Galilean relativity. In his view, the latter makes it manifest that both space and time are "desubjectized mental forms." On this basis he acknowledged that, admittedly, the "reality" that classical (Newtonian or relativistic) physics describes is independent, in the sense that measurements do not interfere with phenomena. But—he claimed—it cannot be independent of their spatio-temporal "shaping up."

Should we follow him that far? As for me, I am reluctant to do so. First, concerning facts, I observe that, while Galileo strictly limited the number of intrinsic properties, still he did keep a few of them. And that to those he kept he imparted both an essential role and a status radically different from that of mere human representations. That is to say, concerning them he took up, as already noted, the philosophical standpoint of a realist.

Second, concerning the concatenation of ideas, I claim that, here as elsewhere, mere indications should not be confused with proofs. It is true that in classical mechanics it is pointless to think of the center of mass of an isolated body as being endowed in some absolute sense with a definite position within a space also considered absolute. I willingly grant that, in this theory, only relative positions of bodies are scientifically meaningful. And that therefore it is technically legitimate to think of space as merely being a condition rendering experience possible, as being, in Petitot's words, a "desubjectized mental form." Nay, it can even be claimed that such an approach is a productive one in that it invites us to discard misleading preconceptions and focuses our attention on Galilean relativity which, through the Noether theorem,[2] reveals the primary role of some invariant quantities.

However, two quite basic points remain true. One of them is that within classical mechanics the *relative* positions (as well as the *relative* motions) of the material points (or objects) are normally said to be real, in the most commonsense (philosophers would say "naive") acceptation of the word. And the other one is that, correlatively, identifying space as a "desubjectized mental form" is a mental move that is in no way compulsory. If some realist Newton follower chooses to think of space as an ultimate arena existing per se and considers that every material point has in this space, at any time, a given position and a given velocity—depending on those of the others only through forces that decrease with distance—we are unable to prove him wrong. We cannot show him that this conception has implications that are at variance with observation. For indeed it has none. First of all, it is obvious that such ontological hypotheses do not in the least imply coming back to Aristotelian physics. But what here is most significant is that they are in no way *inconsistent* with Galilean relativity. For they are, of course, compatible with the three basic laws of Newtonian mechanics, of which indeed it is a consequence that *knowing* whether or not an isolated material point is at rest is impossible. There is no need to stress that, in the opinion of a convinced realist, this inability to ascertain such a fact is but a feature of our nature as human beings, and that claiming to derive from it consequences concerning the ultimate laws of the Universe is just nonsense.

In other words, it is true that Petitot's reasoning is both interesting and, to an appreciable extent, enlightening. It does shed light on the problem under study. On the other hand—as is often the case in the philosophical

---

[2] This far-reaching theorem of mathematical physics links invariance with conservation. Essentially it shows that whenever the "laws of forces" are invariant upon some transformation (translation, rotation, etc.) some physical quantity (momentum, orbital momentum, etc.) is, correspondingly, conserved.

realm—it does not reach so far as to convince an opponent that he is wrong. In view of the apparent plausibility of uncritical realism, to be persuaded such an opponent demands more. The mere impossibility of *proving* that space is real does not loosen his faith in realism.[3] For that end to be reached genuine "arguments against," that is, factual data hardly (or totally) incompatible with the realist standpoint, are needed. Now, in my view, only a critical analysis of twentieth-century physics is liable to produce such reasons. We shall see how.

2-4 A Gradual Overstepping

Compared to Cartesian near realism, the more flexible Galilean ontology already showed the beginnings of an evolution. Soon after it, there appeared, in physics, the notion of force and even that of forces at a distance, which as is well known immediately created perplexity. But at that time the notion in question seemed to merely be a property of the objects. The existence, at such and such a place, of an object gives rise at a distance (so it was claimed) to a force. If the object gets removed the force therefore just disappears. For this notion the nineteenth century substituted in part the one of fields, conceived of as autonomous elements of reality. The electromagnetic field may exist independently of electric "sources," that is, so to say, "in vacuum" (a "vacuum" that, as a consequence, is no longer "void").

That much already constituted a significant overstepping of the "familiar concepts" framework. And it was pointed out above that one would be at pains to find one of comparable importance in any science other than physics (with the exception, of course, of mathematics). Admittedly these other sciences teem with complex notions, but these are secondary ones. They were constructed by combining simpler notions, ultimately derived from the realism of the accident. Also to be noted is the fact that the fields in question still preserve some aspects of the "world of things." Sometimes they are said to be comparable to gelatins of some sort, and such an image is not altogether misleading since, at any time, these fields have at least, at any point in space, a well defined (scalar, vectorial, or tensorial) value. In this respect they stand in sharp contrast with a notion that appeared later, that of the wave function of a system of particles (that

---

[3] Nothing indeed is more surprising than the difference between the spontaneous, intuitive standpoints of philosophers and the general public in this respect. While the idea that space is but a "mental form" sounds fantastically audacious and, in fact, practically unbelievable to all "laymen" (scientists included), philosophers consider it to be quite a natural possibility. So natural that, when they critically analyze Kant's conception, it is one of the elements of it that they hardly consider worth discussing.

of an atom, for example). For a wave function is, in fact, a function of as many x,y,z coordinate triplets as there are particles in the system, so that to speak of the value it "has" at a given point would just simply be meaningless. The "world of things" here truly fades out.

But, some time before the advent of quantum mechanics (to which the wave function notion belongs) relativity theory or, to be more exact, the two relativity theories, the "special" and the "general," appeared. And it may be asserted that it was within the realm of these theories that "universal thing-ism" received the first decisive blow. But on the other hand, it goes without saying that such a "thing-ism" is but a most naive version of realism. So that when all is said and done it may be considered that realism—even in its "objectivist" version—emerged as good as unhurt from the relativist revolution. The reason is that the relativity theories are, in their essence, in no way "antirealist." Both of them may be given a realist interpretation. In them, admittedly, the "real" entities are not objects any more. They are (as already noted) "events" (and, in the second one "geometries"). The point is that, even though, in these theories, many statements get a meaning only through reference to "the observers," events and geometries can be thought of independently of such references. This is why it was claimed above that, on the whole, "realism of the events" is a refined version of the realism of the accidents.

In particular, it should be kept in mind that between Newtonian mechanics and relativity theory there is, somehow, continuity, and that, within the latter, things and events are closely related (since the former are just sequences of the latter). In view of these facts, it should be clear that adopting relativity theory remains, in itself, compatible with the aim of building up a description of phenomena grounded on the hypothesis that things exist and are describable by means of realist concepts. However, such a conception (or "world view") meets with difficulties coming from a very different source and that are most considerable. The next section is an introduction to them.

2-5 Trajectories and Misleading "Pieces of Evidence"

In the debates for and against realism what, within the scientific community, long turned the scales in favor of (physical or objectivist, or etc.) realism was the fact that explaining visible, complex features by means of invisible simple ones was generally successful. Here "simple" means "describable by means of clear, distinct ideas." So that it is—still now— quite often thought (and even considered obvious!) that assuming that the objects theories label by names really exist can only be a help in research. Along these lines some epistemologists consider, for instance, that to claim that any electron exists by itself—with such and such known or unknown individual properties—still is the best way we have of understanding phenomena involving electrons.

It is quite important to know that this is not in the least true, that, systematized in this way, such a view not only does not help at all but is even quite likely to mislead us. Thus, for example, the idea that each one of the electrons in an atom is individually in one definite quantum state (lies on one definite "orbit") is just simply erroneous. (According to the only operationally nonmisleading picture we have, every one of them lies simultaneously on all the "allowed" orbits.) In other words, there are situations in which the vocabulary we use—and in particular such words as "electron," "particle," and so on—is suggestive of "pieces of evidence" that are, finally, but erroneous ones.

Now, what is more, there even exist situations in which such misleading pieces of evidence do not originate in a defective vocabulary but are apparently straightforward interpretations of what we plainly see. Let us consider the already mentioned example of a bubble chamber and the traces that appear in it. Essentially, a bubble chamber is just a box with transparent walls containing a liquid in which a gas has been dissolved. Just as is the case in fizzy mineral water bottles, the gas in question has a tendency to form bubbles. Let us then assume that some small electrically charged object passes through the chamber. It will excite the atoms lying near its trajectory and it is in the vicinity of these atoms that bubbles will form. A whole (visible) string of small bubbles will thereby be created and we shall therefore see a trace of the trajectory of the object.

This being the case, let us examine the way in which such a bubble chamber is made use of in research bearing, for example, on cosmic rays. We know that cosmic radiation comes from interstellar space, so that it falls more or less vertically on the Earth. To study it we set a bubble chamber in working order and wait. We soon notice that more or less vertical traces do indeed appear in it. And we then say: "Now we know what cosmic radiation is. It is of a corpuscular nature. It consists of, so to speak, a "rain" of small corpuscles falling here and there." And we add "This experiment obviously shows that, before any such corpuscle entered the chamber it already had a definite trajectory, which we could not see because of the lack of appropriate instruments but the continuation of which generated the trace we observe within the chamber." We have the feeling that sheer commonsense dictates such conclusions. We think that everything takes place just as when, seeing a white track emerging from a cloud in a blue sky, we say that a jet plane is there that, before it became visible, flew inside the cloud on a certain trajectory we could not see.

Now, believe it or not, this explanation, trivially obvious as it seems to be, is not the appropriate one. It suffers from the same defect as the one that ascribes the alternation of day and night to an assumed rotation of the Sun around the Earth. Just like that one, it relies on a theory—classical mechanics in the present case—that, while qualitatively accounting for the fact at hand, still has the ruling-out defect of being at variance with very many other facts. It must therefore be set aside—just as the geocentric theory was set aside—and replaced by a theory that, in the field under study, does account for *all* of the facts. Such a theory exists. It is quantum mechanics. And according to quantum mechanics there are just simply no trajectories.

How is that? No trajectories! But look here—people will exclaim—what about the observed traces? Isn't the claim that trajectories do not exist blatantly at odds with the fact that, undoubtedly, we see traces? No, it is not. Quantum mechanics does indeed yield an explanation of the latter. And the explanation (Mott 1929) is extremely simple, even though it is surprising.

To grasp its nature, we must remember that at the root of quantum mechanics there are most general predictive rules informing us about what will be observed. The wording of these rules involves terms ("particles," "waves," etc.) that reflect human experience. But, for reasons the next section will make clear, care is taken not to make use in them of too figurative words and, in particular, of the terms "corpuscular" and wavelike" when radiation is referred to. When some radiation (cosmic or otherwise) interacts with an atomic object (here one atom of the liquid) there is a certain probability, whose value quantum mechanics yields, that this object will get excited (more precisely, will appear as being excited). And this readily causes bubbles to appear, as already noted, in the vicinity of this object.

Now, let us consider the question: "What is the probability that two atoms of the liquid be simultaneously excited (still in the above-stated sense) by the radiation under study?" Here again, the quantum mechanical formulas yield the answer. The said probability is extremely small whenever the two atoms are not (not even approximately) aligned in the direction along which the radiation is propagated (here the vertical one). And it is quite appreciable when they are. This result is easily generalized to three, four, . . . , N atoms. Consequently, even in cases in which there is, on the whole, a large probability that many atoms are excited, the probability that those that are should not be aligned is very small. In other words, in an overwhelming majority of cases the simultaneously excited atoms will lie (in our example) along a vertical line and therefore a vertical trace will be observed.

As we see, the theory correctly accounts for the observed phenomena, and it does so without referring in any way to the "classical" view that the incident particle has at any time a well-defined position and thus travels along a definite trajectory.

So, in the microscopic domain[4] our attempt at analyzing observation by means of "clear and distinct ideas" (in the Cartesian sense) turned out to be a failure. What we think such simple reasoning obviously shows, it, in fact, does *not* show at all. Obviously therefore, relying on the use of commonsense concepts and modes of arguing is in this field very risky.

2-6 On the Existence or Nonexistence of Hidden Things: Particles and Dirac's Sea

It may have sounded odd that, in the example of cosmic radiation just given, use of the word "corpuscle" was warned against. It is true that—like everything that has to do with quantum physics, light for example—this radiation does show corpuscular aspects. In the cosmic radiation case the latter are even those that manifest themselves in the most conspicuous way. But, quite generally, it is not without good reasons that in such contexts the uncompromising word "aspect" is, as a rule, made use of. In fact, in order to get rid of difficulties well known to experts in the field, this word must even be taken in the weak sense "aspects for us." The reason is that, in this way, no implicit assumption concerning the existence of *properties* (the property of *being* a particle, the property of *being* a wave, and so on) is actually made. For example, if we merely refer to the "corpuscular aspects" of, say, light, the risk is small that what we say will be thought to imply that a photon possesses a trajectory (and the same can be said concerning all types of particles). Now, this is appropriate since, in fact, such a notion nowhere appears in the theory[5] and

---

[4] In physics, objects and phenomena of subatomic, atomic, or molecular size are said to be "microscopic." Larger ones are termed "macroscopic."

[5] Even though some quantum mechanical computation rules involve a notion of "path" that superficially resembles it. These rules are closely linked with the already mentioned Feynman theory. As regards prediction of observational outcomes, this theory is, as we know, equivalent to quantum field theory, even though it involves entirely different computing procedures, the principle of which is to be described in sections 2-7 and 9-6. Here let it just be noted that it parts just as much from familiar concepts since it introduces the notion of particles being propagated toward the past. Some elements of the formulas it uses have received names with a realist flavor ("virtual particle," for example) but we shall see quite soon that this is but an appearance. And, more generally, we shall find (section 9-6) that a realist interpretation of this theory would meet with the same obstacles as those encountered

assuming that a photon necessarily has a trajectory would render explanations of such trivial facts as the formation of fringes in a Young-type experiment very intricate and clumsy. To this, the fact that, contrary to photons, the cosmic radiation quanta have a rest mass (they are protons) changes essentially nothing. It still is quantum mechanics—that is, the use of wave functions, operators, and so on—that makes it possible to quantitatively investigate the interactions of such quanta, be they termed "particles" or not. And, to repeat, it so happens that the notion of real, continuous trajectories is not compatible with such a treatment. This is the point that should be kept in mind.

But—it might be objected—the foregoing reasoning rests on a philosophical conception according to which the hard core of the theory is the set of its observational predictions. Could we not switch to a theory that would describe what *takes place*, and would thereby restore the trajectory notion? To put it bluntly: is it not because physicists are bent on speaking, not of being but just of knowledge, that, concerning particles, they are led to discard the notion of trajectories?

It must be granted that, for a time, this remark had some truth in it. It would not be utterly wrong to claim that within the so-called "Copenhagen School" the notion of trajectories was partly set aside for this reason (more precisely on the grounds that what cannot be known is meaningless). But today we know more. For indeed some physicists quite explicitly made a point of overstepping the limits of just the prediction of observations, and some did, in fact, succeed in building up ontologically interpretable theories. Louis de Broglie (1928), in particular, did put forward a theory, rediscovered later by David Bohm (1952), that essentially recovered the quantum mechanical predictions while being expressed in an objectivist language. And in some such theories (this one included) the notion of trajectories does admittedly reappear. For example, in some of Bohm's articles[6] quite instructive figures may be found, quantitatively showing that guidance of the particles by the wave results in fringes being formed on the screen.[7] On the other hand, more complex phenomena may also be considered. Imagine, for example, that, within the emitting device, two particles are, each time, emitted together by an atomic source. And suppose that, each time, only one of the two enters the rest of the experimental arrangement and makes an impact on the screen. The formulas of the theory in question then show that the trajectory of the latter depends in a crucial

---

by ordinary quantum mechanics (the principles of which it espouses). The so-called "superstring theory" is an extension of the theory in question.

   [6] See, e.g., Bohm and Hiley (1993).

   [7] The trajectory of every particle is calculated from theory. It is then observed that, under the action of the wave, these trajectories come to group in such a way that, on the screen, fringes are formed.

way on what happens to the other particle, however far away it happens to be then. And that, because of this, fringes do not appear.[8] This is just one of the numerous consequences of nonlocality, getting rid of which by changing theories is impossible, as Bell's theorem will show us. Finally therefore, even within the theoretical models that a priori seem best suited for accommodating the notion of trajectories, the latter is somehow immersed within a Big Whole in which it well-nigh loses significance.

In view of all this, a judgment such as "the theory does not claim there are no trajectories, it merely states it does not deal with them" must be, at the very least, seriously qualified. "Normally" the theory does not deal with trajectories; this is quite true. But when questioned on the subject it does not keep silent; far from it! It has ready answers and these, as we just saw, are rather baffling.

That much being said, the outcomes of the foregoing section and the comments just made thereon invite us to widen the scope of our query. What the example just considered made us question was the idea that it is useful—or even just appropriate—to think of particles, or objects in general, as having at all times a whole set of contingent properties such as position, velocity, etc., attached to them. But a priori we might well wonder whether we should go so far as to extend such doubts to the very concept of existence, to the notion of the existence of particles and, more generally, objects. We might reflect that even if it is true that the objects in question cannot be thought of as possessing contingent *properties*, still they should be considered as, at least, having an individual permanent *existence*.

There is no denying the great—at least psychological—force of the arguments that speak in favor of the latter view. Hardly questionable is the experience we have that things do last, and their constituents even more. That their ultimate constituents should have permanent existence may thus seem obvious. Moreover, this view is akin to the notion of invariance, which has such a considerable role in science. On the other hand, the unquestionable experimental fact that, within high-energy events, particle creation is observed cannot be ignored either. One of the main successes of Dirac's theory was that it managed to reconcile these two apparently contradictory requirements (of course, its other, even more impressive, success was that it *predicted* such creations!). In fact it does not account for all types of particle "creation" but merely for that of fermions.[9] One of

---

[8] But similar, though more complex, effects can be observed. They involve both particles at once, even though they are quite distant from one another.

[9] The particles that, concerning the constitution of what we use to call "matter," are the most significant ones—quarks, electrons, etc.—have features in common in view of which they are collectively called "fermions." The other ones (photons, pions, etc.) are called "bosons."

its basic (and experimentally verified) feature is that fermions are always "created" in pairs, that is—more precisely—together with an "antifermion" of their own species (an electron together with a positron etc.).[10] In its broad lines it consists in assuming the existence of a whole "sea" of invisible and nonlocalized fermions, and in considering the "creation" of such a pair not to be genuinely real, to merely be an appearance, due to one of these fermions suddenly emerging from the said "sea." For such an occurrence creates a "hole" in the "sea" and this hole, which of course shows features (electric charge, etc.) opposite to those of the fermion that "jumped out," is nothing else than the corresponding antifermion.

As we see, this (venerable) theory does not undermine the notion of everlasting existence of each individual fermion. But it preserves this notion at a "conceptual price" that may be termed exorbitant. For indeed it is at least as difficult to "really believe" in the existence of the Dirac sea as to give credence to any of the most fanciful stories mythology ever spread around. This is one of the reasons that, in the theoretical physicist's "toolbox," the Dirac theory was soon replaced, for most purposes, by quantum field theory. The latter theory is quite appreciably more general (it also applies to bosons, it readily accounts for the lack of individuality of the particles, etc.) and, being more abstract, it demands less from sheer "credulity." With it, the mind does not have to struggle with such a disconcerting picture as that of the Dirac sea (still today, however, physicists have no qualms about making use of Dirac's theory whenever they deem it suitable for their calculating purposes).

With regard to the question at hand, that of the existence of particles as lasting entities, this shifting from Dirac's theory to quantum field theory constitutes quite a crucial change. For indeed the glimpse we took at the latter theory in chapter 1 showed us that, contrary to Dirac's, it identifies the existence of a particle as just a state, that is, when all is said and done, merely a property. Hence, concerning particles, there is according to quantum field theory *no* conceptual separation to be drawn between the two notions of property and existence. We already observed that such a standpoint correctly accounts for the "creation," upon collisions, of particles (bosons included) that did not previously exist, that, indeed, did not then exist any more than is the case concerning the position of a cosmic particle before it becomes manifest through a trace in the bubble chamber.

Admittedly, this conceptual jump is one quite difficult to make, especially for persons for whom the postulate that every particle enjoys individual existence is something like a necessary truth. But the latter will have to duly take into account a most serious mismatch between this

[10] As, in substance, Dirac predicted, as we know.

postulate and the experimental data, which, quite on the contrary, demonstrate a complete lack of individuality of particles of the same species lying in the same quantum state.[11] And they will have to note that, conversely, quantum field theory automatically accounts for this lack of individuality since, in it, physical states are characterized by just the *number of particles* they involve. The physicists rightly see in this a confirmation of the pertinence of the theory in question and we, who here are interested in concepts, must as well see in the mentioned facts a testimony to the truth of the views that these facts corroborate.

As we saw, the said views prevent us from considering that a particle exists per se. The only "thing" to which, in view of the foregoing, we might think of attributing such an existence is, as previously noted, the "something"—not multifarious but as good as unique—that changes state when particles get "created."[12] And even that "something" shows features rendering such an interpretation somewhat risky. Under such conditions it is but natural that most physicists are reluctant to engage in "existential" speculations of such a kind and choose to limit themselves to the task of finding formulas making it possible to predict what will be observed. Indeed, it will appear from what follows that such a standpoint is not devoid of some wisdom.

Still, it is quite understandable that it should be felt as a kind of renunciation and, as we already know, as soon as quantum mechanics appeared on the scene a minority of physicists expressed, on that behalf, a deep dissatisfaction with it. One of the most critical was Einstein. What Einstein expected from physics, as noted above, was that it should describe Reality as it actually is. In view of the nature of the quantum mechanical axioms he could not but notice that quantum mechanics does not genuinely aim at such a result. And he found it wanting for this reason. It is true that he himself did not put forward an alternative theory. But, as we saw, other physicists did. Unfortunately, the conceptual clarity of these theories is very much counterbalanced by quite serious inconveniences. In the Broglie-Bohm theory, for example (called "the pilot-wave theory" when taken in de Broglie's version), accounting for the production of particles in collision phenomena proves extremely difficult. Indeed, it requires as much as giving up the view that the ultimate constituents of Reality are of the nature of particles. A similar remark is in order concerning the related question of quantum indistinguishability and its observ-

---

[11] Referred to here are the experimental validations of quantum statistical mechanics, which show that Boltzmann statistics is invalid and must be replaced by either Bose-Einstein or Fermi-Dirac statistics (according to whether the particles involved are bosons or fermions).

[12] Its technical name is "the state vector within Fock's space."

able effects. In view of the fact that these theories originated in requirements of a conceptual rather than "strictly physical" nature, such a dramatic change in the concepts they make use of is, with them, particularly disturbing. The fact that these changes are necessary therefore seriously jeopardizes the credibility of the theories in question and gives the impression that, after all, they are just ad hoc and artificial.

The example of the traces as well as the considerations developed in this section were put forward in order to convey a first and rough apprehension of a point that will appear much more clearly and distinctly in later chapters of the book. The point in question is that bringing attempts at finding realist descriptions of the quantum world to a successful conclusion is considerably more difficult than would have been expected at first sight. Farther on we shall discover in this respect that, due to insufficient knowledge of contemporary physics, many epistemologists delude themselves concerning the usefulness of the notion of "hidden things."

2-7 A "Fabricated" Ontology

Among physicists such delusions hardly appear, not at any rate, among those of them who elected the basic physical theories as their special field of study. For these people do know that, in view of the experimental data nowadays at our disposal, such conceptions as near realism and realism of the accidents are quite strictly untenable. Besides, we already noted that their main objective is to construct widely applicable mathematical models, adequately predicting all the data that the extreme acuteness of contemporary instruments of observation makes it possible to verify. And we also saw that, as a counterpart, they generally leave such constructions quite wide "open" in so far as the nature of the involved concepts is concerned. In fact, this holds true to such an extent that, quite often, even a "practitioner"—trained in the art of apprehending the concatenation of equations and deducing the implications they may have—is at a loss when he tries to disentangle what, in these theories, refers to genuine ideas from what is just a description of clever methods of calculating. To tell the truth, reading the corresponding texts generally leads to the conclusion that their real subject is the description of methods (one of the words most often used in them), that the—numerous and brilliant—ideas they put forward almost always concern calculation procedures, and that, in them, even the words that seem to refer to concepts in fact merely serve to designate some recurrent elements of these methods.

This state of affairs has lasted for at least two generations and has had a curious and somewhat disconcerting effect. Within the physicists' community it generated something of the kind of a pseudo-ontology. This is not the proper place to enter into quantitative details. Let it just be mentioned that this sham ontology spontaneously emerged on the basis of the findings of just one (but outstandingly brilliant!) physicist, the theorist Richard Feynman. Outwardly, the formalism Feynman elaborated—and on which more will be said in chapter 9—quite appreciably differs from the quantum field theory one. In it, for example, there is no question any more of a strange "something" whose state changes account for apparent particle "creation." But the two theories merely predict observational results, and they are quite strictly equivalent in this respect. The advantage of the former is of a purely computational nature. However, this advantage is enormous, due to the fact that, in Feynman's formalism, most complex calculations may be pictured visually. More precisely, these calculations involve long mathematical formulas that are essentially products of several complex factors. But it so happens that the way the latter combine with one another may be represented by means of graphs, constituted of straight line segments, each one of which corresponds to one of the said factors. Feynman had the simple but most effective idea of mentally associating with each such segment the motion of a particle that either is identified with one of those actually taking part in the process under study or is just a fictitious one. For example, if the graph has the form of the letter H the "horizontal" segment in the middle, which has no free ends, corresponds in the formula to a factor the nature of which is that of a denominator. It is mentally associated with a fictitious particle called "virtual." And as for the other segments, those with free ends, they are mentally associated with the two particles whose "collision" the calculation is meant to study.

This procedure, which amounts to replacing a long and abstract formula by a picture, summons up in view of the planned operation the (considerable) visualization abilities of the brain. It thereby makes the calculations easier to such an extent that it enables physicists to perform highly complex ones that could never have been done without this help. Under such conditions, it is not surprising that the physicists in question should, on this basis, have invented something akin to a pseudo-ontology. It consists in, for example, interpreting the H diagram by saying: "One of the incident particles (one 'leg' of the H) proceeds to a certain point, where it emits a virtual particle (the horizontal segment) and continues on its way, while the other one (the other 'leg') absorbs the virtual particle." Note that the physicist who says this does not mean it completely literally (no more than we do ourselves when we say the Sun "rises" or "sets"). If you press him a little on this point he will readily concede that, to some extent, this is but a "way of speaking." He will add that what the graph yields is not an itinerary, or even the probability of such and such an itinerary, that it just represents a piece of information related to the event the formula deals with. He will stress, moreover, that the information in question is not the probability of the event, as a realist would have it, but a "probability amplitude." And if you press him more he will grant that this is a typically quantum notion, the interpretation of which is ticklish. But all the same, the half-allegorical talk the physicists resort to when making use of Feynman graphs is so convenient and efficient that we always feel tempted to raise the words appearing in it to the level of names designating things. I mean, we have to force ourselves not to consider the various segments in such a graph to be actual trajectories of genuine particles.

Such is the fabricated ontology this section deals with. Knowing that virtual particles are but symbols for denominators present in formulas yielding calculation recipes, we may deplore that thinkers of great ability devoted so much effort to the task of understanding to what "category of beings" virtual particles belong. But it must be said in their defense that the physicists' current language could not but incite them to quite seriously consider this question.

2-8 Indications for What Follows

The emergence, in physics, of the fabricated ontology described above forcefully illustrates the kind of tension to which the mind of a physicist interested in the analysis of ideas readily gets subjected. On the one hand, physics is an empirical science. We think of it as essentially being a synthesis of our communicable experience. But, on the other hand we also are naturally inclined toward physical, or even ontological, realism, with the consequence that, at least at some times in our existence, we feel somehow compelled to *interpret* the said synthesis, to consider it as being a description of a reality that is both independent of us and knowable by us (since, by assumption, we describe it). But alas! Nowadays the very outcomes of physics make such a standpoint almost untenable. Consequently, some present-day physicists set to the fore the view that their science is a description of experience and push very much into the background the idea that it describes "reality out there," while others, more impressed by the "obviousness of realism," take up the opposite viewpoint in spite of the difficulties it implies, as we found out. The purpose of the present section is to show that in fact our "reasoning framework" depends very much on which one of these two approaches we choose to take up. Three of the thereby generated differences are especially to be noted because of the major role they will have in what follows.

The first one has to do with *objectivity*. The conditions a statement must obey in order to be objective are, within the realm of the realist approach, much more stringent than they are in the other one. This point is developed in section 4-1-1.

The second difference concerns *explanation*. In the realist conception an explanation normally refers to the existence of some objects (in a broad sense of the word, meaning particles, fields, and so on) and the properties of the said objects. And the explanation is considered suitable even if, as is usually the case, it makes use of counterfactuality. In particular, when it refers to properties possessed by unobserved objects (inferred by means of known laws from those observed in other circumstances, or just postulated). Clearly, a supporter of the alternative standpoint (that physics is but a summary of human experience) has less justification for building up explanations along such lines. This does not mean that the said alternative standpoint is incompatible with explanations of any form. In it, genuine explanations are indeed given. But they get built up by referring to *general laws*. Within this realm, to explain is tantamount to showing that the phenomenon to be explained takes place in conformity with such laws.

The third difference has to do with the quantum mechanical *principle of completeness*. This principle, forcefully claimed to hold true by the discoverers of quantum mechanics, stipulates that this theory is a complete description of reality. Upholders of the "realist" interpretation of physics make the said principle explicit by stating that the wave function (alias the "state vector") of a quantum object incorporates all the numbers that, according to physical realism, correspond to the structure and dynamical properties of the object. In other words, according to them the principle states that supplementary (often called "hidden") variables do not exist. This is what may be termed the "strong version" of completeness or, for short, just "strong completeness." Clearly, upholders of the conception that physics is but a synthetic description of our knowledge cannot, logically, endorse such a formulation. The reason is that, according to this conception, to positively assert that some entity assumed strictly "unreachable" does not physically exist is just as open to criticism as to claim that the entity in question physically exists. However, here again this does not force them to give up any notion of (quantum) completeness. But they must define it differently. In fact, the physicist Henry Stapp gave the clearest definition along these lines. According to him completeness is then just the assumption that "no theoretical construction can yield experimentally verifiable predictions about atomic phenomena that cannot be extracted from a quantum theoretical description" (Stapp 1972). This may be called "weak completeness." Note that weak completeness makes it possible to keep—if deemed advisable—the "hard core" of the completeness notion without taking a priori a stand concerning the existence or nonexistence of supplementary variables.[13]

This approach—which is fully in accordance with the "open realism" standpoint—is the one that we shall take up when examining quantum mechanics. In a spirit of "enlightened agnosticism" we shall therefore refrain from barring the possibilities (i) that the structure of Reality can be imagined and (ii) that it involves parameters (or "variables") not present within the quantum mechanical formalism (hence "supplementary" to it). This means that, concerning the models involving such parameters, our position will be twofold. On the one side, we shall willingly grant that, even if their empirical content is vanishingly poor, those among them that have no false consequences *may* be correct; that they are "compatible with truth." But on the other hand, by virtue of the weak completeness principle, we shall consider that no physical experiment will ever make it possible to know whether they are true or not. At first sight, such impossibility may seem to imply that, for analyzing the conceptual implications of quantum mechanics, the said standpoint is utterly unprofitable. But we shall see in the course of the following chapters that, on the contrary, it is, in some instances, enlightening.

---

[13] The "crypto-ontological" nature of the strong completeness principle is well illustrated by the fact that in some experiments (for example, those concerning distant correlations, described in the next chapter) it would entail without any more ado the existence of supraluminal action at a distance. Such a facile disproving of a basic physical law strongly suggests that strong completeness should not be raised to the status of a principle a priori.

CHAPTER 3

NONSEPARABILITY AND BELL'S THEOREM

§ะ

3-1 Correlation at-a-Distance: Bell's Theorem

WE SAW ALREADY that quantum field theory casts serious doubts on the multitudinist conception of Nature. We noted that, contrary to the latter, it strongly suggests a kind of holism. However, as we also saw, this is but an indication. The existence of two other alternative formulations may not be ignored. Hence, it is now time that we should take into account facts that, within a realist approach, render holism independent, not only of quantum field theory but indeed of any theory whatsoever. Obviously, this point is of considerable philosophical importance and fully deserves our attention.

The facts in question are essentially correlation phenomena. To begin with, let us therefore consider the correlation notion in detail. Let us first have a look at a simple example. Suppose every time the telephone rings at my home it also rings at my neighbor's. If I observed such a phenomenon I should certainly try to find out what its cause is. And—which is significant for what follows!—I would not be prompted into this query by just the pragmatic desire to cure a problem. Even if it did not bother me in the least, still, intellectually I would feel, somehow, the need to find out what explains the correlation. Sure enough, in practice I might rely on induction. I might say to myself that something that already happened many times is likely to happen anew. But I would not rest content with just simply this. I could not consider that such an "observation predicting rule" constitutes, by itself, the explanation. When faced with some steadily reoccurring correlation phenomenon the human mind a priori longs for its actual causes to be found. Of course, it does not assume that, necessarily, one of the observed phenomena is the cause of the other one. In a great many types of correlation-at-a-distance phenomena this, obviously, is not what happens. But in such cases we inquire for a cause common to both and rest content only when the latter is found.

In physics—be it classical or quantum—such correlation phenomena are quite common and in an overwhelming majority of cases they are satisfactorily explained this way: It is immediately observed that events of which two sequences exhibit a correlation have a common cause at their source. In view of the analyses of ideas that will follow, we shall find it convenient to have a very simple model of events of such a kind at our disposal. First, let us think of a pair of darts such as used in the well-known game. Let gravity be ignored and let us assume that, initially, these two darts are lying side by side—at a place that we call "the source"—and point in the same direction. Suppose now that, due to the triggering of some tiny spring placed between them, the two darts get separated, one of them moving to the left and the other one to the right, both of them keeping their common, initial orientation. Finally, assume that initially, at the source, instead of just one pair there are a great many such pairs, oriented in all possible directions, all of which "burst," one after the other, in the just described manner.

Under such conditions a strict correlation-at-a-distance obviously exists between the directions in which the two darts in one and the same pair are pointing, since these directions coincide. Such correlation could easily be observed by means of measuring instruments disposed symmetrically with respect to the source at some distance from it, and that would detect, first the orientations of the two components of the first pair, then those of the components of the second one, and so on. It is even true that, for securing a strict correlation between registered results, instruments detecting the precise orientation of a dart are not strictly needed. It suffices to use instruments every specimen of which defines a direction in space and detects the sign of the projection on the latter of the orientation of a dart. If the left and right-hand-side instruments are oriented in the same direction it is clear that the pairs of registered signs corresponding to one and the same pair will always be either two + or two − signs. And, consequently, the statistical correlation registered on the ensemble of all the pairs will be a strict, positive one (never any + − nor any − +). As for the cause of the correlation in question it obviously consists in what takes place at the source, in the fact that, initially, the two darts composing one pair are parallel to one another.

For reasons we shall see, investigations were recently made bearing on some correlation phenomena that look similar to these but involve photons in lieu of darts, the two photons of a pair being created together upon downward transition of an atom from an excited state to its ground state. The measurements bore on their "polarization vector" (which had there a role roughly similar to that of the orientations of the darts).[1] The

---

[1] In order to understand the reasoning set out here no real knowledge of what "polarization" is is actually necessary. It is enough to note that a polarization measurement along any given axis (perpendicular to the propagation direction) may only yield one out of two results, conventionally labeled + and − (associated with a linear polarization respectively parallel or perpendicular to the axis in question). Incidentally, for the same purpose there is also no need to actually know what a wave function is.

experiments were independently carried out by several teams around the world,[2] but, concerning overall reliability (statistical preciseness and so on), the first really crucial ones were those made by the Alain Aspect group, at the Orsay Optical Institute (Aspect et al. 1982a, b]. For brevity the set of all such experiments will therefore be designated in what follows by the generic name "Aspect-type experiments."

In these experiments the situation is qualitatively much the same as in the imaginary one concerning darts. In them also, the photons originate in pairs from a common source (the emitting atom), so that the quest for a cause of the observed correlation between polarization measurement results is quite naturally directed toward what takes place at this source. On the other hand, since the involved systems are photons it is natural to look for an explanation grounded in our knowledge concerning photons. Hence, since the latter are "quantum systems," we must consider the notions that, in quantum mechanics, have the basic role insofar as observational prediction is concerned, that is, the "wave functions" of the systems.[3] Here, however, comes a surprise. In the example with the darts, two darts were initially prepared together in a special way at the source (so as to constitute one of the above-considered pairs). But this mode of preparation, of course, did not prevent them from having, at all times, individual properties, a knowledge of which, if possessed, would make it possible to predict the results of future observations on these objects. In particular, each one of them had a definite orientation (though it was the same for both) and it was this orientation that caused the result (+ or −) recorded by the corresponding instrument. On the contrary, in the example of two photons coming from the same atom and "correlated" it is not true that each one of them is describable, in principle, by a wave function of its own, that would induce the + or the − result. In such configurations, what is the case is that, mathematically, such individual wave functions simply do not exist. No other wave function than the one of the pair can be introduced. In such situations the pair wave function is said to be *nonseparable* or, alternatively, one speaks of an *entanglement* of particle wave functions.

Such a state of affairs in no way undermines the efficiency of quantum mechanics as regards its use for mathematically predicting what correlation phenomena (between polarization measurement outcomes, say) are to be observed on such photon pairs by means of appropriate instruments. (Just as in the dart case, these instruments are positioned far away both from the source and from one another.) Indeed, for finding out what will

---

[2] Particularly noteworthy, in this connection, are the pioneering works of Clauser (Freedman and Clauser 1972) and Fry (Fry and Thompson 1976).

[3] Most fortunately the technical subtleties that, to some extent, distinguish photons from other quantum particles may here be ignored.

then be observed we even have a choice between two computing methods, one of them descriptive and the other one predictive.

Suppose that the right-hand-side instrument is nearer the source than the left-hand-side one, so that the right-hand-side photon interacts with it (thus undergoing measurement) before the left-hand-side one arrives on the left-hand-side instrument. The descriptive method then consists in making use of a certain computational rule of the formalism stating that, as soon as this measurement takes place, the wave function of the pair must be "reduced," with the result that the left-hand-side photon then *gets* an individual, well-specified wave function. This wave function yields the probabilities of occurrence of the various possible outcomes of measurements bearing on the said photon, conditioned on the outcome of the right-hand-side one. The correlation predictions thus obtained agree entirely with the results of the Aspect-type experiments. And note that such an agreement does not merely hold good when the two axes along which the polarization gets measured are parallel, as was the case in the imagined dart experiment. It holds true generally, that is, whatever the relative orientation of these axes happens to be, and quantitatively, even though the correlation is then no longer a strict one, in general.

The just sketched computational procedure may be termed "descriptive" or "realist," for it is possible to mentally form an image of some elements of reality underlying it. This image consists in considering that the wave functions are "real stuff," so that their "reductions" are also fully real phenomena. Within such a conception, before the advent of the first measurement, nonseparability is itself, obviously, a real attribute of the physical system constituted of our two photons. However, this description is quite disconcerting in that, as we just saw, the measurement performed on the right-hand-side photon (at, say, point A) has an immediate long-distance effect. This is just the fact that the left-hand-side photon, then distant from A, gets a wave function—hence an element of reality— that it did not previously have (and that will determine the result of the measurement it will undergo or at least the probability thereof). Such a view is at variance with the basic rule of special relativity that no influence travels faster than light.

The other method at our disposal consists in not introducing at any stage any wave function other than just the one of the pair. To the mathematical expression of this wave function some calculation rules are directly applied that, according to the formalism, yield the joint probability of observing a given pair of results on the left and right instruments, and this for any pair of orientations of the latter that one may choose. This calculation thus yields correlation predictions and (fortunately!) the latter are identical to those obtained by means of the descriptive method (a confirmation of the fact that quantum mechanics is a consistent theory).

Note that, in this method, no faster-than-light influence explicitly appears. But this circumstance is tightly linked with the fact that the method in question is purely *predictive* and lends itself to no interpretative picture.[4] As we just saw, it just consists in applying a set of observational predictive rules (the quantum mechanical ones) concerning which it has been found that up to now they correctly predicted what was observed. In other words, it is grounded on just induction, quite apart from any reference (not even implicit) to a realist interpretation liable to support the latter and make it somewhat plausible. At first sight, it may therefore be wondered whether it yields a genuine explanation of the observed correlation. In fact, this may be considered a first illustration of the difference, noted in section 2-8, between explanation as understood by the objectivist realists and explanation as conceived of by the upholders of the view that physics merely describes human experience. Unquestionably, within the realm of their professional activity most theoretical physicists will unhesitatingly assert that such a reference to the quantum rules does constitute a genuine explanation. They will claim that much because, spontaneously and implicitly, they stick, in such contexts, to the second of the conceptions just recalled, and because they know that the quantum rules are far-reaching generalizations, the discovery of which has to be considered an overwhelming success. By contrast, in the telephone example above the statement that, in the past, correlation was observed in every instance has no such virtue. This difference justifies the fact that our telephone consumer's claim "since the correlation was always observed in the past it will be observed in the future" is not viewed by anybody as constituting an explanation, whereas the use of the quantum rules is taken to be one. It remains true however that both of them merely refer to predictive rules.

Pursuing the "Conceptual Quest"

On the other hand, instinctively we are, to repeat, objectivist realists. "Explanations" just grounded in observational predictive rules hardly meet our expectation. We get easily convinced that we should look for better

---

[4] The fact should be kept in mind that, in quantum mechanics, a statistical ensemble of pairs described by "the pair wave function" cannot be identified with a *mixture* of pairs whose orientations in space are well defined, differ from one pair to the others, and are statistically distributed. Relative to most correlation measurements these two ensembles lead to quantitatively different outcome predictions.

ones, that should fulfill the requisites of the realism of the accidents or in which, at least, what "explains" would "really exist." And our claim seems all the more grounded as, in the corresponding classical case, the explanation of the correlation observed between the measurement outcomes is precisely of such a type. This we saw on the dart model, for, there, the correlation between the observed outcomes proceeds indeed from a correlation established at the source between the orientations of actually existing darts. Note, moreover, that, in this model, the fact that each instrument registers + signs and − signs implies that initially the dart pairs are not in all aspects identical. Some of them must differ from the other ones with respect to orientation. Quite naturally, this remark leads to the view that in the photon pair case the observed correlation might conceivably be given some similar explanation. It suggests that, although all these pairs have the same wave function, still they might not be strictly identical. Just as in the classical case some variables would exist—here they are called "hidden"—that, right at the start, would actually differentiate these pairs from one another, just as—obviously—their various orientations differentiate, at the source, the dart pairs from one another. And the subsequent behavior of the two photons of one pair would be determined—or at least influenced—by the values these variables have on this particular pair.

Qualitatively an explanation of this sort is attractive because of its simplicity. But it is of course necessary that quantitatively it should not disagree with any of the observable facts. Moreover, if it is demanded that it be consistent with the rule that no superluminal influence exists it is of course necessary that the said rule should not be violated by the very nature of the variables in question. In other words, it is necessary that these variables should be local, that is, relative, at any time, to some well-specified place.

At this stage in our quest we have to take into account a theorem already alluded to and that proves to have a basic role in any attempt at interpreting contemporary fundamental physics. It is called the Bell theorem. To make use of it right now, we need not resort to its full explicit formulation, nor to the whole array of the notions it involves. It is enough that we should take into account one of the main points it establishes, which is that no local hidden variable theory is capable of reproducing in detail all the quantum mechanical predictions that the Aspect-type experiments actually showed to be correct.[5] Since, to repeat, the experiments in question did corroborate all of these predictions, they prove, in view of the theorem, that no local hidden variable theory is compatible with the

[5] For completeness' sake let it be mentioned here that it is John Bell's theoretical discovery that gave rise to the Aspect-type experiments, not the reverse.

data. In other words, they disprove the local hidden variable explanation that looked so attractive at first sight. It is this very disproof that constitutes *the* significant point here. It is important to realize that this is in no way a rejection grounded on some a priori philosophical conception but, quite on the contrary, a disproof based on plain facts.

Clearly, this result is of a considerable generality. Within the realm of a world view consistent with objectivist realism, neither one of the two alternative assumptions that wave functions are real or that local hidden variables exist makes it possible for us to keep to our instinctive way of understanding correlation at a distance.

Yes but—some will ask—perhaps it still could be saved, at the price of somewhat watering down a few of our intuitive claims about particles and so on. To this question the answer is "no." We shall see that, according to its explicit formulation, the Bell theorem implies that the outcome of the Aspect-type experiments is incompatible with a conception of Reality involving a "locality" principle (a principle roughly stating that influences[6] are not propagated at infinite speed). Hence, reasonable as it may look, this principle is violated in Nature (at least if this word is taken in its usual sense). It is this violation that is called *nonlocality* when it is considered all by itself, independently of any theory. But of course its strong link with quantum theory can hardly be ignored. Let us remember our analysis of the two-photon experiment. There we introduced the nonseparability notion. In its broadest sense nonseparabilty implies that we must either accept nonlocality (a precise definition of which appears below) or, as quantum mechanics forcefully suggests, agree to change our concepts to a greater extent than we should care to imagine, indeed, to the extent of giving up the view that human-independent Reality is embedded in a normally structured space (existing independently of ourselves) and made up of parts the mutual interactions between which decrease as distance increases.

First Philosophical Comments

1. In the first chapter and, again, at the beginning of this one we noted that the quantum mechanical formalism seriously questions such conceptions of Reality as the "multitudinist" one. In connection with this, let it be stressed that in this field it is important not to rely on the words we read. Almost all of those that the said formalism makes use of (think of the words "particle," "state" etc.) were taken from classical science (otherwise, how could scientists have expressed themselves?) and there-

---

[6] The notion of influence is subtler than it seems. Concerning it, see sections 3-2-2 (Remark 6), 3-3-1, and 14-7.

fore implicitly carry with them the notions of locality and separability. However, the turn they thus impart to our thinking is quite misleading. When, beyond the words, we consider the quantum formalism itself, we see that it entails not only identifying particle creations as state transitions but also giving up classical, Bolzmann statistics. It implies (as already noted) replacing the latter by either the Bose-Einstein or the Fermi-Dirac statistics, both of which make it impossible to think of the particles as each having its own individuality. It also implies giving up the very notion of a probability, for a particle, to *be* at such and such a place, and replacing it by the probability we have of *finding* the particle there. Clearly, all these ideas are either barely or just not compatible with the notions of multiplicity and locality. In view of all this it is clear that the faultless achievements of quantum mechanics in predicting observations already constitute a very strong indication in favor of the idea that multitudinism and locality are inadequate notions when a description of independent Reality is aimed at.

2. On the other hand, an indication, be it even quite a strong one, is not tantamount to a proof. And the philosophers who, concerning such questions, speak of an "underdetermination of theory by experiment" are entirely right in doing so. Besides, in the past it not infrequently happened that a great theory, rich in well-verified observational predictions, got superseded by another theory, implying the same consequences along with new ones, and grounded on altogether different basic concepts. Consequently it is entirely conceivable that quantum mechanics should, one day, suffer the same fate, that it should be replaced by some other theory based on different concepts and in which, therefore, the conceptual difficulties specific to quantum mechanics will not exist. In view of the great philosophical importance of this remark let it be especially emphasized here that, due to the above-mentioned material (the Bell theorem supplemented by the Aspect-type experiments) locality violation is now a direct consequence of experimental data. It is fully independent of quantum mechanical considerations, and will therefore survive, even if quantum mechanics gets superseded.

Note, however, that conceptions that lie within the realm of physical realism but do not assume locality remain tenable. When discussing the notion of trajectory we already encountered such theories and we shall soon study some of their features.

3-2 Locality and the Bell Theorem

The foregoing section introduced us to the problems connected with locality. Knowing what their nature is, we must now engage in a more detailed and precise conceptual analysis of them.

3-2-1 *The Locality Notion*

In the foregoing pages we noted more than once our natural bent—rational or instinctive, never mind here—in favor of physical realism, that is,
for the idea that physics should, somehow or other, describe "reality as it
really is." For short, this conception may be called "the realist vision."
In fact, we instinctively tend to consider not only that, underlying our
experience, there is a reality that does not depend on it, but also that this
notion accords with some basic views we deem obvious. We therefore
entertain the idea that the latter somehow inform us in advance of some
quite general features common to the structures, the details of which physics will reveal.

One of the basic views in question is the principle of "divisibility by
thought." Admittedly, it is not inborn in the human species. Primitive
societies have a vision that might rather be termed "holistic." But it is
quite unquestionable that, from the times of Galileo and Descartes on,
the growth of science was largely grounded on a rejection of the said
vision. In fact, it greatly relied on its exact opposite, that is, on the principle—explicitly stated by Descartes—according to which problems and the
objects they deal with have to be divided by thought into simple elements
before the whole is reconstructed. And indeed, throughout the last centuries the success of this method was so impressive that, quite normally, we
came to consider it an indication that reality itself is essentially composed
of parts. In other words, we developed the multitudinist world-view.
Quite normally, therefore, the many among us who take up the thesis that
physics is a description of reality picture the latter to themselves as being
composed of more or less localized objects and "local" fields.[7]

It is true that, at this stage, this conception is still rather qualitative
and vague. But it so happened that with the help of the (classical) special
relativity theory, and by means of the corresponding notions of events
and lightcones, it was possible (for a time) to express it in a much more
precise way. This followed from the fact that, in Maxwell's theory for
instance, the values of the electric and magnetic fields within any finite
region R of space-time are determined by those these fields have within
the backward lightcone[8] of R, and, even more simply, by those they have
within a limited (hence "localized") region V completely closing the backward lightcone in question (see fig. 3-1). It follows that if the fields in V
are known it is possible to predict everything that will take place in R and,

---

[7] Functions whose space variables are the coordinates of but one point are called *local*.
Functions that depend on several points at the same time are termed *nonlocal*.

[8] The *backward lightcone* of a region R is the set of all the space-time points from which
moving bodies (photons included) liable to reach R could depart.

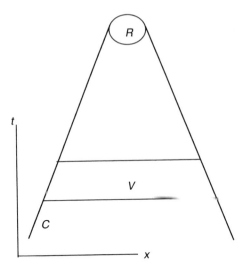

Fig. 3-1. A space-time region R, its backward lightcone C, and a space-time region V that fully closes C. In order to make a two-dimensional representation possible the three space dimensions were here reduced to just one, labeled x. The space and the time units are assumed to be such that if the two straight lines picturing the cone represented rectilinear motions of objects the said objects would move with the velocity of light.

correlatively, that the predictions in question can in no way be modified by additional information originating from observations made in regions spatially separated[9] from R.

Since V is bounded, just like R, the preceding sentence may be considered as expressing a "classical locality principle." Admittedly, such a formulation is rather far from corresponding to what we intuitively have in mind when we utter such words as "localization" or "locality," since, when macroscopic distances and time lags come into the picture, the region V has, on the human scale, a most considerable extension. But, in compensation, the said formulation has the advantage of being strict. It involves no approximation.

This "classical locality principle" was, in classical physics, unimpeachable. However, it relied on determinism and, nowadays, we know that determinism cannot be taken as an axiom. This does not mean that the principle should be altogether given up but it implies that it should be reformulated. The guideline idea of John Bell's reformulation of it was

---

[9] Two space-time regions are said to be *spatially separated* from one another when each one lies entirely outside the lightcone of the other.

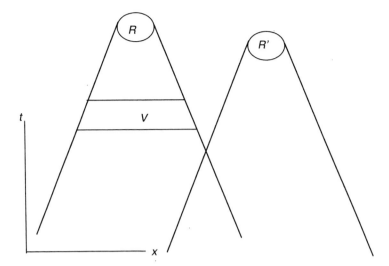

Fig. 3-2. Two space-time regions R and R' spatially separated from one another, their backward lightcones, and a space-time region V fulfilling the condition stated in the *Locality Principle* (see main text).

that, in order to preserve the substance of this locality notion within the realm of an indeterminist theory (such as quantum mechanics) we should express it as follows. We should assume that, as in the deterministic case, if we knew and took account of all the parameters (hidden or not) relative to the elements of reality lying in V, no additional information from observations made in space-time regions spatially separated from R could modify our predictions concerning what will happen in R. Let this reformulation be called the *locality hypothesis*, or *principle*.[10,11] Its precise wording is as follows (see fig. 3-2).

Locality Principle (or "Hypothesis"). A theory obeys locality if the probabilities of events taking place in a given space-time region R are not modified by specification of what events take place in a space-time region R' spatially separated from R *when what happens in the backward*

[10] John Bell (whose mode of presentation is taken up here [Bell 1987b]) called this principle "local causality" because it generalizes Einsteinian causality to probabilistic theories. Philosophers may be surprised by the appearance of the word "causality" in the name of a hypothesis involving probabilities. The explanation is given in section 14-3.

[11] Not to be confused with another, purely formal principle to which, in quantum field theory, this name is sometimes, somewhat unwarrantedly, given. This other principle simply states that the observables relative to a region R are compatible with all those relative to a region R' spatially separated from R.

lightcone of R is already sufficiently specified: for example through a full specification of the events in a space-time region *V* that completely shields off from *R* the overlap of the backward lightcones of *R* and *R'*.

GENERAL COMMENT

The italicized phrase is essential of course, for we obviously expect that information relative to an event in *R'* should modify a probability concerning *R* that would just (or partly) refer to our ignorance. For instance, in the dart example (in which we assume we know nothing of the initial orientation of the pair) we take the probability of a + outcome on the right to be 1/2 when we do not know the outcome on the left, whereas we take it to be either 1 or 0, as the case may be, when we know it. But, on the contrary, it seems quite normal to consider that such information should not alter a probability defined as intrinsic, that is, relative to a situation in which all the parameters that according to a basic rule of relativity theory might have an effect on *R* have specified values.

COMMENT FOR THE USE OF QUANTUM PHYSICISTS

In quantum mechanics, a realist who adhered to the principle of strong completeness (no hidden variables, see section 2-8) might well wonder whether the locality hypothesis is at all meaningful. For indeed this hypothesis centers on the "event" notion and it is clear that the latter, evoking as it does what takes place at one point, does not fit well with the view that the (as a rule poorly localized) wave function is one and sole constituent of reality. But on the other hand note that within a realist world view incorporating strong completeness the very existence of wave functions with extended support, and even more that of nonseparable wave functions (such as the photon pair one in section 3-1), rule out at one stroke any notion of locality, or, at any rate, any notion of locality related to what we have in mind when we utter the word "local."

In view of the latter remark, a realist anxious to "save" locality might then think of replacing strong completeness by the more pragmatic, Stapp-like notion "weak completeness" (section 2-8). For with such a substitution the idea becomes (at first sight) tenable that the "true reality" of the physical systems is neither their wave function nor the self-adjoint operators associated with their observables, that it is a set of parameters, called "additional" or "hidden," whose values at any time actually constitute the "real events." A priori it could then be hoped that the experimentally observed correlation phenomena (e.g., the photon pair ones) might receive an explanation fitting with the general spirit of the realist vision because it would rest on relationships established between such parameters right at the source (as in the dart case). But the Bell theorem applies to such a model and disproves this hypothesis. The latter remark shows that, as long as the Bell theorem is not taken into account, the locality hypothesis cannot be said to be inconsistent with what, in quantum mechanics, we know for sure to be correct, namely, the set of its predictive rules. Only this theorem and the corresponding experiments do really disprove the hypothesis in question.

REMARK

Since, generally speaking, the locality principle involves the notion of unknown events really taking place, it is clear that it normally implies that realism (and even objectivist realism) holds true, at least in some of its aspects. The question as to which of these aspects are relevant concerning this is a somewhat ticklish one. It is considered in section 3-3-4 below.

Definition.   To designate a type of reality assumed to obey locality the expression *local Reality* will sometimes be used.

3-2-2 Bell's Theorem and Nonlocality

We already had a glance at this theorem (Bell 1964, 1987b). It is a proof that, *independently of any theoretical formalism whatsoever, the conjunction of the locality hypothesis (implying realism) with that of "free choice of experiment" implies that some inequalities bearing on measurement outcomes (called the Bell inequalities) must hold true*. The importance of this theorem is due to the fact that these inequalities are *violated*, both by the quantum mechanical predictions bearing on the results of some definite measurements and by the outcomes of the Aspect-type experiments. The measurements in question concern, as we know, correlation-at-a-distance effects.

Note that the premises of the theorem are largely modeled on commonsense. It is therefore not surprising that, in the overwhelming majority of cases, the results of correlation-at-a-distance experiments do obey the Bell inequalities. But in molecular, atomic, and subatomic physics the quantum observational predictive rules show, as we just noted, that there are experiments in which the said inequalities should be violated. And the Aspect-type experiments confirm that this is indeed the case. The theorem then implies that one at least of its premises must be given up. Unfortunately, they all seem secure. But since those of reality and of a free choice of the experiments to be performed are viewed by most physicists as being, in a way, basic truths, it is locality that quite generally appears as being the one that should be dropped. Nonlocality is said to hold.

Some remarks will make more precise the conceptual tenor of the theorem.

REMARK 1

The condition that the choice of the experiments is taken to be a free one means that the experimentalist must be thought to be able to choose them at will, without being unconsciously forced to one or the other choice by some hidden determinism. This condition has an important role in the proof of the theorem. It is often left implicit because of its apparent obviousness. Here it is explicitly stated. But let it be observed that, when all is said and done, it appears as constituting the very condition of the possibility of any empirical science.

As for the proof itself, it involves some calculations. It is given here in appendix 1.

REMARK 2

It is appropriate to realize that, in the form in which it is presented here, the Bell theorem does truly posit objectivist realism (see section 1-2) or something quite akin to it. Indeed, it assumes locality, the definition of which involves specification of events taking place in the backward lightcone of R, and these, like all events, are characterized by real numbers. Consequently, the theorem does assume that the instantaneous, microscopic objective state of any physical system is specified by a sufficiently large (continuous or discontinuous) set of knowable or unknowable parameters that are real numbers. (In the dart example they are those that specify the initial orientation of the pair.) And the theorem also assumes that the fact that these parameters have such and such values has observable consequences (here these consequences take the form of measurement outcome probabilities). As previously noted, philosophical systems could presumably be conceived of that would postulate realism without fitting into the above-defined framework. The Bell theorem could not be applied to them (at least, not in the sense in which it is here understood ; see section 3-3-3).

REMARK 3

It follows from all the foregoing material that the Bell theorem is basically *negative*. By itself, it shows neither that the notion of "human-independent Reality" is meaningful, nor that such a Reality possesses such and such features. It merely shows that factual data prevent us from picturing this Reality to ourselves in such and such naive ways.

But, in compensation, so to speak, Bell's theorem does not depend on some theory or other. This asset we already noted. Even if, one day, quantum theory gets replaced by some "better" one, the conclusions stemming from the theorem will remain valid, since they are independent from this theory.

REMARK 4

Concerning the Bell theorem, a misunderstanding often occurs. It consists in believing that the inequalities it involves bear on quantities necessarily attached—as dynamical properties—to the micro-objects themselves, in imagining, for example, that, before impinging on the instrument it is heading for, each one of the two photons carries a polarization vector and that the Bell inequalities somehow bear on some features (components or whatnot) of these vectors. If this were the case, the theorem would lay itself open to several types of criticisms, such as "What lets you think that each photon possesses, all by itself, a definite polarization vector?" Or, in a more incisive style, "By thus assuming that each photon has a definite polarization vector (or a momentum or etc.) you are making quite a strong implicit hypothesis that dramatically restricts the bearing of the theorem." It should be realized that such criticisms totally miss the mark for, to repeat, Bell's theorem does not bear on quantities assumedly carried by the involved particles (here the photons). In fact, it is by virtue of a mere convenience of language that photons or other particles are mentioned in its connection. As the proof described in Part A of appendix 1 clearly shows, the numbers that appear in the Bell inequalities (and which are called "measurement results" to go along with common use) are mere *observation* results. They are—free from any strict interpretation in terms of "values" possessed by "physical entities"—the numbers of clicks registered by the detecting instruments.[12]

More generally, let it be noted here that the—fully justified!—surprise aroused by the Bell theorem gave rise to a rich critical examination aimed at detecting some implicit assumption supposedly used for proving it (so that discarding it would invalidate the theorem). Of course, it is an excellent thing that such investigations should be made. But this is only true provided their authors do not themselves lack the critical mind they suspect the contributors to the theorem not to have adequately made use of.

---

[12] The conditions of validity of the Bell theorem have been the subject of many extensive, critical analyses. The interested reader will find some of the main relevant references in appendix 1.

Unfortunately up to now this condition has not been fulfilled. Sometimes the conjectured "hidden hypothesis" was quite a subtle one, at other times it was one any "man in the street" could have thought of. But, until now, a careful study of such papers always revealed that the suspected "hidden hypothesis" was, in fact, by no means used in the proof. In the same vein, authors sometimes suggest that for locality to be recovered it would suffice that the measurement results (on each photon) should be determined merely statistically. They thereby overlook the fact that the Bell theorem covers such cases, as we just saw.

These are but examples, mentioned here merely in order to illustrate the extreme generality of Bell's theorem. It should indeed be realized that this theorem has nothing in common with ideas closely linked to a particular technique or theory, and that may therefore be disproved rather easily, just by questioning the validity of the latter.

REMARK 5

Of course, the violation of locality brought to light by the Bell theorem (and the corresponding experimental data) is not equivalent to the mathematical nonseparability we met with in section 3-1 (the nonseparability of the pair wave function). It should rather be said that the latter has set us on the track of the former. Still, as already noted, they are related. Qualitatively, this is clearly seen on the foregoing example of two photons that were created together and separated. For if these two photons (or, more generally, any two particles) do still constitute but one "reality" as nonseparability suggests (within the framework of a realistic interpretation of the wave function), it is not surprising that this should be reflected in the existence of instantaneous influences of one on the other (more precisely, "of what happens at the place where one measurement takes place on what happens at the place of the other one"). For maximal accuracy it should, however, be pointed out that nonlocality was defined— hence superluminal influences suspected—only within the realm of objectivist realism, that is, by taking it for granted that things somehow lie within a really existing space-time. If this standpoint is given up it becomes meaningless to speak of locality—and hence also of nonlocality— whereas, on the contrary, viewed as a structural property of the quantum formalism, the nonseparability notion remains valid.

REMARK 6

The notion of influence is quite akin to that of cause. Now, for well-known reasons (to be considered in section 14-2) the latter notion raises complex conceptual problems to which that of "influence" is of course not immune. Hence it is important to note that neither one of these concepts appears in the foregoing definition of the locality principle; and that therefore the fact that experiment violates this principle cannot be ascribed to some incorrect use of the concepts in question. When all is said and done we shall see, however, that it is almost impossible not to interpret the said violation as revealing the existence of superluminal influences of some sort (see sections 3-3-1 and 14-7 below).

3-2-3 The Supplementary Theorem

To grasp the true philosophical implications of the foregoing developments it is essential that the quite crucial point this section deals with should be duly taken into account. Let it be called here "the supplementary theorem" ("supplementary" to the Bell theorem, of course). It was established independently by P. H. Eberhard (1978) and G. C. Ghirardi, A. Rimini, and T. Weber (1980).[13] It is a rather straightforward consequence of the predictive quantum laws, that is, of what, in quantum mechanics, cannot be questioned. It states that *insofar as the Bell theorem is interpreted in terms of superluminal influences, these influences carry neither matter, nor energy, nor any usable signal.*

If we decided (as a radical idealist would perhaps feel inclined to) to restrict the meaning of the expression "influences at-a-distance" to "transmission of usable signals," then, clearly, we should have to infer from the theorem that the "influences" seemingly following from the Bell theorem do not really exist. But within the framework of objectivist realism, defining a physical phenomenon merely by referring to some human abilities or lack thereof is obviously inadequate. And, within the framework in question, a possibility indeed remains of defining, if not influences-at-a-distance in complete generality, at least, in special situations, cases in which such influences are meaningfully said to exist. In fact, we shall check in some detail below (section 14-7) that, in situations in which the involved events take place in two spatially separate regions, such cases correspond to those in which the locality principle is violated.

REMARK 1

This remark is useful for making the content and bearing of the theorem quite explicit. As we saw, the violation of locality implies that the probability of the outcome of a left-hand-side measurement depends on the

---

[13] The present author had previously mentioned it as a most likely conjecture (d'Espagnat 1975, note 30); in this connection see also (d'Espagnat 1976) section 11-6.

outcome of the right-hand-side one (in a way that a reference to "common causes" does not explain). The present remark consists in observing that the said violation does *not* imply that the probability in question depends on which physical quantity the right-hand-side experimentalist chose to measure. And the supplementary theorem shows that, in fact it is independent of it. It only depends on the outcome of this measurement. The phrases "outcome dependence" and "parameter independence" are often made use of for referring to this quite significant distinction.

REMARK 2

Regarding semantics, it is useful to know that the authors engaged in such studies do not, all of them, make use of the same words. In particular, some choose to define locality as merely being the fulfillment of the "parameter independence" requirement. This enables them to claim that the violation of the Bell inequalities is not a violation of locality. Admittedly, any definition is a matter of convention. It seems nevertheless that, at least in the eye of any would-be realist, outcome dependence constitutes a significant violation of any notion the word locality may be attached to without its sense being seriously altered. This is why, in accordance with most authors, we made use here of the word "locality" for designating the principle we labeled under this name.

REMARK 3

Special relativity sets it as a basic law that the light velocity is an upper limit of all possible signal velocities. Now, within the realm of physical realism, that is, of a conception in which the formulation of a physical law should not refer in any way to the abilities of mankind, the word "signal" can hardly mean anything else than "influence." Consequently, as long as special relativity is understood within the framework of physical realism the last sentence in Remark 6 of section 3-2-2 somehow implies that the Bell theorem entails a violation of the said basic relativity law. Of course, this would constitute a considerable and most unwelcome perspective change. But, on the other hand, in the next chapter we shall see that quantum mechanics strongly suggests interpreting physics along the general lines of instrumentalism[14] or operationalism, rather than along those of realism. Within the realm of such an interpretation—centered as it is on the human abilities of feeling and acting—it is, to repeat, difficult

---

[14] In epistemology the word *instrumentalism* essentially means a conception according to which scientific theories are fundamentally considered to be instruments used for predicting what will be observed.

to even precisely define the notion of influence-at-a-distance without referring to that of a signal. It follows that the very notion of "influences-at-a-distance not making it possible to send any signal" becomes obscure. We then see that, provided relativity theory be, like quantum mechanics, understood in an instrumentalist sense, the supplementary theorem makes it possible to reconcile its most basic principle with the facts and data reported in this chapter.

3-2-4 Conceptual Bearings of These Results

Bell's and the supplementary theorems are to be regarded differently according to whether we believe or not in the universal validity of the quantum mechanical predictive laws. (The respective degrees of plausibility of these two alternative possibilities will be examined below.)

If we negate such a universality, that is, if we consider that the quantum laws exclusively bear on—roughly speaking—the "microscopic world," we have no a priori reason to believe that nonlocality has any bearing on macroscopic phenomena. The reason is that, as we noted, the premises of Bell's theorem are modeled on commonsense and that commonsense just mirrors our experience of macroscopic phenomena. Hence, concerning at least an overwhelming majority of the latter, the premises in question must automatically be fulfilled, so that there is no reason to expect that the Bell inequalities should be violated. In other words, within the hypothesis presently under study there is no reason to even *suspect* that, among the purely macroscopic phenomena, some should obey nonseparability (superluminal signals for example). True, this is not to say that such phenomena are impossible. For instance, still within the same general hypothesis it is conceivable that, on the borderline between the "micro" and the "macro" worlds certain phenomena strictly obeying neither the classical nor the quantum laws should exist, and should be such that nonseparability (of the wave functions) would apply to them while the supplementary theorem would not. However, it is clear that this hypothesis, backed up by no evidence whatsoever, is totally artificial, and hence radically implausible according to our present knowledge.

If, on the contrary, we believe that the predictive quantum laws enjoy universal validity, we must consider the violation of locality to be universally true. But in return, the supplementary theorem is also to be viewed as having universal validity, so that, in this case also, no superluminal signaling is possible. In other words, within this hypothesis a theory conceived of within a realist framework and matching the facts does imply a violation of locality (which it is hard not to think of in terms of supraluminal influences) but this violation (and these influences) cannot be used.

Hence, they cannot be directly detected. More generally speaking, in any such theory Reality obeys nonseparability while the observed phenomena involve separate "perceived objects." Such a state of affairs brings to light a point that, for physicists—owing to the attractiveness the universality notion has for them (we shall see why, on this, they differ from the bulk of philosophers)—is important. This point is that if, within the realm of the universality hypothesis, we assume realism (*open* realism, of course), we cannot identify the set of phenomena with the thus postulated "basic Reality." This imparts plausibility to the idea that because (in particular) of our limitations at perceiving, we are ourselves called into play by the phenomena. Once we have realized this, we are less surprised by the fact that the quantum mechanical textbooks formulate some physical laws (i.e., some general descriptions of phenomena) by (implicitly) involving us (more precisely, by involving "measurements" in a basic way).

If physics has to be described this way, then it obviously cannot be stated exclusively by means of the objectivist language defined in chapter 1. With respect to the world view—and the conception of our own relationship to the world—suggested by classical physics, this, philosophically speaking, is a kind of revolution. There are conceivable escapes from it. They are the ontologically interpretable nonlocal theories the existence of which has been mentioned already. But, considering what we noted relative to Bohm's, we may easily guess that this path also will prove uneasy.

REMARKS ON TELEPORTATION AND QUANTUM CRYPTOGRAPHY

The (somewhat debatable) name teleportation (Bennett et al. 1993) was given (by its discoverers, who granted that they borrowed it from science fiction) to a theoretical procedure loosely connected with the subjects under investigation here. In spite of what its name suggests, the procedure in question involves no transfer of anything akin to a material object. It does imply some transfer of information, but no superluminal one, so that it is entirely compatible with the "supplementary theorem." In fact, as the quoted authors point out, teleportation is the solution of a problem that, in classical physics, would raise no difficulty whatsoever since it just consists in reproducing at a distance—but at infraluminal speed—the state of some given physical system. For example, we might have to do with a vector having some given direction and length (a nail, say), that we would be asked to reproduce elsewhere. In classical physics the operation is easy. It suffices to measure the said direction and length and reproduce them at the other place. But in quantum physics things are different. If the vector in question is, say, a spin, we cannot perform all of the required measurements (we can measure any one of its components but in so doing we disturb the others). What the quoted authors found was a procedure making it possible to elude the difficulty, albeit at the price of destroying the model state (for quantum states cannot be cloned). This work is most interesting to the experts, for it sheds light on some features of quantum theory that are far from being obvious. Unfortunately, in some periodicals the word "teleportation" gave rise to considerations wandering very far away from the truth.

By contrast, the violation of the locality hypothesis led other authors (Bennett and Brassard 1989; Eckert 1991) to the discovery of some unbreakable theoretical cryptographic procedures, based on quantum mechanical principles. It is quite possible that, in the field of advanced scientific technology, such ideas will generate quite important developments.

3-3 Discussion and Philosophical Implications

The content of the foregoing pages is likely to raise some questions. Those of them that, from a philosophical point of view, seem most interesting are examined in this section.[15]

3-3-1 What if These Correlation Phenomena Were but Trivial?

And what if all these uncouth terms holism, nonseparability, and so on, set forth here as carriers of novel ideas, in fact just covered up trivial correlation phenomena? The idea was put forward. And it must be granted that, on the surface, the supplementary theorem seems to support it. For it seems true that all the indications that may conceivably reveal the existence of some influence at-a-distance are—by definition, one might say—of the nature of an exchange of "signals." If the correlation phenomenon under study does not offer, at least theoretically, a possibility of such an exchange, is it not true that its specificity may be questioned? An example meant to support this view was already discussed in *Veiled Reality* (d'Espagnat 1995). Let it here be considered in more detail.

The example is based on the following device. An extremely powerful lighthouse is assumed to stand at the center of a circular wall whose radius is so large that the light spot produced on the wall by the rotating beam moves on it with a velocity greater than the velocity of light. For what follows it is even convenient to assume that, by acting on the lighthouse projector, it is possible to stop it and set it in motion at will (with fixed angular velocity), and that the lighthouse keeper does this on request.

---

[15] A reader aiming at getting an overall view of the problems at hand before tackling a finer analysis of their constituents may well, on first reading, pass this whole section.

Concerning this thought experiment a few points should be noted. The first one is that the two events "point A on the wall is lighted" and "point B on the wall is lighted" are correlated. Each time the first one (call it A') takes place, the other one, B', takes place a moment afterward. The second one is that B' takes place so quickly after A' that, in order to successively coincide, like the spot, with these two events any object would have to travel with supraluminal velocity. Let us say that the correlation between A' and B' is "almost instantaneous." The third one is that this whole experimental arrangement does not make it possible to send a usable signal (an order, for example) from A to B. The reason is that normally the person standing at A does not control the spot motion. Admittedly she could decide to set the whole device into action all by herself. But for that she would have to send the lighthouse keeper—by radio, say—a message that cannot be instantaneous. And it has been shown that indeed the signal she thus sends to B cannot travel faster than light. And finally, the last point to be noted is that obviously this whole experiment involves phenomena entirely subject to classical physics and that everything that is observed in it is trivially explained it terms of the said classical physics. In other words, in this experiment the novel notions of nonseparability, etc., are entirely irrelevant.

The objection here under review is the claim that such is also the case concerning the Aspect-type experiments. Indeed, "almost instantaneous" correlation phenomena are observed also within the latter (between the two measurement events corresponding to one and the same pair). And, also in them, these phenomena do not make it possible to send usable signals. It is true that, in the latter case, the physical picture suggested by the theory is that of some almost instantaneous *influence* (remember the procedure, sketched in section 3.1, according to which the first measurement reduces the pair wave function, thus imparting a wave function to the second photon). But—the upholders of the objection in question would claim—this is merely a subjective interpretation suggested by a computing method, and what is objective is merely the observed correlation. Hence, even though the Aspect-type experiments and the lighthouse are analyzed by means of technically quite different computing methods, there is no point in considering that the first one brings into play great general notions qualitatively differing from those used for describing the second one. Basically, so goes the argument of the people who, by means of the comparison considered here, pretend to show that such words as nonlocality, etc., merely introduce false problems.

Though stimulating at first sight, this objection is, when all is said and done, not valid; for, between the lighthouse model and the Aspect-type experiments there are crucial differences. The main one is that, in the spot case, the correlation observed between the temporary illuminations of two points such as *A* and *B* is obviously due to a set of causes common to both (the variables that, at any time, determine the projector position). In fact, the very structure of the device implies that, for somebody who, hypothetically, would know the *exact* values of these variables, the probability (equal to 0 or 1 in this classical physics case) for point *B* to be lighted at some given time would not be modified by any additional information on what took place at *A*. Consequently, in the lighthouse experiment locality is not violated. By contrast—and this is the crucial point—locality *is* violated in the Aspect-type experiments since the Bell inequalities (which it entails) are violated there. This shows that locality violation prohibits explaining some correlation phenomena in the ways in which, up to now, correlation phenomena had always been explained, that is, either by direct action of one event on the other or through common causes lying at some source of both. Unquestionably, therefore, the violation in question, revealed by the Aspect-type experiments, is something new. And in view of the content of section 3-2-3 we may even say: something it seems natural to interpret in terms of superluminal influences.

3-3-2 Bell's Theorem Does not Postulate Hidden Variables

We already mentioned the existence of hidden variable theories that exactly reproduce the observable predictions yielded by standard (or "textbook") quantum mechanics. In section 9-3-2 we shall see why these theories remained very much apart from the mainstream of theoretical physics development. Some of the real entities they bring into play are nonlocal, which is consistent with the Bell proof that any such theory is necessarily nonlocal.

Informed as they were of the existence of such theories, some physicists working in other fields quite naturally thought at first that Bell's theorem merely applied to such questionable hidden variable theories, and correlatively that the locality violation entailed by this theorem merely concerned the theories in question; that, with respect to the interpretation of *standard* quantum mechanics, locality in Bell's sense (which, to repeat, supposes the main features of objectivist realism) remained valid. In fact, such a standpoint is untenable as we may see by considering once more the section 3-1 example with photon pairs. And it is so independently of whether or not we adhere to the "strong version" of completeness (section 2-8). For, think of one photon pair. Obviously, under suitable experimental conditions only the parameters, call them q, having, directly or indirectly, something to do with it are relevant for determining the objective probability p that on the left, say, a + result should appear. Now, let us take up first the strong completeness hypothesis. According to it the pair wave function constitutes, all by itself, the whole reality of this pair. Among all the parameters relative to events contained in the backward lightcone of the left-hand-side measurement, the set of the relevant ones, the q's, thus boils down to the ones that, at the source, specified the wave function. Now, via the wave function these parameters yield for p a well-defined value p_0 ($p_0 = 1/2$, in the example). And since, as we just saw, there are no other (unknown) relevant parameters within this lightcone, it is impossible to conceive of an objective intrinsic probability p that would differ from p_0 by virtue of being conditioned on more refined data. In other words, p_0 is an instance of the (objective intrinsic) probabilities the locality hypothesis refers to. And therefore, if locality were true, no information concerning the outcome of the right-hand-side measurement could change it. In fact, however, such information *does* (trivially!) change it (in the photon pair example it changes it to 1 if the said outcome is + and to 0 if it is −). We must therefore grant that locality is violated.

If we do not adhere to the "strong version" of completeness this reasoning does not hold good since, as we noted, according to "weak completeness" there may exist supplementary (hidden) variables. Hence p may well differ from p_0. On the other hand, Bell's theorem of course applies, in particular, to this case. And, in view of the experimental results, it shows that locality is violated (more precisely, that "outcome independence" is violated; but above we agreed to consider this to be a violation of locality).

Note that in all this reasoning what is central is the assumption of realism. It is that physics, far from "boiling down to mere recipes," is taken to be about some "real stuff," underlying our experience. Within the strong completeness hypothesis such "real stuff" can only be the wave function (or more sophisticated mathematical beings replacing it). Within the weak completeness hypothesis it can be the wave function and something else, or just this "something else," the wave function merely being a tool. As formulated above, Bell's theorem is, in a way, so general that it applies as soon as the notion of such "real stuff" is taken seriously, whatever the nature of the said real stuff may be.

3-3-3 Does the Bell Theorem Necessarily Presuppose Realism?

Apparently, of all possible notions "reality" is the clearest, and no opinion is more firmly grounded than our firm belief that we experience it. At all times, nevertheless, philosophers considered the question of the relationship between (mind-independent) reality and experience to be both ticklish and crucial. In fact, it could even be claimed that it lies at the very core of all philosophical queries. For some thirty years now, Bell's theorem has been contributing to this field, making it more complex and enlarging still more the already very large set of possible vistas concerning it. Here it is impossible to systematically go into the subtleties of the new domains of investigation thereby opened. This section is just meant to give a few indications on the subject.

The question whether or not Bell's theorem presupposes some form of realism is one of those that are not easily answered, for the Bell inequalities were proven in various manners, based on various sets of premises. Some of the latter explicitly define locality within the framework of a realist point of view. Such is the one described in sections 3-2-1 and 3-2-2. But not all of them do. Those that do not yield proofs concerning which it has been claimed that indeed the locality hypothesis they are based on is free from any—even just implicit—reality postulate. If this is actually the case and these proofs hold, the link between the Bell theorem (the violation of the Bell inequalities) and the philosophical problem "does physics describe reality or does it merely describe experience?" is obviously somewhat weakened.[16] For the violation in question should then be viewed as showing that among the notions we use for interpreting experience one of them, even more deeply ingrained in our mind than that of "reality," is wrong.

So, what is the case? To try and shed some light on the subject let us note that, schematically, the proofs in question are of two kinds.

Those of the first kind—put forward by Stapp (1977), Eberhard (1977, 1978) and others—rely on counterfactuality and essentially nothing else. More precisely they do without the notion of "objects," which, as we saw in section 1-2 under the heading "Realism of the Accidents," is a basic component of the realist picture. Their fundamental concept, called *counterfactual definiteness,* was given different formulations. All of them, however, were based on the idea that if, at space-time point A, a given event was observed, it would have been observed as well if, at another space-time point B spatially separated from A,[17] some event differing from the one actually observed had been observed. At first sight this idea seems to be both obviously true and free from any links with realism since it only mentions observations. But the extensive debates[18] to which it gave rise led to a much more cautious opinion (which seems to be supported by most, albeit not all, participants). This view is that, because of the way the idea is made use of for deriving the inequalities (in particular, since it

---

[16] However, it goes without saying that the possible validity of these proofs has no effect on the validity of the above-discussed one, described in appendix 1. Whether or not these proofs hold good, the combination of realism and the above-defined locality notion, therefore, goes on being falsified by the data.

[17] Or, more generally, such that no signal can be sent from B to A.

[18] See in particular Clauser and Shimony (1978), Shimony and Stein (2001), and Stapp's reply (2001).

has to be iterated; counterfactuals of counterfactuals come into play) the type of counterfactuality there made use of presupposes some sort of implicit underlying realist representation.

The proof here given in appendix 1, section B constitutes together with possible variants the second set of proofs just alluded to.[19] These proofs only apply to the, rather special, cases of strict correlation phenomena, such as those discussed in the dart example and in idealized versions of the Aspect-type experiments. (Incidentally, note that, due to several types of technical constraints, such idealized versions cannot actually be experimentally tested.) By considering the structure of this proof we might come to think that, though initially developed within the realm of objectivist realism, its validity in fact does not depend on that hypothesis. However, this conjecture would not be correct. In the derivation realism, in fact, comes into play, just due to the fact that, for the proof to go through, we have to attribute by thought the predetermination of the left-hand-side outcomes (inferred by induction as explained in the said appendix) to parameters pertaining to some *entities* (photons in the present case). If we did not we would be at a loss to introduce the notion (crucial in the proof) of subensembles of elements on each one of which *both* the outcome of a measurement of *A and* that of a measurement of *B* are predetermined.

To sum up, the above shows that the question forming the title of this subsection is a debatable one but that when all is said and done it seems it should be answered positively. And at any rate it seems altogether unlikely that the Bell inequalities should be derivable within a strictly operationalist approach to physics. Hence it appears that such a purely operationalist conception of this science is automatically immune (without any further assumption being needed) to any conceivable criticism based on the violation of the Bell inequalities. This will be implicitly useful in later sections of the book.

3-3-4 To What Extent Does Bell's Theorem Concern Mind-Independent Reality?

Still, the foregoing analysis leaves questions open. They have to do with the objections that philosophers with a bent toward Kantianism or neo-Kantianism may raise against the very idea of discussing questions (such as locality, etc.) concerning reality-as-it-really-is. The main one is well known. It consists in noting that, to know whether or not a representation

---

[19] Its leading idea goes back to Wigner (1970). I tried to make its premises more precise and more general (d'Espagnat, 1979a,b).

formed on the basis of observed phenomena correctly expresses some feature of mind-independent reality, it would be necessary to go into a comparison procedure; to compare representation and reality. This, they
stress, is quite impossible since we have access only to the phenomena,
not to reality-per-se. It is on the basis of this remark that the philosophers
in question (whose ideas will be further commented on in chapter 13)
claim that science can in no way inform us concerning the reality in question. Their reasoning does not undermine the validity of knowledge and
in particular of classical physics and sciences other than physics since the
latter may be considered to bear, not on reality-per-se but just on "empirical reality," that is, on what experience yields, directly or indirectly, access
to (see chapter 4). But in the problem under study the situation looks
different. The reason is that, in the foregoing sections, nonlocality appeared to be tightly linked to what was there called the "realist vision."
Whence the question: "Is it consistent, is it even conceivable that a theoretical development susceptible of experimental testing, hence bearing on
phenomena, should be basically grounded on an essentially ontological
assumption such as realism?" Obviously, like any other experiment, those
of the Aspect-type merely involve phenomena. Is it at all thinkable that
their content should reach beyond phenomena? That they should inform
us concerning a reality conceived of as primary with respect to the latter,
in other terms (as some would say), that they should have metaphysical
consequences?

The answer to this enigma is to be found in the fact (pointed out in
section 1-2) that any "realist" world-view (in the philosophical sense of
the epithet) is, in fact, made up of two elements. One of them is the (vague)
assumption that experience somehow enables us to "say something"
about reality-per-se; and, as we just noted, such an assumption is not
provable. The other one is a *representation* of the said reality, build up
on the basis of some general features of the observed phenomena that the
thinker with a realist turn of mind tends to identify with elements of mind-
independent reality. When this thinker is an objectivist realist, these features, as we saw, are the remarkable stability of some groups of impressions, named "objects," positions and forms of objects, numerical values
of these quantities, etc., on the one hand, and counterfactuality on the
other hand. The reason why the just mentioned twofold structure of objectivist realism does indeed solve our enigma comes from the following
fact. If we sum up what we saw above concerning Bell's theorem and its
proof, we observe that, of the two elements composing the realist world-
view, only the second one—the *representation*—actually comes in. In
other words, what is made use of in the proof in question is not the "metaphysical" concept "reality-per-se." It is the *representation* elaborated by
our mind on the basis of the phenomena. It is counterfactuality together

with the fact that, within the said representation, mind-independent reality is represented by a set of real numbers. It is true that we know neither the value most of these numbers have nor even the physical nature of the physical quantities they are measures of. But what the Bell theorem shows is that, whatever this nature and these values are, if such and such general conditions (stated in the locality hypothesis) are imposed on a function (namely, the conditional probability that the event occurs) of which the said values are the variables, this implies verifiable consequences.

Now in this nothing violates the great law of philosophical idealism (and Kantianism!) according to which only an idea may be compared to an idea. The crucial step that justifies this remark was, in fact, taken right at the start, for it consisted in going over from an elusive, noumenal "reality-per-se" to some clearly understandable *idea*. And this was done within our very definition of objectivist realism, when we made the point that reality-per-se can be represented by a set of knowable or unknowable parameters and that counterfactuality applies. The rest is but a matter of calculating and comparing results with experimental data. To put it differently, it is true that the events lying within region *V* of figure 3-2 are assumed real in some ontological sense. But this is not what is actually made use of in the proof. What *is* used there is just the idea that these events are adequately pictured by one among an infinity of possible *representations* of the type "sets of numbers" (and that counterfactuality holds true). Admittedly, we do not know which one of these representations is the correct one. We cannot even know with what probability any one of them is. But the beauty of the Bell theorem is just that, within it, this unattainable knowledge is not needed. The theorem shows that, whichever representation is the true one—nay, even, with whatever probability any one of them is—as soon as the locality condition is imposed on it, assuming free experimental choice, testable consequences (the Bell inequalities) follow. Consequently, it should not come as a surprise that the above-stated philosophical objection does not, in fact, apply to the Bell theorem.

A question rather tightly linked with the above one is whether or not the essentially negative notion of nonlocality may, as was suggested (Lévy-Leblond 1997), be replaced by some positive one. The answer is that in fact it is not without good reasons that the negative mode is made use of. As shown by the foregoing analysis, Bell's theorem does not infer from the phenomena the existence of some property that, transcending the said phenomena, would be ascribable to mind-independent reality. It merely shows that if we build up too naive a representation of the latter (the one corresponding to locality) we get results that experiment falsifies. Aiming at changing this essentially negative statement into a positive one might well result in a description of some alleged property (that would then be an altogether fundamental one) of mind-independent reality. For the above stated reasons, such a move would not be justifiable. It would be a valid one only within the realm of some special ontological model such as the Broglie-Bohm one. Whoever does not adhere to a model of such a type should keep to the negative formulation.

In fact we should even take a step further. To see why, let us first remember that the idea according to which experience yields some information on "reality-as-it-really-is" (alias "mind-independent reality") is nowadays generally considered a mere conjecture but that as such it is legitimate. We shall see below that it is even plausible, at least in its most "open" (i.e., less "descriptive") version. However, let us note once again that it is not provable. And we should realize that, strictly speaking, this lack of any possible proof renders questionable not only the positive *but also the negative* pieces of information we may, in such a field, try to infer from the data. In other words, within the problem at hand suppose we claimed that the available data *prove* mind-independent reality is not "local" (in the sense of section 3-2-1). To this claim it could be retorted that, since the idea that experience yields information on mind-independent reality is just a conjecture, it may well be erroneous, and that it is therefore quite possible that scientific experiments in general, and those of the Aspect type in particular, inform us merely about the phenomenal representation we build up of reality and not at all, either positively *or negatively*, about reality "as it really is."

Such an objection would be fully justified. To repeat, the Bell theorem does indeed bear on the *representation* we form of mind-independent reality. Since, within the realm of plain objectivist realism (within which Bell reasoned), mind-independent reality and its representation are one and the same thing, within such a view the theorem does of course bear on mind-independent reality (alias reality-per-se). On the other hand, within the conceptions that reject such identification that much, of course, cannot be claimed. Nevertheless it *may* be claimed that even within them Bell's theorem is indirectly meaningful with respect to mind-independent reality. The point here is that obviously, to be at all worth taking into account, a description should not be totally arbitrary. Now, in view of the results of the Aspect-type experiments, claiming that reality-per-se is ultimately composed of localized objects is admissible only if it is posited that the said reality-per-se is totally disconnected from experience. But then, why should we take this particular description of it into consideration at all, since the only reasons we had for doing so were that for a time we believed it to agree with our experience and have now found that this was a delusion? Indeed, to keep it after having found that much would be a totally arbitrary move. It would therefore run counter to the just stated rule, the one that even though, admittedly, tentative pictures of reality-per-se may be built up on considerations extremely remote from scientific experience, still totally arbitrary such pictures are not acceptable. Hence, to put it in a nutshell, any tentative description of reality-per-se involving locality is either erroneous (if we believe that experience informs us on the said reality) or senseless (if we believe that it does not).

REMARK

As already noted (section 1-2) the element of objectivist realism that we called *representation* seems to consist of notions coming from our ancestral experience, found to be outstandingly useful and slowly raised therefore to the level of "obvious truths" during the course of humanization. Considered from this angle the developments reported here may be understood as follows. Little by little, during prehistoric times, human beings worked out, on the basis of experience, a representation of reality. And they had good reasons to believe that at least some of the general features they imparted to it—the idea of objects, counterfactuality, etc.—were "true," that is, (in an implicitly operational sense) in accordance with *all conceivable* experiments. The discovery of nonlocality is then nothing else than our finding out—thanks to the Bell theorem and the corresponding experiments—that, strictly speaking, this is not the case. Viewed along these lines nonlocality appears free from any "metaphysical admixture." But we have to grant that this is at the price of a considerable conceptual upheaval.

3-3-5 More Concerning Counterfactuality

As noted above, our prehistoric ancestors handed down to us the counterfactuality notion, at least in its intuitive form, as an inherent component of their commonsense, macroscopic realism. Thus indeed, nowadays counterfactuality is considered to be a necessary component of any "nonesoteric" form of the latter, of, in particular, the realism of the accidents and, more generally, physical realism. This "evolutionary fact" has a few consequences worth mentioning.

1. It justifies the prominent role counterfactuality has in some at least of the proofs of Bell's theorem; such as the one reported in appendix 1, section B, involving photon pairs enjoying the property postulated there of strict correlation-at-a-distance. In this proof indeed, it is first shown that when the polarization of the right-hand-side photon of such a pair along some definite direction is measured, the result of a measurement that might conceivably be made of the polarization of the left-hand-side

photon of the same pair along the same direction is predetermined. And in a second step it is inferred that this result would also have been predetermined if the measurement on the far away right-hand-side photon had not been performed. Obviously in this second step, along with locality counterfactuality (as defined in section 1-2, at the end of the "Realism of the Accident" item) is also made use of. Within the framework of the proof described in section A of the same appendix, the role of this notion is less obvious. It may, however, be considered that counterfactualit intervenes there through the very fact that conditional probabilities are made use of.

2. It implies that even theories that are formulated in apparently realist terms cannot be considered realist in some sense akin to the commonsense one if they violate counterfactuality. Such is, for example the case concerning the "consistent histories" theory put forward by Gell-Mann and Hartle (1989) and popularized in Gell-Mann's book *The Quark and the Jaguar* (1994) (see appendix 2 for details).

3. It removes an apparent paradox bearing on the reasons that made us bring the influence-at-a-distance notion into the picture. Nowadays, in science, notions are normally introduced by implicitly (or explicitly) referring to some conceivable possibilities of observation or (human) action. And concerning the notion of influence, the necessity of distinguishing it from that of correlation would seem to imply that the reference should be to action. But here, as already noted, we have to do precisely with a state of affairs in which such a reference does not work since the influences whose existence we have been led to consider do not correspond to any possibility of action. So, why, after all, did we make use of the word "influence" in this connection? Taking into account the evolutionary process mentioned above—I mean the fact that, along and in connection with realism, counterfactuality has become an inherent component of our "normal" world-view—is a way of answering this question. Indeed, we may quite naturally claim that, by slowly and approximately extrapolating common experience (i.e., experience of macro-objects) our prehistoric ancestors and ourselves, after having tacitly built up the counterfactuality notion, finally raised it to the level of a principle. That we sort of inserted it into our very conception of reality, nay, almost, even, within our logic. Under such conditions, its violation within some correlation-at-a-distance experiments could not but be called "influence," even though it does not enable us to act. Such considerations show that there are words—"influence" being one of them—that fit our ancestral, macroscopic experience in a most adequate manner but the handling of which is, within our "microscopic" experience, unavoidably defective.

3-3-6 Nonlocality versus the Principle of Analysis

In the so-called "hard sciences" the principle of analysis, roughly sketched in section 1-2, is nothing else than the "divisiblity by thought" rule already encountered above. It may be stated as follows.

Any extended physical system—whether composed of particles, or fields, or both—may be thought of as composed of parts that in theory are knowable and are localized in mutually distinct (though often adjacent) regions. Moreover, if the forces that connect these parts with one another are known, then a complete knowledge of the values of the physical quantities belonging to each one of the parts yields ipso facto the knowledge of the whole system.

Unquestionably, in classical physics as well as in all the sciences outside physics such an idea has proved correct. It is true that within classical physics there are quantities, such as the potential energy of a two-body system, that, by definition, are "nonlocal" since they cannot be attributed to either one of the two bodies. But these quantities are derived ones. Their existence does not falsify the foregoing statement since knowledge of the forces and the relative positions of the two bodies makes it possible to calculate them. Moreover, in practice the principle at hand obviously remains, even now, of crucial importance since the whole of classical physics seems to obey it. And yet we shall presently verify that, as already suggested in chapter 1, because of the facts reported above in this book the principle in question can no longer be set at the root of an acceptable ontology.

In fact, as long as we try to ground our ontology on the wave function notion (as long as we believe the wave functions are elements of physical reality) the impossibility in question is easily seen. For let us take up again the photon pair example of chapter 1 (incidentally, in it, the pair could just as well be composed of particles other than photons). In order that the principle should be obeyed it would be necessary that, just as in the potential energy case, we should be able to calculate, at least in principle, the pair wave function on the basis of properties belonging to each photon separately. But then, *which* properties? Their individual wave functions? No, since, as we saw, in the situation considered the latter just simply do not exist. Other properties? Within the quantum mechanical formalism none exists that could play this role. In fact, the pair wave function is obtained using calculation methods that cannot be appropriately described here. Suffice it to note that, conceptually, they proceed from an altogether different approach since, in them, the wave function is inferred from the production mode of the pair.

For salvaging divisibility by thought we could then imagine shifting the role of "representative of reality" from the wave function to some other element within the quantum formalism. Along these lines some authors relied, at one time, on "mathematical beings" that happen to be, in the example, ascribable to each one of the two photons separately and that, in their jargon, physicists call "density matrices." But this attempt failed for the reason that, according to the said quantum formalism, even a complete description of these density matrices and of the forces does not make it possible to account for some correlation effects that the pair wave function shows should exist, and that are, indeed, observed.[20]

In the same spirit we might also, of course, imagine not to take the "conceptual" indications stemming from quantum mechanics completely at face value. According to this other line of thought there is no reason to demand that all the notions that "fit reality" should be elements of the set of quantum mechanical predictive tools. With perhaps the exception of very strict, diehard positivists, most of us would indeed find it a priori legitimate to search for some representation of reality not exclusively based on the elements in question, and of which it would just be demanded that it should not lead to predictions contrary to the quantum, experimentally verified, ones. Is it possible to build up along these lines some universal theory consistent with the above-stated principle of analysis, or, in other words, with the (spatial) divisibility by thought?

It is likely that this was, at least in part, the hope entertained by the discoverers of the hidden variable theories. And, at first sight their enterprise may well appear successful. Consider, for example, the well-known "pilot-wave" model due to Louis de Broglie and David Bohm, which adequately reproduces all the quantum observational predictions, at least within the nonrelativistic realm. It rests on the view that physical reality basically consists of a gigantic swarm of corpuscles, each one of which has at any time a well-defined position and a well-defined velocity. In this, clearly, it is fully in accordance with the divisibility by thought idea. On the other, hand however, let us not forget that, in the model, these corpuscles are assumed driven by a "wave" which is, just like the corpuscles, a basic element of reality but is *not* divisible by thought into localized elements.[21] All things considered, it would thus be a mistake to think that the type of reality the model describes is compatible with divisibility by thought. Before the discovery of the Bell theorem it could still be hoped

---

[20] As far back as 1965, in *Conceptions de la physique contemporaine* I drew attention to this point. The interested reader may find details in d'Espagnat (1965, 1976a, 1995).

[21] Instead of a "pilot wave" Bohm used the notion of a non decreasing, i.e., nonlocal, "quantum potential." But, especially in later texts Bohm and Hiley (1993) emphasized that, along with this nonlocality, their theory also implies a quality of wholeness linked to the global character of the many-body wave function.

that this was just a peculiar, unfortunate feature of the said model. But the violation of the locality hypothesis demonstrated by the theorem has shown that this is not the case. Any theory aimed at describing "reality as it really is" and compatible with the experimental data is unavoidably incompatible with the principle of analysis, as soon as the latter is understood, as explained above, according to the mode of spatiality.

REMARK 1

The Broglie-Bohm model has another distinctive feature. Contrary to standard quantum mechanics it is essentially deterministic. The question of determinism is not quite as important as that of locality, so that in this book its study is postponed to section 14-5. But this is not to say that it is not significant. Of course it is! Indeed, when quantum mechanics was discovered the fact that it is not deterministic was the one that, by far, was found the most striking. Correlatively, when the Broglie-Bohm model was discovered, the fact that by introducing hidden variables it restored determinism was very much set at the forefront, while, by contrast, the nonlocality of the model, although Bohm stressed it, did not attract much attention. Now, it is a fact that, at first sight, the notion of hidden variables and particularly that of hidden particle positions seems to convey the idea of locality. This explains that Bell's theorem, which disproves all realist *local* theories and, among them, hidden variable ones, was (and still is) interpreted by some to mean that *all* hidden variable theories are disproved. And since it seems impossible to restore determinism without introducing hidden variables it was (and sometimes still is) mistakenly inferred that determinism is strictly refuted. This reasoning is wrong, as shown, in particular, by the fact that actually the predictions from the (deterministic!) Broglie-Bohm model *agree* with the quantum mechanical ones. Moreover, it had the pernicious effect of concealing the real nature of what, combined with Bell's theorem, the Aspect-type results actually prove, namely, nonlocality. As for determinism, it is true that the results in question do contribute to making it implausible. But, to repeat, strictly speaking they do not disprove it (and this is why Bell himself could go on considering the Broglie-Bohm model to be consistent and interesting, even after he had discovered his theorem).

REMARK 2

With respect to the nonseparability—or nonlocality—of the pilot wave or of some other wave functions it should be remembered that, within the language convention here made use of, an "ordinary" wave (on water, in air, etc.) is always "local"; and that, similarly, the classical electric and magnetic fields are local. As we know, these entities are so styled (even though they normally are nonzero over a wide range of values of their arguments) because they are functions of, besides time, but the three coordinates x,y,z of *one* point in space. On the contrary, the pilot wave, like the wave functions of particle pairs introduced in section 3-1, is a function of, besides time, $3N$ variables, N being the number of involved particles.[22] This fact is worth noticing since it shows how misleading it would be to consider nonlocality to be a feature of classical electromagnetic theory or any other classical field theory.

3-3-7 Nonseparability and the Atomic Theory

Challenging the atomic theory may look surprising. Is it not proven that the atomic hypothesis accounts for an enormous amount of facts, in a considerable variety of fields? Were atoms not counted? Were they not, more recently, seen? And, what is more, manipulated?

And yet, such a challenging is appropriate. Not, of course in view of expressing reservations concerning the usefulness of the atomic—or, more generally, corpuscular—representation. There are immense domains within which, in order to describe experiments and their results, the notions of atoms, particles, molecules, and so on have to be introduced. It remains true, however, that usefulness never constitutes a proof, or even a convincing indication, of existence. In order to clearly realize this, let us imagine a person who has thoroughly assimilated general relativity theory, with matter-induced space-time curvature and so on, and is fully convinced that this theory correctly describes reality. If this person decides to build a house, he or she will have to manipulate bricks and other heavy objects. Hence she will, repeatedly and painfully, experience the notion of heaviness. In other words, the usefulness of the concept of a gravitational force pulling downward will become quite patent to her. Nevertheless, if she is as faithful to Einstein's message as we assumed her to be, this will not (and should not!) convince her that something like a gravitational force "really" exists.

The apologue should also remind us of the "underdetermination of theory by experiment" already alluded to. Plentiful as may be the set of the operations (including scientific experiments) that the atomic theory enables us to perform and to interpret, the idea remains conceivable that some other entirely different theory would do as well in that respect.

---

[22] It is said to be an N-point function, or equivalently a one-point function but defined in the abstact $3N$-dimensional space called configuration space.

Such objections—some may claim—cannot be raised against the plain fact that it is nowadays possible to "see" individual atoms (thanks to the "tunnel effect microscope"); nay, even to move them at will. Well, it is true that these possibilities are now available. However, it cannot be claimed that they really show that atoms "exist" independently from the rather special circumstances at work in such experiments (where the atoms are those at the surface of some crystal lattice). And, much more generally it must be observed that no strictly *certain* apprehension of an object ever takes place. The so-called "optical illusion" phenomenon always remains a possibility, as the commonplace "broken stick" experiment trivially shows. And as for the possibility of manipulating objects, it also does not prove that the objects in question "really exist." This clearly follows from the fact that, in the broken stick experiment, although, according to our normal way of thinking, the "break" of the half-immersed stick does not really exist, we manipulate it most easily, just by gently raising or lowering the stick.

Admittedly, remarks of this type would rightly be termed mere chicanery if, from a scientific point of view, multitudinism, that is, the philosophical thesis that everything is composed of a myriad of tiny objects, could be taken up as a universal, conceptual framework. In other words, if it were one in which the various sciences could naturally get inserted. But, through its very existence, quantum mechanics shows that such is not the case. Indeed, if one among its features catches the eyes at first sight, it assuredly is the fact that the "wave functions" used in it are "spread." That, for example, the one of the electrons of an atomic or molecular system appreciably differs from zero over either the whole volume of the system or at least quite a substantial part of it. Moreover, this wave function is, as a rule, nonfactorizable, which means that, just as in the example of the photon pairs considered in section 3-1, it would be meaningless to speak of the wave function of each individual electron. In other words, the function in question is one, not of just three coordinates, as is any function in ordinary space, but of $3N$ variables, N being the number of involved electrons. Clearly, it is right at the outset that the mathematical ingredients of this discipline suggest a conception quite at variance with the picture (characterizing the atomic philosophy) of several corpuscles each one of which occupies, at any time, within three-dimensional space, a place distinct from those of the others.

To such observations one might be tempted to object that at least the corpuscles that are "free" (not bound to other corpuscles) do exist individually, each one being at any time at some definite place, and that this is clear since, in bubble chambers, we see their individual traces. Alas! As we noted above, in the last resort this proof is not one either. For, indeed, the account quantum mechanics yields of the formation of these traces is the only one that fits into a general theory and it makes use neither of the notion of trajectory nor even of that of an individual "something" the marks of which are registered.

It is true that the underdetermination of theory by experiment that was used above as an argument for questioning the atomic theory seems to be a double-edged one. Somebody having kept aloof from contemporary physics might well think of using it in order, on the contrary, to weaken the doubts expressed here concerning atomic theory. "It is impossible" she would claim "to a priori bar the hypothesis that, in the future, a theory will be discovered that will differ from quantum mechanics but will yield the same observational predictions while merely bringing into play localized corpuscles linked by ordinary forces."

Indeed, in the literature concerning problems of this type such an argument is often put forward. And of course it can be made irrefutable. To this end it suffices to take the phrase "it is impossible to a priori bar" as signifying "we cannot bar, etc., as long as we remain within the realm of, exclusively, general philosophical ideas." Otherwise said: "as long as we refrain from making ourselves acquainted with some precise quantitative data." Unfortunately this amounts to wearing blinders, for indeed the data in question may well entail constraints that falsify the conclusion. And in the present instance we know this is just precisely the case, by virtue of Bell's theorem.

To what extent is philosophical atomic theory thereby refuted? To form an idea of the answer it is best to consider again the most remarkable pilot-wave model due to Louis de Broglie and generalized by David Bohm, already mentioned in section 3-3-6 and elsewhere. As noted there, in it we do find the idea of a large number of corpuscles existing by themselves, each one occupying at any time a well-defined place, and moving around. Admittedly, this is the guiding idea of the atomic picture. But it is considerably watered down there; indeed, to the extent of being deprived of what, philosophically speaking, constituted its whole interest and significance, namely, the idea that everything that exists is composed this way. For, according to the model, the pilot wave also exists, as we have seen. And, as we noted, it is not divisible by thought. It constitutes a great whole, definable only in a $3N$-dimensional space, N being the corpuscle number. Such an existence "per se" of an essentially nonlocal entity is what, in the model, explains the above-described well-correlated-at-a-distance behavior of photons that are created together. It is therefore quite an essential element of the said model, and because of it the model is, in the last resort, nonatomistic. The complete generality of Bell's theorem implies that this conclusion is general and holds good concerning any model that fits into objectivist realism and the observable predictions of which are not at variance with quantum mechanics. In view of the theorem, any such model is bound to imply an essential nonlocality obviously quite incompatible with philosophical atomic theory.

It is true that, in the pilot-wave model, what is actually perceived is the set of the corpuscles, that is, the variables usually termed "hidden" (a misleading name, it must be admitted), not in the least the pilot wave. In other words, it is proper to say that the latter merely "pulls the strings," without ever appearing on the scene. From this it follows that the persons for whom the said model is the paradigmatic example of all ontologically interpretable theories sometimes tend, in their philosophical arguments, to forget about this nonseparable pilot wave and claim that everything is made of atoms. Needless to say, particularly concerning philosophical reasoning, oversimplifying matters in this way is unacceptable! Consequently, all that may be claimed by somebody anxious to save something of the ontological line of thought that led to the atomic world view is that there are models (e.g., the pilot wave one) that *are* ontologically interpretable and in which *some* entities (the corpuscles) are endowed with properties fitting philosophical atomism.

We shall come back to the atomic conception when we study materialism (section 12-3).

3-3-8 Remark on the Role of the Pilot Wave

It is, if not nonseparability proper, at least the fact that the wave function is spatially extended that explains why the Broglie-Bohm model is compatible with the outcome of the Young fringe experiment. Of course, in this model any corpuscle passes through but one slit. But it is acted on by the pilot wave, which does pass through both slits at the same time and which, beyond the diaphragm, is therefore not the same according to whether the other slit is shut or open. The state of affairs at a place that (on the atomic scale) is far from the corpuscle trajectory thus has, on the behavior of the latter, an influence that may be termed "nonlocal" (although no entanglement is involved).

Note that, in this model, it would just be absurd to take the presence, at some time *t,* of a corpuscle within one particular slit as a pretext for attributing to it a wave function that at *t* would be nonzero only within that slit, and to ground predictions on it. In the model, predictions exclusively depend on the pilot wave as defined by the overall experimental conditions (source, diaphragm, slits, and so on). More generally, while, in the model, a corpuscle has at all times definite properties, its future depends not in any way on the latter but merely on the overall wave function. This is what explains why some conceptual difficulties specific to the "standard" form of quantum mechanics and with which we shall get acquainted later (section 8-1) are not present in the considered model.

CHAPTER 4

OBJECTIVITY AND EMPIRICAL REALITY

৪৬

4-1 Strong Objectivity and Weak Objectivity (Alias Intersubjectivity)

O UR ORIGINAL program is well on the way. The need to go beyond the familiar concept framework was demonstrated in chapter 2 and the necessity of going over from multitudinism to a more holistic view of whatever is meant by "Reality" was established in chapter 3. What remains to be considered and will hold our attention for a longer time is the philosophically even trickier question whether or not the objectivist language is of universal use in physics. In other words, it is the question whether or not physical realism may be kept, at least as a universal "everything takes place as if." Of course, these three subjects come within the scope of general epistemology. As regards the first two it is nevertheless on the basis of facts, not exclusively of ideas, that we examined them. Abiding by this method, we shall do the same with respect to this one.

4-1-1 Quantum Interference and "Weak" Objectivity

In the quantum domain the most significant fact is, by far, the existence of interference effects involving particles. The phenomenon is well known. Consider a diaphragm with two parallel slits.[1] If a particle beam passes through this device and impinges on a screen, so-called "interference fringes" appear on the latter, which could not be explained if the particles behaved according to classical physics and hence seem to reveal that they somehow possess wavelike features. Because of its similarity with Young's original experiment (bearing on light) this phenomenon is often called the "Young slit experiment" even when it involves particles other than photons, and we shall take up this convention. Note that for the fringes to appear it is necessary that, on their way, the particles should undergo no appreciable interactions. If, for example, some dense gas were blown

---

[1] Needless to say, all the following descriptions of experiments are considerably schematized, in order to bring out their most significant aspects.

between the diaphragm and the screen, with the consequence that very many particles would hit molecules of that gas, the fringes would fade. At high density, that is, if such interactions became the rule, it is expected that the observed behavior of the particles would simulate the one they would have if they obeyed classical physics. Their impacts on the screen would gather within two blobs corresponding to each one of the two "pencils" (of light if the particles are photons) issuing from the two slits.[2]

The formation of such interference phenomena is a massive, bulky fact, the conceptual bearings of which are all the more considerable as it is but one out of many equally uncommon facts, all of which are well accounted for by an extremely general mathematical formalism called quantum mechanics. However, the words "accounted for" should, as we already know, be understood here in a well-specified and somewhat restrictive sense. Quantum mechanics synthesizes these facts, encompasses them in its equations, and yields rules for predicting them. But the light it sheds on them does not reach beyond that point. If we are anxious to better understand "what takes place" it is up to us—along with many other interested people!—to try to find out the answer.

One important point to be noted first is that, in such experiments, no "fraction of a particle" was ever observed. On the screen the impacts are always those of "full fledged" particles. If we try to capture one—by means of some measurement or other—"as it goes past," what we shall observe will always exhibit the features of a "whole" particle. And yet, if at all times they kept their status of genuine "particles," that is, of essentially localized entities, each one of them should pass through but one slit and hence fall on the screen at a place roughly opposite that slit, which implies that when all the particles in the beam have finally reached the screen their impacts on it should group within the two already mentioned blobs. Still, as we noted, in the no-gas case, we see fringes. To try and remove the enigma, two approaches may be considered.

To be grasped, the first one requires—at least at the stage at which we now are—no conceptual effort whatsoever. It consists in granting that, indeed, the particles preserve their status of particles even when not observed and hence that each one of them passes through one slit only, but in assuming that they are driven by a special force—termed a *field* or *wave*—that bends their trajectories in some peculiar way. It is then conceivable that this force should somehow depend on the existence of the

---

[2] Up to now the "with gas" scheme is but a thought experiment. It serves to describe observational predictions that unambiguously follow from the quantum predictive rules when the "particle" is assumed to interact with the environment, here symbolized by the gas. For more details and references to actually performed experiments (with neutrons, etc.) see, for example, Giulini et al. (1996), p. 67.

other slit, the one the particle does not pass through, and the conceptual difficulty is thereby removed. This, as we know, is just the guiding idea of the Broglie-Bohm model (the quantitative features of which this is not a proper place to report on).

The other approach seems stranger at first sight, for it is grounded on a "great truth" that is of no use in everyday life and of which, therefore, we hardly ever think. This is simply the fact that the domain of validity of a notion may well be limited. Here, admittedly, the particle concept adequately fits our experience concerning emissions, impacts, and more generally what "takes place" when some direct or indirect observation is involved. But the point is that this does not guarantee that the concept should be valid generally, and in particular in between observations. Within the realm of this second approach, which centers physics on experience more than on any "real stuff" notion, the use of classical concepts—the one of *particle* (that implies locality), those of *velocity*, *field*, *force*, and so on—is not at all banned. But it does not partake of any kind of rational necessity and is not, therefore, imperative irrespective of circumstances. In the present case it must be considered that, when we have to do with the simpler of the two experiments considered above, the one without gas, the particle concept applies at the point of impact on the screen, possibly at the source, but not in between. As regards the question "what takes place in the interval?" it must be approached with caution. Having prepared the particle beam he or she makes use of, the physicist has pieces of information concerning it. These she synthesizes in the form of a mathematical symbol—call it D—conventionally termed the "initial wave function" by some and the "initial state vector" by others. Moreover, she has at her disposal a general differential equation (the Schrödinger one), that enables her to derive from D, for any later time t, a new "symbol" D_t called the "time-dependent wave function" (or "state vector").[3] And she also disposes of a general rule called the *Born rule* that, applied to D_t, yields probabilities. Not really probabilities for the particle to *be present* at such and such a place at such and such a time, since the idea that a "particle" is always localized independently of our knowledge is not assumed; just probabilities for the particle to *be found* at such and such places and times if it is looked for—then and there—by means of some appropriate device. In our experiment, the screen is such a device. The density of the impacts

[3] As physics students know, in practice the calculation is even simpler. The particles are prepared with a well-defined momentum, and this means that their wave function can be approximated by a plane wave extending at any time over the whole space, which renders the computation of D_t fully trivial and, in fact, unnecessary. On the other hand, the approximation in question is but a computational simplification procedure void of conceptual significance. In a concept-oriented analysis such as this one, introducing and discussing it in the main text would have needlessly weighted down the latter.

we observe at any point of it must therefore be proportional to the probability thus calculated at that point. And indeed it is. The prediction thus computed is in full agreement with the observations relative to the fringes, their existence, the distances between them, etc.

Since D_t synthesizes all the information he or she has, the physicist is often tempted to consider that it is a physical reality, to assert that, between emission and impact, the particles "are" wave functions or, more briefly, waves. But in view of the foregoing such a claim is open to criticism. For it implies assuming that something not directly accessible to observation may be described as if it were, that is, may be given a familiar name—wave—of which we are not strictly sure that its domain of validity covers such cases. The attitude of naive uncontrolled objectification such a language implies is not really in the spirit of the considered standpoint.

The two approaches just described have been investigated by successive generations of physicists. Above (chapters 2 and 3), the reasons were briefly indicated why, nowadays, the first one is in favor with but a small minority of them. Still, we shall come back to it later on (and we shall find it has some interest nevertheless). But, for the time being, let us consider the second one. Within it, it is interesting to investigate what the result is of applying the foregoing remarks to the "thought experiment" of Young fringes with gas. When, as we noted, the gas is dense enough it is expected that the interactions between the particles and the gas molecules should make the interference fringes vanish. In other words, it is expected that in the area between the diaphragm and the screen the particles should, as far as appearances (or, in other words, "phenomena") are concerned, behave more or less as they would if they obeyed classical physics. Someone who, having momentarily forgotten how the beam was formed, merely took into account the particle behavior within that area, might therefore be tempted to attribute to them classical properties such as positions, velocities, etc. Within what conceptual limits are such ways of thinking valid? The question is a delicate but far-reaching one, as we shall see in a moment.

On this subject the first point to be noted is that, strictly speaking, even under the considered assumption the "way of thinking" just described (attributing, under the specified conditions, classical properties to the "particles") would not be really acceptable. It is true that, when some dense enough gas is present, the possibility of observing interference fringes on the screen is expected to vanish. And, more generally, that the same is expected to take place concerning the possibilities of observing any kind of interference phenomena merely involving the particles composing the beam. But if the particles then *really* became classical, these possibilities are not the only ones that would vanish. The same would hold true concerning those of a similar nature but involving, along with the said particles, the gas molecules with which they interacted. Now, technically observing quantum phenomena of such a complex nature is extremely difficult and it is hard to imagine how it could actually be done. But it so happens that experiments fundamentally similar to that one are at least conceivable, and that, concerning them, the (generalized) Born rule predicts that phenomena similar to the interference ones (for a while, call them "phenomena B") will appear. This theoretical consideration should suffice to prevent a physicist informed of the beam preparation from thinking that, when there is enough gas, after having gone through the diaphragm each one of the initial particles is "really" in one or the other of the two "pencils" defined by the two slits.

We saw that, by virtue of the rules and equations of the quantum formalism, the mathematical symbol D_t calculated from D yields the probability that, if we place somewhere a measurement device enabling us to possibly observe a particle, we get the outcome "particle observed." Of course, since the above statement looks complicated, we feel tempted to make it simpler and just say that D_t yields "the probability for the particle to *be*" at that place. But it turns out that this change would be a mistake. Not only would the allegedly simpler formulation be alien, as already noted, to the spirit of the considered approach, indeed, it would actually be faulty. In the no-gas case the mistake would be quite patent since the above simplified statement (the one between quotation marks), if it is at all meaningful, implies that at any time the particle is with some probability at some definite place, even if it is not observed. This means that, at any time, it is intrinsically localized (even if this localization is not predetermined by the initial conditions). But this would bring us back to our "first approach" and—for lack of the very special guiding mechanism that is an inherent element of the latter—would prevent the fringes from being formed. Note, moreover, that the oversimplification would prove to truly be one also when the gas is present. For it is incompatible with the existence of the phenomena called above "phenomena B," which admittedly are complicated and elusive ones but still should in principle be observable.

From this a most important lesson is to be drawn. It concerns the notion of scientific objectivity. Scientific statements are termed objective in contradistinction to those bearing on opinions, tastes, and so on, called "subjective." And many scientists seem to think that this is all there is to say on the matter. But this is not entirely true for, in science, the word "objective" indistinctly qualifies statements that, in fact, are expressed in significantly different ways. Some have a form that makes it possible to consider that they directly inform us of some attributes of the things under study. They do not refer (not even implicitly) to the community of the people who know—or get informed of—their content. Let them be called *strongly objective* statements. But others are not of this type. They explicitly or implicitly refer to some human procedure (of observation, for example). Such statements are quite appropriately called objective since, by definition, they are true for everybody. But their form (or context) makes it impossible to take them to be descriptions of how the things actually are. Let them be termed *weakly objective*.[4] Of course science swarms with assertions quite generally considered strongly objective. In classical physics, practically all statements—at least all of the basic ones—are of this type, so that there was a time when it could reasonably be conjectured that *all* scientific statements might be set in that form. But to this conjecture the content of the foregoing paragraphs nowadays constitutes an outstanding counterexample. For indeed, is there, within contemporary physics, anything more basic than the quantum *Born rule*, which, in its most general version, yields the probabilities that if such or such measurement is made on a system prepared in such and such a way, such and such a result is read? Now, this is manifestly a "rule of the game," stated in weakly objective terms. And, on the particular example of the quantity "position," we just saw that, in the "particle" case, it cannot be reformulated in strongly objective terms.

Finally we therefore see that, unless we choose to leave the main stream of contemporary theoretical physics and swing to one of the ontologically interpretable models (whose difficulties we mentioned), we have to grant that weak objectivity is one of the prominent features of contemporary physics. Let it be precisely defined.

Weak Objectivity.   A statement is "weakly objective" when it implies (directly or indirectly) the notion of an observer but is of such a form (or occurs in such a context) that it implicitly claims to be true for any observer whatsoever.

4-1-2 The Traces Example

The quantum theory, reported in section 2-5, of the traces that particles generate in cloud chambers—and in photographic emulsions as well— offers a nice illustration of weak objectivity and of the role of *observa-*

---

[4] Bringing in new expressions—or making use of common ones in a new sense—is always a delicate move, to be taken only when the formulation of what is to be stated is thereby truly simplified. But I do really think that this is the case here. On the other hand, the choice of the adjectives "strong" and "weak" could admittedly be questioned on semantic grounds. Let it be clear that it no more implies a judgment of value than is the case concerning "strong" and "weak" convergence in mathematics. In fact, the choice of the epithets is of no significance. The significant point is that, when two notions differ, their names should somehow be distinguished.

tional prediction in quantum physics. At first sight, as we noted, these alignments of bubbles seem comparable to the white trails produced by a jet plane in a blue sky. Hence, when the "particle" source is external to the chamber (or the emulsion), we not only attribute to each "particle" inside this device a well-defined trajectory (coinciding with the trace) but also do not hesitate to continue the latter to the rear, by thought, up to the particle source. So that, if we have to do with cosmic rays, as in the section 2-5 example, we think of them as objects traveling through space with something like a rectilinear, uniform motion, and therefore occupying, at least approximately, a well-defined place at any time. However, as we also noted, such a picture does not fit with quantum mechanics (nor, incidentally, with the Broglie-Bohm model, in which the corpuscles continually undergo deviations dictated by the whole-universe wave function). And anyhow it is clearly inadequate since, in the case in which a diaphragm with two nearby slits is placed between the source and the observational device, the picture in question would imply that the particle, having a trajectory, necessarily passes through but one slit, which obviously (the Broglie-Bohm mechanism not being assumed) would be contrary to what we know from the Young slit experiment. The true explanation of the observed alignments is not, therefore, to be looked for within the realm of such ideas, great as may be the force with which our intuition puts them forward. It essentially lies in the fact that, when the initial conditions are sufficiently known quantum mechanics makes it possible to predict what will be observed.[5] It does this, as we know, by introducing mathematical symbols that were given names (wave function, state vector, etc.) and many of which evoke some picture. But, to repeat, the pictures thus called up are unreliable ones and play no role in the calculations. What quantum mechanics in fact yields are merely the probabilities that, for a given initial flow, microblobs will be observed at such and such places within the device. And, as already noted, the probabilities concerning the cases of the blobs being aligned along the general direction of motion are considerably larger than those relative to any other configuration. In other words, what quantum mechanics predicts is just that, within the device, we shall see alignments (of microblobs or bubbles) consistent with what we actually observe and naively interpret as being "traces." This justifies the quotation marks we placed above around the word "particle." Here as in the foregoing section the word "particle" is manifestly just a name, borrowed for convenience sake from the vocabulary of the physics of yore but to which we should take care not to impart the meaning it then had. And we also find on this example that a theory

---

[5] Or to know the probability that such and such result will be observed. Here the word that is significant for our purpose is not "probability." It is "observed."

merely aimed at predicting observations may quite efficiently compete with a descriptive one with respect to the feeling of understanding it may generate.

4-1-3 Comments

In view of the deliberately philosophical trend of this book, special precautions were taken here that do not, as a rule, appear in quantum mechanical textbooks. In fact, the latter mention the notion of "domains of validity" only quite exceptionally. And practically none of them ever points out that it is conceivable that our basic conceptual equipment should not be complete "right at the start." That, in other words, cases may exist, that are simply not describable, even in principle, by means of the array of our familiar concepts. Concerning the interference phenomena these oversights have the consequence that the readers of these textbooks instinctively tend to fill in (so to say) the blanks. That is, they tend to think of the particles by means of well-known notions, even at times when, by assumption, they are not observable. This leads many of them to take at face value expressions such as "wave function," "state," and "state vector," and consider therefore that between the production and observation times the particles either "are waves" or at least "are in a state" adequately represented by the corresponding state vector. True, this is a language that is convenient, and even practically necessary whenever calculations are to be done, since it exempts us from forming long, roundabout sentences that would weigh down the procedure. But it must be stressed that, conceptually, taking it at face value makes the measurement problem (to be discussed later) even less tractable. Admittedly, to some extent at least the use of this language makes it possible to do without the "weak objectivity" notion. For example, in the context of the Young slit experiment some note (as we did above) that the interaction that takes place at a certain time between the screen and the particle is, in a way, tantamount to a "measurement" performed by the former on the latter. They then claim that—as all measurements do—this event "reduces the wave" that, immediately before it took place, the particle "was," that it renders it pointlike. And that therefore the probability the formalism yields is indeed a probability that, just after the interaction, the particle should "be" at the considered place. However, as we see, this restoring of strong objectivity is only achieved at the price of introducing at least two most unpalatable ideas: the one of a sudden "reduction" of a physical wave and the one that *some* interactions should be identified with a special phenomenon, termed "measurement," the nature of which is left undefined.[6] Clearly, these are "ontological incongruities" that we should avoid if we can.

In this spirit some theoretical physicists, upholders of the "orthodox" quantum mechanical approach, for a time laid their hopes in formulations of quantum mechanics that differ from the above-described one while yielding the same observational predictions. One of them (a most useful one, by the way) is called the "Heisenberg representation." Its specificity is that it represents the physical quantities by means of "time-dependent self-adjoint operators." Another one—which took up, in fact, different forms—consists in modifying logic in some ways. The investigations carried out in these fields were of considerable interest in that they increased our knowledge of the logic-mathematical structure of the formalism. It must, however, be observed that, for reasons to be made explicit later (sections 9-5 and 9-6, "Remark") they do not, by themselves, open a way toward restoring strong objectivity.

On the other hand it is noticeable that—despite such a persistent problem of consistency, the existence of which most theorists willingly grant—many physicists cling to a realist, descriptive standpoint that renders the weak objectivity concept quite unpalatable to them. Incidentally, let it be granted that in favor of their view they have arguments of a general, philosophical nature that cannot and should not be just simply brushed away. In fact, to "decide" whether or not objectivist realism—basically aimed at describing "what really is"—is *the* minimal demand fundamental knowledge should satisfy is truly a crucial debate. Some do think it is. As for me, I tend to consider, when all is said and done, that it is not. However, this is a problem that—while debatable by physical means as we shall see—essentially pertains to the philosophical realm. Investigating it here would draw us aside from the observational physics data the information developed in this chapter bears on. Let us therefore postpone its examination for a moment. The next chapter is entirely devoted to it.[7]

---

[6] Traditionally, at this stage, the notion of an "irreversible act of amplification" is alluded to, and sometimes it is even viewed as being the *physical cause* of the "reduction" in question. As Wheeler (1984) appropriately stressed, a difficulty with such an approach lies in the (formidable) question once asked by Wigner: "What does amplification mean?" (or, more generally, "what does irreversibility mean?"). Can it be defined as a purely physical process, that is, without any reference being made to our apprehension abilities? Wheeler's answer was a definite "no," for, as long as we do not look at them, processes that we should unquestionably consider irreversible can in fact—or could, in principle—be reversed, and prevented thereby from "coming into being." In substance, he inferred therefrom that events should be explained on the basis of *meaning*, not the reverse. I think this argument is conclusive (for more details on this see my *Reality and the Physicist* [d'Espagnat 1989]).

[7] Within the approach centered on the weak objectivity notion it is clear that we cannot (hence should not) speak of any action of the consciousness of the (individual) experimentalist—held to be a "reality-per-se"—on the wave function, considered to be another "reality-

4-1-4 Miscellaneous Remarks

1. The weak objectivity of quantum mechanics is quite remarkably brought to light by the following statement, due to Niels Bohr:

> The description of atomic phenomena has in these respects a perfectly objective character, in the sense that no explicit reference is made to any individual observer and that therefore . . . no ambiguity is involved in the communication of observation" (Bohr 1958).

Identified, as in the above quotation it explicitly is, with intersubjective agreement, this objectivity obviously differs from strong objectivity.[8] And already in previous works I called it "weak" in order to distinguish it from the latter.[9]

2. On the other hand, it should be noted that the intersubjectivity considered here has hardly anything to do with the partial, or "relative," intersubjectivity that philosophers and sociologists are mostly interested in. What *they* have in mind is an intersubjective agreement bearing on perceptions, that is, on already very much interpreted sensations (here is

---

per-se." In recent articles or books the fact that no such action needs be assumed is often stressed, and rightly so. On the other hand, the reason sometimes put forward is that it is merely *indistinguishability* that makes the difference between cases in which there are interference effects and hence no "reduction," and cases in which reduction takes place. The claim then is that there are no interference effects whenever it is in principle possible to know which path the particle took, and there are whenever this is impossible in principle. A difficulty with this standpoint is that it is ambiguous. Taken literally, the statement that in such and such circumstances "we cannot know" which path the particle took implies that, even in these circumstances, the particle did take a definite path (that we do not know). It is then difficult to understand why interference effects nevertheless take place, and the fact that they do appear precisely in the cases in which *we* cannot know "which path" must be considered a mysterious coincidence. Alternatively, the statement in hand may be thought to be something more basic than just a criterion that happens to work. It may be understood to be a strict application of the operationalist axiom that, whenever an assertion is such that, in principle, we cannot know whether it is right or wrong, this assertion is meaningless. This second standpoint seems more reasonable than the first one. But it is clearly centered on the notion of knowledge, viewed as being a primeval one. Hence we should realize that it goes along with recognition of the fact that quantum mechanics is essentially focused on knowledge, that is, weakly objective. According to it, physics is a description of but *empirical* reality (section 4-2-3 below).

[8] As stressed below, its domain of validity is larger. All the strongly objective statements can be couched in a weakly objective form (by referring to the corresponding observations), whereas the opposite in obviously untrue.

[9] That Bohr identified objectivity with intersubjectivity is a fact that the quotation above makes crystal clear. In view of this, one cannot fail to be surprised by the large number of his commentators, including quite competent ones, who merely half-agree on this, and only with ambiguous words. It would seem they could not resign themselves to the ominous fact that Bohr was not a realist.

a flower, a heavenly body emerging from the sea, a ghost, etc.). And they rightly stress that such collective agreements largely proceed from a community of knowledge or ideas and cannot therefore be universal. On the contrary, weakly objective statements keep very close to the raw, unanalyzed sensations. They are of the type "in such and such circumstances a luminous sensation is felt." Such statements do not refer to any culture. The fact that a theory should be grounded on them thus in no way implies that it should merely reflect one of the latter. It may be just as universal as an ontologically interpretable theory. All the same, there may exist different types of weakly objective statements. One of our tasks will be to define them.

3. The fact that "orthodox" quantum mechanics (corresponding to what was called the "second approach" in section 4-1-2) involves axioms, such as the Born rule, that are but weakly objective makes it possible to evaluate the pertinence of an analogy made use of in section 1-1. The question was whether or not the "conceptual revolution" generated by contemporary physics may be compared in importance to the one that resulted in the sixteenth century from the advent of Copernicus' astronomical theory. In scientific circles the bent is rather toward a negative answer. Many scientists feel inclined to consider that, on this point, we should remain cool headed, their argument being that classical physics may be regarded as an approximation of the present-day one, which obviously the geocentric theory is not with respect to the heliocentric one. However, this argument is most often put forward by physical realists and, coming from them, it is not really acceptable. It is true (we shall verify this in detail in chapter 8) that, from a purely operational point of view, the classical physics phenomena are adequately explained on the basis of quantum physics, with the help of suitable approximations. But, for the persons in question, to accept to consider science as a merely operational predictive tool obviously would not constitute a self-consistent attitude. As a matter of principle, they think of it as being a slow, gradual approach to reality as it really is. However, considered under this light (that is, as a strongly objective description) classical physics in no way appears an approximation to "orthodox" (textbook) quantum mechanics since, short of falling into the above noted "ontological incongruities," the latter must be held to be but weakly objective. If these scientists' realist standpoint is adopted, the argument therefore fails. And it follows that, logically, whoever has a natural tendency to share their conceptual outlook should consider the prevalence of "orthodox" quantum mechanics to be—potentially at least—a conceptual revolution comparable in scope to the Copernican one.

4. It is not infrequently (and quite rightly) stressed that the (orthodox) quantum formalism is predictive rather than descriptive. But an additional point should be stated. As the analysis of the Young slit experiment makes

clear, the formalism in question is not predictive (probabilitywise) of *events*. It is predictive (probabilitywise) of *observations*.[10] Correlatively, its main innovation with respect to classical mechanics does not lie in the fact that it calls in intrinsic probabilities but in the fact that its probabilistic statements are but weakly objective. This point is all the more to be stressed as commentators, including most competent ones, seldom even mention it. The question is to be considered again in section 14-5.

5. As a sequel to Remark 3 of section 3-2-3 it is worth noting that the fact that the quantum formalism is but observationally predictive alleviates a difficulty about reconciling nonlocality (apparently implying supraluminal influences) with the special relativity law that no supraluminal influences occur. The point is that any theory, or just assertion, can obviously be expressed in terms of observational predictive rules (instead of saying "this rag is red" we may always say that if you, or I, or anyone else, looked at this rag we would have the sensation "red"). And relativity theory is, of course, no exception. Now, quantum formalism plays a prominent role in physics. This may incite us to formulate not only it but also all the rest of our knowledge in observationally predictive terms. But then the notion of influence boils down to that of signal, in a sense ultimately referring to human actions. Consequently, when such a "weakly objective" conception of scientific knowledge is universally adhered to, the very concept of an influence that cannot carry any signal whatsoever becomes unscientific. Clearly, then, the "supplementary theorem" removes the difficulty, since, according to it, what, in a realist approach, was (in view of the Bell theorem) quite naturally called a "supraluminal influence" does not carry any signal and is therefore, in this new approach, no longer an influence.

6. From a philosophical point of view the notion of weak objectivity is nothing new. Right from the start, but mainly since Kant's time, philosophers have been stressing the precariousness of any pretension at describing—beyond all possible human representations—"reality as it really is, up to its contingent features." As we saw, they point out that such a claim—which can easily be read between the lines of many scientific texts—cannot be backed up by any *proof* of its validity, and they assert that what science yields is but a representation. They might therefore believe that, for a physicist, giving up all attempts at forcing physics into

---

[10] This is true notwithstanding the fact that in many cases we normally hypostatize such observation outcomes and (conventionally) take them to be actually occurring events. This is legitimate for we then think of them within the framework of the (human-centered) empirical reality concept (see section 4-2-3). As a consequence of this the notion "event" is meaningful only within the realm of the latter. Correspondingly, since the wave functions are here taken to be just tools serving for observational prediction and are not even attributed *empirical* reality, their changes may not be considered to constitute, by themselves, events.

the strong objectivity mould is essentially tantamount to coming round to a view that has been known by them for a long time. However, such a judgment would not be entirely correct, for it so happens that philosophers underestimate the bearing of their own observation. Indeed they seem to take it for granted that the human representation that science yields should necessarily be expressible by means of the realist *language,* and that therefore everything takes place *as if* what physics describes were reality-per-se.[11] In this field, what contemporary physics teaches us—in so far as its interpretation in terms of weak objectivity forces itself upon us because of its efficiency—is that we must go farther and give up this "as if" standpoint, or at least its universality.

4-2 The Measurement Problem and Empirical Reality

4-2-1 A Remark Concerning "Complementarity"

Above, a sentence from Bohr was quoted that clearly shows that his own notion of objectivity coincided with the one we termed "weak." It is true nevertheless that, before the time when he wrote this, Bohr had introduced his famous "complementarity" notion, and that some authors thought a kind of revived, original realism might be construed on that basis.

The main difficulty at this point is that Bohr himself never gave any explicit definition of what he meant by "complementarity." But what is sure is that this word, wherever it appears in his writings, is not to be taken in its usual sense. When, in ordinary life, we speak of two complementary descriptions of the same object, we mean that these descriptions are partial, but in no way opposed to one another, that they are as two photographs taken of the same object, one full face and the other one in profile, so that they can be combined and that, quite naturally, this yields a more detailed knowledge of the object. Now—as is most well known—this commonplace idea is not the one that Bohr referred to. According to him it is admittedly true that, while no set of classical concepts entirely covers up the thing to be described, still some of them yield information of a certain kind, and may be used, together with other ones (called *compatible* with them) for building up a representation of the object. But the picture thus obtained is valid only within the realm of well-defined

---

[11] Some of them turned this into a principle. According to Kant the basic concepts that we may meaningfully use in our reasoning all belong to a fixed set (that of the "categories of understanding"), which more or less coincides with the set of concepts that the realism of the accidents makes use of.

experimental conditions, corresponding to the possibility of performing joint measurements of the physical quantities these concepts represent. It is incompatible—contradictory—with the one that, under different experimental conditions, we should build up of the same object. In other words, it is impossible to combine the two pictures in order to form a description more detailed that the ones yielded by each one separately. If we wanted to go along with the photographic allegory we might think of comparing this situation with the case when we take a picture of a landscape showing several planes using a wide-angle camera. If we focus on the foreground, the background appears fuzzy, and conversely. Such an analogy would, however, be defective and even, in the last resort, misleading. For it implicitly assumes the simultaneous existence—independently of the photographer's decisions—of all the objects lying in these different planes and of their forms. Whereas, on the contrary, within the Bohr approach the forms in question—nay, the objects themselves—depend on the experimental conditions.

Such is Bohr's complementarity. In view of the emphasis it lays on pictures we might be tempted to interpret it as a kind of return to realism. Not, obviously, to naive realism (refuted by the notion of conflicting pictures) but, perhaps, to dialectical materialism and its central theme of "contradictions within the things." The expression "regional ontologies"—borrowed from Husserl by François Lurçat (1990) for analyzing Bohrian complementarity—may also suggest the idea. But still, we should take care not to mix up different things. Upon reflecting on Bohr's views we must always keep in mind the essential role they impart to the experimental conditions, to, in Bohr's words, the "conditions which define the possible types of predictions regarding the future behavior of the system" (Bohr, 1935). For indeed, according to Bohr, these conditions "constitute an inherent element of the description of any phenomenon to which the term 'physical reality' can be properly attached" (*loc. cit.*). Now the experimentalist chooses these conditions. They therefore are "in some way, human" as Lurçat himself granted (*loc. cit.*). The "physical reality" Bohr had in mind cannot therefore be in any way identified with a "human-independent" or "per se" reality. Unquestionably it is but an *empirical* or *contextualistic* reality. And the corresponding objectivity can be but weak.

Some time ago, interesting theoretical developments by Roland Omnès (1988) were considered by him to constitute a refinement of Bohr's ideas. They have to do with the notion of "partial logics" and its use in reformulating quantum mechanics. Like the (related) theories of Griffiths (1984) and Gell-Mann and Hartle (1989), the Omnès one is expressed in a language with a realist overtone, in which such notions as those of events, sequences of events, etc., are systematically made use of. And indeed it seems likely that, at the start, these four authors either explicitly aimed at restoring a form of realism or at least thought such a standpoint could harmlessly be kept as, so to speak, a general background to their theory. But as such constructions took shape—and were refined in the light of some criticism or other—it became more and more clear that such a compatibility could not be easily upheld.[12] In particular, while, as we saw (sections 1-2 and 3-3-5), the counterfactual definition procedure constitutes an inherent element of conventional realism, Omnès' approach implies—just like Bohr's—standpoints that are definitely at variance with it. Such is, for example the one about a property of a system ceasing to be true if and when we remove the instrument with which the system was about to interact. Such conceptual oddities appropriately illustrate the fact that in such matters the scientific language almost unavoidably misleads those who are not on their guard. This is due to the fact that the scientist has, of course, to express himself, that he can only do this by means of the language he has, and that this language, for trivial and obvious reasons, is mostly made up of terms coming from the realism of the accidents.

In fact, as he himself finally acknowledged (see quotation in section 4-1-4) it was essentially within the framework of *weak* objectivity that (at least concerning the micro-objects) Bohr conducted his thought. And as a consequence, within the physicists' community the complementarity notion got but a mixed reception. Often formally praised, it was in fact quite rarely used. And, in particular, the theorists who worked on the measurement problem practically never had recourse to it. For reasons that will become apparent, we shall here follow their example.

4-2-2 Measurements and Quantum Superposition Effects

When investigating quantum measurements two points have to be taken into account. One of them is that (as already noted) the hypothesis of the universal validity of quantum mechanics rests on very strong arguments, the force of which will become more and more apparent as we go along. The other one is the fact that a realist, descriptive approach is favored, as we also saw, by most physicists. And that, within its realm, since the instruments, their pointers included, are made up of atoms, they must themselves be quantum systems (like, by the way, their environment). Surprisingly however, the quantum nature of the measurement instruments raises most serious theoretical difficulties that are still a subject of inquiry.

---

[12] These works were commented on in detail in *Veiled Reality* (d'Espagnat, 1995).

They will be analyzed in some detail in chapters 8 and 9.[13] Here I shall merely sketch the main lines of the one that is at the root of all others. Its importance proceeds from the fact that it arises as a direct and almost ineluctable consequence of what constitutes the mainspring of the whole quantum formalism, namely, the *superposition effect* (alias *superposition principle*).

In order to sketch the basic idea of the latter without mathematics, let it just be mentioned that, within the "descriptivist" approach under study, a physical system—let us think of an electron—may be in various states (of motion for example), that the formalism describes these states by means of appropriate mathematical symbols, *a,b,c,* say, conventionally called "state vectors," and that it may happen (this has no analogue either in classical physics or in everyday experience) that one of these symbols is the sum of two or several other ones. *c*, for example, may be equal to *a+b*, *a*, *b*, and *c* describing,[14] to repeat, three possible states of an electron.[15] *c* is then said to be a "quantum superposition" of *a* and *b*. If the measuring instruments, including of course their "pointers," are just quantum systems it is to be expected that the same should sometimes happen to their states. For example, we can imagine a measurement that, when performed on an electron in state *a*, sets the pointer in (quantum) state *A* and when performed on an electron in state *b* sets the pointer in (quantum) state *B*. And a question then arises: "What happens to the pointer when we decide to bring into play the same measurement procedure but, this time, on an electron lying in the state described by the symbol *c = a+b* ?" Quite unruffled, the formalism placidly answers: "This operation sets the pointer in the state *A+B*, called a quantum superposition of *A* and *B*." (In order to keep notations simple we here assume the electron gets absorbed within the "matter" the pointer is made of, an assumption we shall soon cast off.)

Now—alas!—not only does this answer fail to match our experience but we are even at a loss to grasp what it may physically mean. First, it does not match our experience since, when the measurement in question is actually made on an ensemble of systems all lying in state *c*, what we observe is clear and neat: sometimes we see *A* and sometimes *B*. And, second, we do not grasp what it means physically for, to what physical state of the pointer—to which one of its possible positions on the dial— could the *A+B* symbol correspond? Clearly, to none!

---

[13] In *Veiled Reality* they are analyzed more systematically but in a way that involves mathematics.

[14] It is on purpose that neither the symbols ψ, ϕ, $|>$ etc. nor the names usually attached to these symbols are used here. For a "layman," all of them are inadequately suggestive and hence generate faulty enigmas and illusionary vistas.

[15] For example, *a*, *b*, and *c* (or, better, $2^{-1/2} c$) may describe three different "spin states," of the electron, along differently oriented axes.

We might be tempted to consider such a "conceptual imbroglio" a good motivation for just simply rejecting the whole formalism. Or at least for judging that the universality hypothesis is unacceptable and that the said formalism is meaningless outside its "domain of efficiency," atomic and subatomic physics. But we would thereby ignore the fact that quantum superposition effects involving macro-objects are indeed indirectly observed in all the—admittedly rare—situations in which theoretical calculations show, both that they should occur and that, with the help of suitable devices, consequences of their forming are observable in practice. How is it possible to reconcile this with the fact we just noted that the pointer is always seen to lie in but one, definite, scale interval?

This is a difficult question, even though the theory has in it a feature to which a solution of the riddle may conceivably be related. This is the fact that, at the quantum level, macroscopic systems strongly interact with their environment. Consequently, when the dimensions of one of the involved systems are "macroscopic" (say, some 10^{-6} cm or larger) its environment cannot be ignored. And it follows that devices that would make it possible to exhibit quantum superposition effects are outstandingly complex, so that assembling them is extremely difficult. In an overwhelming majority of cases, including that of instrument pointers (viewed as quantum systems), this difficulty is tantamount to an utter impossibility. This circumstance is at the root of an important theoretical development, called decoherence theory, that will arrest our attention below. Here, we merely note that it entails what follows. Let us imagine a statistical ensemble of electrons e, prepared in state c. And let us assume that on each one of them a measuring process similar to the one considered above is carried out. This essentially implies that each electron is made to interact with an instrument pointer P, which we now consider to be just a (highly complex) quantum system. Let us assume all these pointers to be in the same quantum state at the start. On the systems S composed of one e, one P, and the environment (including of course the rest of the instrument) we may, by means of auxiliary instruments, measure all sorts of physical quantities. And in a situation such as the one considered above, measuring *some* such quantities would, in theory, yield results incompatible with the view that some pointers lie in state A and the others lie in state B. But decoherence theory shows that (fortunately!) all these quantities are so complex that measuring them is practically impossible. Consequently, by considering that, when the process is over, some pointers *do lie* in state A and the other ones *do lie* in state B we run no risk of being found wrong. In other words, the theory then "saves the appearances."

To neatly grasp, at the level of ideas, the situation we are here facing, the simplest way is to compare it with the one concerning the "Young slit with gas" thought experiment. The two problems are manifestly similar.

In both of them the formalism yields a large set of outcome predictions concerning observations that could *in theory* be made, most of which, as a rule are not actually performed, and among which a choice has to be made. And in both of them the observations in question form two groups. One of them consists of those that are most easily made: impacts on the screen and positions of pointers. The other group consists of observations that are extremely difficult or practically impossible to perform: correlation effects involving both the particles and the gas molecules and quantum superposition effects simultaneously involving the pointer and its environment. In both cases we noted that by disregarding the second group a vision of the world consistent with classical views is made possible. Indeed, this procedure of disregarding makes it possible to imagine, without putting oneself at variance with any data of a purely observational nature, that, between the diaphragm and the screen, each particle is "totally" either within one or within the other of the two beams the slits define. And, similarly, it makes it possible to consistently think of each pointer as being either "genuinely" in state *A* or "genuinely" in state *B*. In both cases these descriptions are entirely consistent with the testimonies of our eyes. The difference between them is merely one of outlook. It lies in the fact that, in the particle case, we quite naturally take into account what we know of the past history of the particle beam. And we therefore consider it unthinkable that the mere presence of a gas between the diaphragm and the screen should force the particles to pass (nay, even, to "have passed"!) through one slit only. On the contrary, in the case of pointers, lacking such an "intellectual railing" we feel almost invincibly led to believe that what we see is what "really is."

On this, however, a little thinking should make us cautious. When—to repeat—I look at some obliquely half-immersed stick, I see it is broken just as distinctly as I see that the sky is blue, or that a pointer lies at some definite place on a dial. And, nevertheless, I judge it is *not* broken![16] I must, therefore, doubt that there are observations that are direct to such a degree that I am assured they reveal reality "as it really is." Consequently, when I see that an instrument pointer lies somewhere on a dial, what I am sure of is merely that I have this visual impression. It is not that the pointer "is really" there, in some absolute, ontological sense.

4-2-3 Empirical Reality

On the other hand, our whole science is construed on such observational data. To take as "unreal" such "massive" facts as the position of a pointer

---

[16] "My reason straightens it," the writer of fables, Jean de La Fontaine, pertinently wrote.

on a dial would amount to a rejection of our whole scientific knowledge and, in the last resort, knowledge in general. Clearly therefore the interpretation of quantum mechanics that forces itself, as it were, upon us consists in taking the above mentioned "impressions" as seriously as we can. In other terms, it is to say that after a measurement process of the foregoing type (and still assumed to bear on a state of type c) has taken place, the pointer "really is" in some definite position (either A or B). The reason why such a claim remains tenable is that, all the same, it is objective, even though it is so only in a *weak* sense. For indeed quantum mechanics casts no doubts on the idea that if you, I, or anyone else looked at the instrument we would see the pointer to be in some definite position. The conceptual difficulty we are in merely comes from the fact that the claim in question fails to be strongly objective. The point is that strong objectivity implies counterfactuality, as we saw. And that in the considered situation some extremely complicated quantum superposition tests are conceivable, that are not done but could presumably, in theory, be done, and whose outcomes might quite well be inconsistent with the claim. Admittedly, such tests are quite infeasible in practice. But it should be clear that, for a realist, this does not remove the conceptual difficulty[17] since this very notion of *practical* infeasibility involves human beings and their aptitudes (or lack thereof). It is therefore quite impossible to make use of it for justifying the consistency of an alleged description of reality-in-itself.

Hence, if we really believe quantum mechanics is universal we cannot escape the conclusion that the statement "the pointer is really etc." is merely *weakly* objective. Correlatively we then have to say that, in it, the expression "is really" is not to be taken in the ontological sense many people would naively attribute to it. Since we nevertheless are quite forced to consider that some idea of reality is there involved, we shall have to make use of a philosophical notion, namely, that of the *phenomenon*. We shall say that what is at stake is "empirical" (or "contextual") reality, a notion to be considered—somewhat in the manner of Kant—as designating a set of *phenomena*, themselves partly defined by referring to the abili-

---

[17] Admittedly there exists a simple and well-known way of getting rid—in words!—of this difficulty. Its adepts first assert that instead of introducing such an abstract word as "reality" they will make use of simple, clear ones, such as "world" or "universe." They then state that they have no bent for brooding over the "ultimate nature" of the universe and are content with investigating such matters as lie within human reach. And they take advantage of this for setting aside considerations on measurements we are quite unable to perform and similar things. This is fine, except for the fact that they thereby leave in abeyance the question whether the physical properties they describe *belong* to the universe or are just elements of *our vision* of the latter. This, however, is *the* question that quantum mechanics raises! Leaving it in abeyance is not an answer.

ties of human beings at acting and perceiving. We already met with such a notion when discussing Bohr's complementarity. Here we note its pertinence concerning the measurement question.

On the other hand, to claim that, after all, physics is a way of accounting for phenomena thus defined rather than a description of reality-in-itself amounts to turning away from (any version of) physical realism. Hence a philosophically crucial question remains open, that of knowing whether or not it is possible to disregard the, apparently most impressive, arguments that are in favor of the said realism. This point will, to repeat, be the subject matter of the next chapter.

REMARK

So it appears that, after all, the weak objectivity notion plays quite a central role in our understanding of the measurement phenomenon. It is therefore important to form an adequate idea of what it is. Its difference from strong objectivity is clear, of course, from its very definition (section 4-1-1). Indeed, with proper specifications it may just as well be termed "intersubjectivity," as we saw. It is, however, worth stressing that such intersubjectivity totally differs from just simple subjectivity.

To clearly grasp this let us turn back to the Young slit experiment (without gas). It is often stated that a condition for fringes to develop on the screen is that no attempt be made at finding though which slit the particle passes. Roughly speaking this statement is correct. But the way it is expressed seems to call in just simple subjectivity and it is somewhat ambiguous in this respect. For example, suppose that, without yet looking at the screen, we try to detect the particles passing though one definite slit, and let us assume that the outcome is negative. No particle is detected. Should we then predict that when we look at the screen we shall see fringes, or, on the contrary, should we predict we shall see none? In fact, what is the case here is that the way the question is asked is not precise enough to allow for an unambiguous answer. Clearly, if the result is due to the fact that our detection instrument was oriented incorrectly and did not interact with the particles the fringes should be there. On the contrary, if the interaction took place and the said result is due to some defect internal to the instrument, no fringe should appear. In other words, what determines the existence or nonexistence of the fringes is not in any way the observer's individual subjectivity. It is the nature of the overall observational device.

Now, should this be termed "strong objectivity"? No, it should not. The arguments we saw in this chapter rule out such an inference. Admittedly a measurement outcome—such as the position of a pointer on a dial—is not dependent on some individual subjectivity. It is public property, so to speak. But it should be kept in mind that in the expression "public property" there is the epithet "public," which refers to a collectivity of *persons*, observers or actors. This, finally, is the essence of the weak objectivity notion. Some of the physicists who deal with quantum mechanics seem not to have quite grasped this point. For them, the very notion of measurement outcome implies the view that the outcome in question is what it is quite independently of the very *existence* of beings able to perceive it. That, in other words, it is objectively real, in the sense of strong objectivity. If they were right it would mean that physical realism is a well-established conception and we should even have to consider that intersubjective agreement contributes to the proof that indeed it is. In the next chapter (section 5-2) we shall, however, see that this is not the case; that the said agreement does not yield such a proof; and that it is therefore consistent to link up measurement outcomes merely to an *empirical* conception of reality.

4-2-4 *Two Types of Weak Objectivity*

The above-mentioned "complicated tests" that, were they feasible, would reveal the "quantum superposition" of the pointer final states are, to repeat, of an incredible complexity. In most cases they are quite impossible to perform (they would involve not only the measured object and the instrument but also just about all the air molecules that came in contact with them, etc.). However, many other simpler tests are conceivable, that actually are not done but that could, in practice, be done (such as testing the pointer position with a subsidiary instrument, etc.). And even in cases of type c the outcomes of such tests would be compatible with the view that the pointer lies on some definite scale interval. In other words the view that the pointer always lies (as we "see" it does) in but one of these intervals is compatible with the counterfactuality condition realism dictates with only one reservation, crucial from a logical point of view but most academic for the physicist at work. This is that some theoretically well-specified but practically infeasible tests should be ruled out of consideration. We shall express this fact by saying that the corresponding statements have a *strongly empirical* truth value, or that they are *empirically* true or false. The objects they bear on (ordinary macroscopic objects, classical fields, and so on) are the elements composing our *empirical* or *contextual* reality, such as we just defined it.

Although we must take them to be but weakly objective, these "empirically true" statements would be considered strongly objective by persons who do not know quantum mechanics or do not consider it universal. In our terminology it is therefore appropriate to distinguish them from statements whose very grammatical form directly reveals that they are weakly objective, namely, those that are explicitly expressed in terms of predictions of observations. (Some of these, as we know, rank among the "axioms" of conventional quantum mechanics.) To distinguish statements of this latter kind from the ones above, we shall say they are *epistemologically* true or false. Here the qualifier "epistemologically" is meant to signify an opposition to ontology stronger than what the words "empirical," "empirically," etc., imply. The root "epistemo" serves to stress the *explicit* reference to human knowledge that statements of this type entail, and that makes it quite manifest that they cannot be ontologically interpreted.

4-3 "Quantum Rules" and "von Neumann's Chain"

As already noted, what most basically differentiates quantum mechanics from classical physics is not (as often believed) the fact that its axioms involve intrinsic probabilities. It is the fact that it is not descriptive but essentially predictive and, more precisely, predictive of outcomes of observations. It can therefore be claimed that its "hard core" reduces to a set of rules.[18] Let us here review the structure of the latter, just to see how simple it really is.

In a first stage, we gather all the impressions we have had at a certain time t_0 concerning what appears to us to be a "physical system." And we code this information by means of a mathematical symbol that, if the information is detailed enough, takes the form of a "wave function" (or "state vector") defined at t_0. In a second stage we compute, by means of the Schrödinger equation or otherwise, what this wave function becomes at the time t of the measurement of the physical quantity we are interested in. And in a third stage we compute, on the basis of the said wave function, the probability that we shall then "have the impression" of reading such or such value on the measuring instrument. The latter calculation is made by means of a formula usually called "the Born rule."

Admittedly, the fact that the rules in question are easily stated does not imply that they are always easy to apply. In particular, expressed as above they immediately raise the question: "What should we include into what we call 'the system'?" In principle, the answer should be "everything on

---

[18] All of what, but *merely* all of what, in these "laws" or "rules" is useful for dealing with the subject matter of this book will be made explicit. As axioms, the said rules are exhaustively stated in some (but not all!) textbooks on quantum mechanics. The interested reader may find them, for instance, in either one of the two references (d'Espagnat 1976, 1995).

which our impression (alias our 'becoming aware of') will bear." This, however, immediately raises a difficulty for we must realize that our "becoming aware of" will, in fact, bear on the measuring instrument and that the latter, being macroscopic, is a highly complex system. Should all the parameters specifying its state be taken into account and incorporated within the initial wave function? Fortunately, inspection of the mathematical formalism shows that in normal practice this is not in the least necessary and that we run no risk of modifying the result if, instead of doing so, we do as if the instrument itself were the observing subject. A fortiori, it follows from this inspection that, when writing down the wave function, we need not include in it parameters specifying the structure of the observer's eye, optical neurons, etc.

In this domain, in fact, the formalism yields more general information. Assume, for example, that we have to do with a composite system composed of a microsystem S (a particle, an atom, etc.), an instrument I "measuring" a quantity G on S, another instrument I' "measuring" the position of the I pointer, still another one I'', doing the same on I', etc., and finally an instrument I_f, the pointer of which is observed. Within classical physics the said observation would directly inform us of the value G had on S before the whole operation took place. In quantum physics the problem is to compute the probability that we shall have the impression of indirectly perceiving, via I_f, such or such a value of G. The formalism shows that, for so doing, we have in theory several procedures at our disposal. We may, for example, include the parameters relative to S, I, and I' within the wave function and treat the macrosystem composed of I'', . . . , I_f as if it were the "perceiving subject." Or we may do the same, but leaving only S and I on the "quantum side" (i.e., within the wave function) and setting I' on the "perceiving" side. Or finally—and this is by far the simplest—we may leave only S on the "quantum side," which amounts to including no instrument "parameter" (alias "variable") whatsoever in the wave function. This shows that we can indeed "do as if" the large, system composed of I, I', I'', . . ., I_f constituted the observing subject. This theoretical equivalence of several modes of calculation was pointed out for the first time by J. von Neumann and is called the "von Neumann chain."

On the other hand, in principle we may of course consider more general questions, such as that of knowing what correlation effect would be observed (by means of appropriate instruments) between some physical quantity pertaining to S and some physical quantity pertaining to I. In such a case it goes without saying that we could not set I on the "classical" (or "perceiving subject") side. We would have to use a wave function involving the parameters (i.e., "variables") of both S and I.

Finally, concerning the quantum rules let us note that, within the convention taken up in this book, the set of them all does include the (generalized) Born rule (which yields probabilities of observation on the basis of knowledge of the wave function) but does not include the (already alluded to) procedure called "reduction" or "collapse" of the wave function. The point is worth mentioning since, in many texts, the two procedures are described together and are not clearly distinguished. In fact, while the latter is often extremely useful, in theory it never is absolutely necessary. In section 3-1 we saw, for example, that, in the simple photon pair case, the procedure involving such a reduction (called there the "descriptive method") may be replaced by a procedure that involves none (the "predictive method"). And with the help of the von Neumann chain notion other examples can most easily be built up. Moreover, we shall see in chapters 8 and 10 that the reduction procedure raises considerable interpretation problems. Such are the reasons why it is not included here inside the set of basic rules.

All the foregoing confirms—if the necessity were felt—that analyzing the conceptual foundations of quantum mechanics reopens the question of realism and makes it more acute than it ever was. It is thus high time we should tackle it.

QUANTUM PHYSICS AND REALISM

§§

5-1 Strong Objectivity and Realism

A S ALREADY NOTED, the view that physical realism is true and that physics should describe phenomena in terms compatible with it is still today very much in favor among scientists and "laymen" alike. This is partly due to the fact that the view in question is in no way a "purely instinctive" one. On the contrary, it reposes on arguments that do admittedly carry weight and to which it is proper to devote attention. Such is the purpose of the present chapter.

5-1-1 Objectivist Realism and Counterfactuality

Some philosophers defined reality as that on which we can act. But this definition may not be taken at face value for it implies that stars are not real. It must therefore be widened, which entails making use of the conditional tense. Not as a definition but at least as a *sufficient criterion* of reality, the objectivist realist thus claims that "is surely real what we could act on or could possibly act on us."

Unquestionably, such a widening looks straightforward and natural. When we think something really exists we all implicitly have some such an idea in mind. Still, it is to be noted that conditional statements belong to the realm of modal logic, which means that they fall outside the domain of *classical* (alias traditional) formal logic. Modal logic is indeed a branch of formal logic that developed relatively recently, and this is even more the case concerning the—related—logic of conditional propositions. This elementary remark is to be viewed as a first indication that objectivist realism, even though it is by far the most widely spread philosophical attitude, still is a conception that is not quite as "elementary and obvious" as it may seem. Let us therefore try to make it more explicit. This may be done by noting that when the objectivist realist claims that a thing—or a property of a thing—is "real," this, in his views, implies the truth of such propositions as "if, relative to it, we performed such and such observational act—which in fact we do not perform—we should observe this or that." The distinctive feature of such assertions lies in the phrase "which in fact we do not perform." And this is why (as we already know) they are called "counterfactual."

Unfortunately (I mean, from the realist's point of view!), in quantum formalism such counterfactual assertions, as a rule, hardly find a place. The said formalism is most powerful as regards answering questions such as "if some measurement M has *actually* been made and its outcome was such and such, what information does this yield concerning the system under study?" But it in no way guarantees that the piece of information thus gained would also hold good if some measurement M' had been made instead of M. And this—surprisingly enough!—is true even in cases in which some apparently convincing arguments show that neither M nor M' could possibly disturb the system.[1] Such is the source of the tension between quantum physics and objectivist realism.

In view of the following discussions let us finally note that, because of the counterfactuality request inherent in this realism, the strongly objective statements characterizing the latter are bound to quite a strict condition of validity. To wit: of all the consequences they would entail supposing the system they bear on were subsequently subjected to any conceivable tests, none should be false.

5-1-2 On the "No-Miracle" Argument

But—some might say—why all this talk? At least in the macroscopic domain, is not the idea of reality the clearest of all those we have? And isn't it obvious that many simple, straightforward arguments substantiate it? Strangely, the answer is "no," as philosophers well know. It is perhaps not useless to briefly show why such seemingly self-evident arguments fail.

Of all the realist arguments, the apparently most convincing one, the one that, in a way, summarizes the others, is known under the name of *inference toward the best explanation* and also under the slightly too incisive but suggestive one of the *no-miracle argument*. It is aimed at justifying the realism of the accidents. Its supporters build upon the unanimously acknowledged fact that if a scientific theory enables us to make a great number of predictions and observation substantiates them all, this constitutes an extremely strong argument in favor of the view that the theory, or some equivalent one, is correct. Of course, our reason for judging this way is that it would be an extraordinary coincidence—a real miracle—if all the predictions from the theory happened to come right just by

---

[1] The so-called EPR (Einstein, Podolsky, Rosen) problem allows for a clear illustration of this point. See, e.g., *Veiled Reality* and, here, section 7-1-1.

chance. Now, the supporters in question point to the fact that immensely many predictions are made at any time by billions of people on the basis of the "theory" that physical things "are really there," have such and such properties, etc. And—they continue—practically all these predictions do come out right. How could we fail to consider this to be a dazzling corroboration of the truth of the "theory"?

However, the argument is but outwardly convincing. In fact the reservation "or some equivalent one" has the effect of making it ineffective. For such a reservation—which, from a logical point of view, is obviously necessary—calls for the question: "Is there, relative to the problem under study, a theory other than objectivist realism and capable of yielding the same 'immensely many' successful observational predictions?" And to this question the answer is "yes." It is now known that, when it is considered to be universal, "standard" quantum mechanics (no hidden variables assumed) is such a theory. Indeed, quantum mechanics is a theory aimed, as we know, at predicting observations, and its "axioms" in no way involve the assumption that physical objects exist by themselves and are endowed with definite forms, positions, etc. But still, it can now be shown[2] that concerning macroscopic objects it yields a set of rules—a "theory"—that is "equivalent" to the commonsense one according to which such objects have, at any time, individuality, forms, positions, and so on. Here of course the word "equivalent" is not to be taken in any ontological or would-be ontological sense. It merely means (and this is quite enough) that both theories lead to the same *predictions of observations*. Hence, contrary to what the no-miracle argument seemed to show, the fact that the predictions in question are successful does not prove that the commonsense theory is the correct one. The "true" theory may, just as well, be quantum physics.

And yet, it so much seems to go without saying that the no-miracle argument is right, that the latter spontaneously comes to mind, even when we are not in the least trying to motivate objectivist realism! Consequently its—just shown—failure at justifying objectivist realism legitimately raises some questions. Let two of them be considered here.

First Question.   Is it possible to demonstrate the said failure in an a priori manner, that is, without referring to the "equivalent theory" that quantum mechanics constitutes; just by means of a purely philosophical reasoning?

Apparently several philosophers consider this to be the case. One of their argument (see, e.g., J. Worrall [1989]) is that the parallelism with a scientific theory is not so strict as it seems. The reason that Newton's

---

[2] See, e.g., Omnès (1994b).

explanation of the planetary orbits looks so convincing is that it explains more than just the said orbits. It also accounts for gravitation, the motion of the Moon, the tides, the return of Halley's comet, etc. In other words, it is checked by independent observations that confirm its validity (for an upholder of Descartes' cosmology, for instance, Halley's comet had no reason to come back at a certain date rather than sooner or later). Whereas, concerning the fact that realism of the accidents "explains" our day-to-day successful predictions we are at a loss to find corroborating facts paralleling these.

The objection is a sensible one but we may consider it not to be fully decisive. It questions the value of the realist explanation but it stops there and puts forward no alternative one. We shall come back to it in section 13-3-3.

Another "purely philosophical" objection to the no-miracle argument rests on the observation that, anyhow, what is observed has to be interpreted (it would be definitely *erroneous* to state that, in the morning, the Sun climbs up the eastern hills and that, in the evening, it plunges into the western sea). Consequently, the only "reality" our realist may meaningfully try to establish is that of the entities that our (intuitive or sophisticated) theories elaborate. However, within the framework of realism—be it "of the accidents" or "Einsteinian"—it so happens that, due to "scientific revolutions" or mere changes of outlook, the said entities are sometimes just simply *replaced* by some other ones. Think again of the substitution of Einstein's general relativity for Newton's theory. Within the realist conception such changes imply dramatic upheavals of our description of the world since an entity that in the former theory was basic and hence supremely real—the gravitation force in the example—suddenly appears nonexistent. It is quite understandable that realists finally came to the point of questioning the cumulative character of science (we shall have a further look at this question in chapter 11). And there are cases in which the situation is still more puzzling. They are those in which the mathematical formalism of the theory—the formalism that makes observational predictions possible—may be expressed in two different forms, involving quite different concepts but still leading to exactly the same predictions. In such cases the persons whom the no-miracle argument turned into realists are in the position of Buridan's ass. They have no rational means of deciding whether it is the set of concepts corresponding to the first or the second theoretical structure that refers to real entities.

To sum up, it may be said that, yes, purely philosophical considerations do already seriously reduce the weight of the no-miracle argument. But, as we just saw, they are essentially negative. Their existence therefore in no way diminishes the relevance, in these matters, of the fact that quantum theory exists and constitutes, in theory, a genuine substitute for the realism of the accidents.

Second Question.   Should the failure of the no-miracle argument at convincingly justifying objectivist realism incite us to give up every kind of realism and any quest for some realist explanation?

The answer is "no," for a reason that the next section will make explicit.

5-1-3 The Need for Open Realism

The philosophical objections just described to objectivist realism implicitly refer to the fact that the realist bases his or her explanation of why our predictions are successful on the notion that the objects are really this way or that way. In other words, the said objections refer to the *descriptive* character of realism and are essentially directed against the "best explanations" that the realists strive to base on such descriptions. But, taken in its widest sense, realism neither reduces to, nor even implies, "power to describe." In the realm of an extension of the foregoing discussion we shall therefore, from now on, not refer any more to the notion of "best *explanation.*" We shall focus instead on the aspect of the prorealism argument that is best expressed by the phrase "it would be a miracle if." Our answer to the "second question" above will then be "no," and will be based on the fact that, expressed this way, the argument in question no longer refers to the notion "description."

Admittedly, the foregoing objections showed us that the no-miracle argument fails to justify the *conventional* versions of realism (realism of the accidents and Einsteinian realism), which rest on the postulate that there are well-specified, describable elements of reality. But from this it does not follow that the said argument is powerless concerning realism in general. Indeed, we noted that concerning ordinary, macroscopic, phenomena quantum mechanics leads to observational predictions equivalent to those obtained on the basis of traditional realism, so that we can dispense with postulating the intrinsic reality of objects and their attributes. But the reason why quantum mechanics implies such predictions is because it admits of laws. Now, be they descriptive or just simply predictive, these laws were discovered,[3] not invented. More precisely, even though it is true that human beings imparted to them a certain form, which could be different from what it is, it could not be *arbitrarily* different. To put it

---

[3] Which of course does not mean that they just emerged from experience. As most rightly stressed by many authors, imagination had to be made use of in order to build them up. But the crucial point is that only those that facts corroborated were kept.

otherwise, there "is something" that tempers the imaginative impulses of our creativity. In electromagnetism, equations would be simpler if the electric field were a scalar instead of being a vector, but objectively this is not the case. Making such a substitution "by fiat" would ruin the agreement between theory and experiment. In other words, the physical laws do not totally depend on us, which means that they also depend on something else. And in fact, radically negating the very existence of this "something else"—claiming that the corresponding notion is meaningless—would immediately arouse quite serious conceptual difficulties[4] relative to the nature and the very existence of the said laws. Finally, therefore, even though the no-miracle "postulate" fails at justifying the conventional forms of realism, it constitutes a valid argument for justifying the idea that the notion of "mind-independent reality" is meaningful, in other words, what we called "open realism."

5-2 Intersubjective Agreement

Along with the no-miracle one, there is another argument that, at first sight, also seems to prove that the realism of the accidents holds true and of which it may also be shown—by referring to quantum mechanics—that in fact it proves nothing of the sort. It is grounded in intersubjective agreement and we shall now scrutinize it. Let it be specified right away that here we shall only be interested in intersubjective agreement concerning the observation of contingent facts. We are interested in the fact that when, for example, a teapot is on a table we all agree in judging that, at that place, there is a teapot.

5-2-1 Philosophical Discussion

The intersubjective agreement thus defined is a fact of experience in which we are a priori inclined to see quite a decisive proof of the validity of objectivist realism. For indeed if Alice and Bob both claim that, at about five o'clock, they saw a teapot on the table, the assumption that there was then really a teapot there seems to be by far the simplest explanation of such a convergence of observations. If (as, apparently, the opponents to realism would have it) the statement "the teapot really exists there" had, for Alice, no other meaning than that of specifying how she mentally organizes some of her sensations (and the same for Bob), then the fact that they have the same sensation at the same time would be most surprising. It

---

[4] An analysis of which will be developed in chapter 17.

would be sort of a miracle, constantly renewed within any form of social life. The argument is so simple that it looks as if it were irrefutable.

And yet, philosophers managed to concoct two objections against it. The first one—call it (1)—rests on the fact that the offered explanation is based on assumed similarities between Alice's mental images and the real world *R* on the one hand, and Bob's mental images and *R* on the other hand. For the explanation to be valid it would therefore be necessary that some cause-effect relationship should be extant between *R* and Alice's mental images (as well, of course, as between *R* and Bob's). However, *R* itself is neither observed nor, actually, observable. It belongs not to the realm of phenomena but, in Kant's words, to the noumenal one. The "teapotlike" explanation thus takes us far beyond what is actually established concerning the domain of validity of the causality concept, since it is only relative to relationships between phenomena that causality can be checked at all.

Basically the other objection—call it (2)—has some relationship with (1), but still it is worth being stated separately. In fact, it is but a transposition of one we already met with in connection with the no-miracle argument. It is grounded in the general idea according to which, for an explanation to genuinely be one—for it not to be purely ad hoc and of the *virtus dormitiva* type—it is necessary that the *explicans* should account for more than just the phenomenon we intend to explain. However, such is not the case here. To "explain" the intersubjective agreement we call forth the notion of a reality assumed primary with respect to the phenomena, whereas such a notion is made use of in no other explanatory proposal, and whereas, in particular, it never intervenes in any of those that science puts forward, which always are from phenomena to phenomena.

These two objections are worth considering but it seems likely that they will not convince everybody. Both are grounded—the first one obviously, the second one indirectly—on the distinction between phenomena and noumena. They therefore do not persuade the objectivist realist who, right from the start, rejects this distinction. Moreover, just as is the case with the philosophical objections to the no-miracle argument, neither one offers any alternative explanation.[5] Consequently, it seems reasonable to consider that on the whole they carry less weight than does the intuition—

---

[5] Strangely enough, few philosophers have ever looked into this problem. In the corresponding literature, only the intersubjective agreement concerning general or mathematical notions is discussed extensively. Admittedly, it is there explained that we all have the notion of a triangle because we are all similarly constituted. But it is not clear how the fact that Alice and Bob are similarly constituted explains why the two of them see a teapot at the same place. Only Husserl seems to have taken an interest in this question (in the fifth of his *Méditations cartésiennes* [1969]). But his analysis is so complex and so involved that it is difficult to consider it decisive.

universally entertained by scientists and laymen alike—according to which our mutual agreement concerning the objects we see obviously proves they are real.

5-2-2 Scientific Discussion

However, from a scientific point of view the problem calls for a more detailed study. With such an end in view let us first make the "basic" realist argument more open to analysis by reformulating it in terms of possibilities of prediction. It then reads as follows.

"If Alice believes in objectivist realism she considers that the teapot is really on the table and that Bob also sees it there. Hence she can predict that when both of them come to meet, their memories of the event (the notes, say, in their notebooks) will coincide, a prediction that then gets verified indeed. Whereas, on the contrary, if she did not believe the objects exist independently of herself, she would have no reason to suspect that Bob also sees a teapot standing on a table. Nothing would therefore incite her to make the prediction in question so that, when they later compare the notes in their notebooks, the observation that they coincide could not but come (to her and Bob) as a surprise. (Objectivist) realism is therefore an ("the") explanation."

In the quantum physicist's opinion the truly "weak point" in this reasoning was in fact missed by both of the two above-discussed philosophical objections to it. It lies in the phrase: "nothing would (therefore) incite her . . . " Admittedly, if Alice disposes neither of objectivist realism nor of a theory equivalent to it with respect to the point under study she will claim that the coincidence is unexplainable. But[6] she will not utter such a claim if she knows of some nonrealist theory predicting the coincidence. Now, here again quantum mechanics, assumed universal, is equal to the task. To show it, let us imagine replacing the teapot by the pointer of the measuring instrument considered in section 4-2-2, and let us assume that Alice and Bob both look at this pointer after the instrument was made to interact with some quantum system initially prepared in the quantum state $c = a+b$ (the notations are those used in the said section). As we know, the quantum formalism directly yields observational predictive rules, without resorting in any way to the "realist conjecture" that the pointer must lie at some definite place. And as we also know, it does not tell us with certainty what will be observed. It just tells us that, when observed, the pointer will be seen to lie either in state A or in state B (and it yields the corresponding probabilities). But in compensation it yields

---

[6] As already noted in *Veiled Reality* (sections 1.4 and 14.9).

with certainty (and this is what counts here) the prediction that Alice and Bob will both see the pointer lying in the same state (i.e., at the same place). Either they will both see it lying in state *A* or they will both see it lying in state *B*. But, whatever computation mode is chosen, the formulas yield a probability zero that one observer sees it in state *A* and the other one in state *B*. In other words, the formalism predicts intersubjective agreement. And (to repeat) it does so without in any way implying that before the measurement took place the pointer "really was" in the observed state. (In section 16-2-1 this result will be made use of in connection with phenomenalism and operationalism.)

5-2-3 Discussing the Discussion

1. It is worthwhile entering into some details concerning the Alice-Bob agreement and the reasons thereof. To this end, let us first remember that there are no "truly direct" observations. Even when engaged in observing a teapot, or an instrument pointer, the observer always makes use of her eyes (and spectacles if she is shortsighted) as auxiliary instruments. And let us note that therefore the problem is conceptually the same—at least under the universality assumption (to be further studied in section 6-5)— whatever the size of the pointer is taken to be. Only, if this pointer is very small, the auxiliary instruments must be chosen appropriately. If it reduces to one electron only (and hence must be quantum mechanically dealt with), Alice and Bob have to make use of instruments suitable for observing the position of an electron (or, say, preferably its momentum, since, for a free electron, momentum is a "constant of motion" whereas position is not). Within the formalism, equations corresponding to such measurements may be written down. And they show that the indications yielded by the two auxiliary instruments necessarily coincide. It is true that in the case in which the "pointer" is an electron to make use of the formalism we have to assume the two measurements not to be strictly simultaneous whereas, in the "real pointer" case Alice and Bob look together. But in fact this does not matter for such observations always last for a finite time so that they may just as well be thought of as being sequences of shorter ones, made by Alice and Bob alternately.

2. Up to this point we have primarily been looking at cases in which what the two observers observe is the position of an object (teapot or pointer). And we saw that "standard" quantum mechanics does predict intersubjective agreement even in "superposition" cases, that is, cases in which, as a consequence of the very quantum laws, the object in question cannot be thought of as lying in any definite position. This does show that, contrary to naive expectations, the fact that we universally observe

intersubjective agreement is not a proof that realism of the accidents holds true. Admittedly, we could wonder whether other situations may be thought of in which, to account for this agreement, there would be no other way than assuming that, already before any observation, the object possesses the observed property. But the formalism is so general that it seems this hypothesis can be discarded without qualms. It may even be added—but there the proof is more complex (Omnès 1994b)—that, concerning macroscopic systems, the said formalism predicts that the evolution in time of what will be observed (by Alice, Bob, or anyone) must reproduce the classical physics predictions.

Conceptually, these conclusions reach much farther than those that philosophers could come to. In the matter in hand the developments of the latter merely aimed at invalidating the explanation of intersubjective agreement that the realist trusted in. They were mute concerning possible alternative ones. In fact, they limited themselves to suggesting that the very notion of explanation might well be obsolete (which looks somewhat like acknowledging a failure). Here, not only is the philosopher's negative conclusion recovered but, in addition, a positive explanation is yielded, at least if it is considered that reference to universal, well-verified predictive laws constitutes an "explanation."

Moreover, we must remember that, for reasons mentioned above, the said laws may not be considered to exclusively depend on us, and that this makes open realism credible. It can therefore be claimed that, when all is said and done, intersubjective agreement contributes to the plausibility and internal consistency of a philosophical standpoint according to which the concept of existence is not radically regarded as being under the dependence of the concept of the human mind. And this, even though—as we saw to be the case—in view of the phenomena quantum mechanics deals with, the said existence can only be that of a "reality" lying far beyond what can be actually described.

In this respect, let us finally remember that, as is well known, one of the difficulties radical idealism (which makes "awareness" primary) has to cope with is the plurality of minds or, in other words, of the I's. This difficulty (called "the arithmetical paradox" by Schrödinger [1958]) essentially consists in that, within the idealism in question, it is hard to discern what accounts for the intersubjective agreement between all these I's (short of introducing the ad hoc notion of some "universal I" having in fact nothing but the name in common with what the pronoun "I" refers to). Since the conception proposed here does not coincide with radical idealism the difficulty in question is, in it, less acute. It cannot be claimed that it altogether vanishes for the impersonal character of the rule yielding probabilities of observation (the Born rule) still looks somewhat enigmatic. However, it is less offensive, for we have long been familiar with the idea that physical laws are universal, so that we are hardly surprised at observing that this universality extends even to those that are merely predictive of observations. (More about this in section 18-5).

5-2-4 The Agreement within the Broglie-Bohm Model: "Bell's Calculation" and Contextuality

The above-described situation stands in contrast with the "lessons" of classical physics since there intersubjective agreement directly reposed on an "observed existence," so to speak, namely, the existence of things. Within the assumption that quantum mechanics is universal the explanation proposed here of the said agreement does still refer, in last resort, to the notion of an existence, namely, that of "Reality." But the link is indirect and inferred rather than observed since, as we saw, a detour via the notion of "law," nay, even of "observational predictive law," proved necessary. It is interesting to examine what the matter looks like in the models—credible or not credible, this is not the question—that are, so to speak, conceptually classical but predictively quantum mechanical, in that they exactly reproduce the quantum observational predictions while being ontologically interpretable. And, of course, the Broglie-Bohm model is ideal for this purpose.

Since, to repeat, this model is, like classical physics, ontologically interpretable, we quite naturally expect that in it, just as in classical physics, intersubjective agreement should *directly* rest on the reality of things. And, in a way, this is the case. But "in a way" only. For indeed, in our current way of thinking the word "things" covers both the notion of "perceptible reality" and that of "localized objects." Now, in the model perceptible reality corresponds to the "supplementary variables" (thus misleadingly called "hidden"). And, according to it, it is indeed the fact that these variables (or at least some of them) have definite values, collectively perceived by us, that generates intersubjective agreement. But, contrary to the "things" of classical physics, these supplementary variables are not "local."[7] And as a result, at least in the case of some configurations, the explanation of the agreement takes up a most unwonted form. Indeed, in such cases it radically differs, as we shall presently see, from the "teapot type" explanation and shows features making it somewhat similar to the one offered—via predictive laws—by "standard" quantum mechanics, and which has been described above.

---

[7] In the sense that what takes place at some point A may well crucially depend on what value some such supplementary variable has at some other point B arbitrarily distant from A.

To build up such a "configuration" let us imagine that Alice and Bob, instead of observing the same object (a teapot or some quantum system) measure different physical quantities on different but strictly correlated quantum objects (say, two particles). And let it be the case that the quantity Alice measures and the one Bob measures are themselves strictly correlated (as in the dart experiment as well as in the simplest form of the Aspect-type experiment) so that the values they get are the same. Of course, what we have in mind concerning intersubjective agreement is that, because of the strict correlation, everything should take place as if the two measurements were made on the same object. We argue: "Both on Alice's and on Bob's side the outcome of the measurement (the number appearing on the dial) is determined by the value the measured physical quantity has on the particle that carries it. And the reason the two outcomes are the same is that (because of the correlation established right at the source) these two values are the same."

Now, quite surprisingly, in the model this is not the correct explanation. This unambiguously follows from a relatively simple—but conceptually most far-reaching—computation made by John Bell in the 1960s and which, for future reference purposes, will here be given the name "Bell's calculation" (Bell 1966, 1987b, chapters 1 and 15; *Veiled Reality*, section 13-3; see also, here, appendix 3).[8] The crucial fact is that, in the Broglie-Bohm model, the particles are driven by the wave function, which is an explicitly nonlocal entity. In some situations[9] this entails—as the said calculation shows—that once Alice has performed her measurement on "her" particle the outcome of the measurement Bob will later perform on "his" particle does not exclusively depend on the set of the "hidden variables" of the latter. Strangely enough, it also depends on various parameters related to the outcome of the measurement made by far-away Alice. In such a situation intersubjective agreement still takes place but, as we see, it involves nonlocality, which means that it is not explained by any classical correlation mechanism.

Note also that, whereas, in the dart case, the agreement in question is directly explained by a *fact*, namely, the common orientation of the two darts at the source,[10] in the case under study here the said agreement

---

[8] (A note for theorists.) It should not come as a surprise that the variable Bell labels x is identified by him with a particle coordinate in his chapter 1 and with the "pointer variable" in his chapter 15. For indeed, in a Stern-Gerlach measurement it is the particle coordinate along an axis chosen parallel to the magnetic field that plays the role of the "pointer variable." The same remark hold concerning the content of *Veiled Reality*, section 13-3.

[9] The one Bell considered is just simply that of a system of two spin-1/2 particles lying in a definite spin state, assuming the two particles move apart as the darts do in the section 3-1 example.

[10] For the physical laws involved are so simple that they are not worth mentioning.

shows up as a consequence of a *law*. For indeed it follows from a rather complex formula which, itself, is a consequence of the postulated mathematical structure of the model. On the conceptual side the said model may therefore be considered an intermediate step, used in forming, stage by stage, a view that may be called an "explanation" of the intersubjective agreement. This gradual elaboration starts with the "commonsense-like" assumption that the said agreement should be due to *contingent* features of the "real stuff outside us." When we decide to apply it to the microscopic realm this conjecture naturally induces us to make use, at least as a "temporary scaffolding," of some ontologically interpretable model, such as the Broglie-Bohm one. But then the surprising result of Bell's calculation suggests to us that, after all, the agreement in question should rather be due not to contingent features (such as correlation at the source) but to some *deep structures* of Reality. And then, as a last step, we may think of just keeping this last idea while removing the scaffolding, considered not very reliable. The end result of this thought process is that we do not consider intersubjective agreement to be a miracle. It is indeed due to some deep structures of mind-independent reality. Only, since all ontologically interpretable models are unreliable, we cannot claim that we can, for sure, describe these structures. At best, all we may claim is that we do know something of them, but, only in that we know the observational predictive laws of which they are the hidden source.

Contextuality

Quite independently of the intersubjective agreement question, Bell's calculation yields the possibility of introducing in a simple way the interesting "contextuality" notion. To this end it suffices to point out that, account being taken of what has been noted above, the outcome of the measurement made by Bob is not determined by the parameters relative to "his" particle and those of his instrument. It crucially depends on what takes place where Alice is, including details relative to how her instrument works.

More generally, we now know[11] that the ontologically interpretable theories can match the experimental data only at the price that the outcome of the measurement of a physical quantity be dependent on whether or not some other quantity[12] is simultaneously measured and on the choice of the latter. This is what is called "contextuality." Nonseparability and contextuality are the two basic features that set limits to the possibility

[11] Thanks to Bell's works (1966) and also, independently, to those of Kochen and Specker (1967).

[12] Compatible with it, of course.

of interpreting the quantum phenomena (in a broad sense) in terms of objectivist realism.

5-2-5 Explaining Agreement by means of Personalized Probabilities

As it appears in the account of the quantum rules given in section 4-3, a quantum probability (as yielded by the Born rule) is not only intrinsic but also radically impersonal. It is the probability *we*—as a community—have of receiving such and such impression. This universality is the reason why we all perceive the same events. It accounts for the fact that if Alice and Bob both look at the pointer of a measuring instrument they see it lying at one and the same place. It also accounts for the fact that if, on two distant objects the pair of which is described by a wave function, Alice and Bob measure two quantities strictly correlated in the quantum sense, the outcomes they will get will indeed be strictly correlated. A priori we may wonder at the existence of intrinsic probabilities that are impersonal to such an extent that they are simultaneously relevant even for several individuals arbitrarily far from one another. Hence there is some point in noting that, in a way,[13] to this impersonal probability rule a "personal" one, that is, one defined individually for each observer, might be substituted. For this it would suffice to adopt an interpretation of measurement the logical possibility of which was pointed out some decades ago[14] and whose specific feature is that it introduces a distinction between the observer's state of mind and his or her state as a physical system. Its guiding idea is just that, upon measurement, the ultimate registering takes place in the former, not in the latter. And consists in that the observer's mind sets itself either in a state E_A corresponding to A or in a state E_B corresponding to B (notations are those of section 4-2-2), the probabilities of these events being those yielded by quantum mechanics.

The model was suggested by the Everett theory (section 8-4) but it only borrows from it the idea that once a mind has tied itself to a "branch" it keeps tied to it. Albert and Loewer (1988, 1992) and, more recently, Zwirn (2000) revived it. It exactly reproduces the quantum mechanical predictions but shows up specific features concerning the intersubjective agreement issue. To most simply grasp what it consists of, let us first remember that systems composed of several subsystems often happen to be in so-called "entangled" states. For example, a system composed of

---

[13] I mean, if we set no a priori limit to the apparent extravagance of the consequences of such a choice.

[14] d'Espagnat (1976), section 23-2.

two particles U and V may lie in an entangled state such as $u_+v_+ + u_-v_-$, where the states u_+ and u_- correspond to two different values U_+ and U_- of a quantity $\mathbb{U}$ belonging to U and the states v_+ and v_- similarly correspond to two values V_+ and V_- of a quantity $\mathbb{V}$ belonging to V. Of course, the model then assumes that if Alice measures $\mathbb{U}$ on U she may get either outcome U_+ or outcome U_-, which, in the model, means her mind may then find itself either in a certain state to be called A_+ or in another state, to be called A_-. And the same, of course, concerning Bob. But the model further assumes that, whatever the outcome Alice thus gets, if Bob measures $\mathbb{V}$ on V his mind may, as well, find itself either in state B_+ or in state B_-.

At first sight, such a hypothesis seems to conflict with the quantum mechanical predictions since, in the considered situation, quantum mechanics predicts a strict correlation. But in fact there is no conflict. For let us assume that after the measurements Alice and Bob find themselves in the states of mind A_+ and B_- respectively, and that Alice asks Bob to let her know what he himself observed. The request has to pass through the channels of sound, voice, possibly telephone, auditory nerves, etc., that is, purely physical organs. Unavoidably it thus takes the form of a measurement performed by Alice, not, directly, on Bob's mind but on his neuronal and registering system. And since in the considered hypothesis, this system is entirely governed by quantum mechanics it may be shown, just by applying the formalism, that the answer Alice will hear will not be "I got outcome V_-" but "I got outcome V_+." Hence also here, inasmuch as it is verifiable, intersubjective agreement holds good. We must grant however that it here takes up a truly unexpected look. Admittedly, poets and moralists alike have long being asserting that each one of us feels in some unique way what touches him or her "at the deepest." Still it is not very easy to get convinced that this profound remark reaches so far as being valid · concerning our perception of teapots or pointers. We know that quantum physics is strange, but is it *that* peculiar? Doubt is permitted.

5-3 Intersubjective Agreement and Empirical Reality

For a long time now philosophers of various schools—Berkeleyans, Kantians, neo-Kantians, etc.—have been making us aware that objectivist realism ("transcendental realism," in their language) may be given up. That such a conceptual move is neither a logical absurdity nor even a break of consistency. They could establish the impossibility of proving that science (and, more generally, knowledge) bears on "what is out there" rather than just simply on *our* experience. And we must therefore grant that *they* fathered the notion of an empirical reality, conceived of as being the whole of what, within reproducible and communicable human experience, may be expressed in an objectivist language. That nonseparability leads us to ideas of such a nature cannot therefore come up as a surprise to them. But nevertheless it must be granted that the situation we are in looks strange and new. This is due to the fact that what, in these philosopher's reasoning, rendered the empirical reality notion palatable was, as we noted, the correlative "as if" idea. It was the idea that introducing such a "relativity of knowledge" notion could be interpreted as being nothing else than just a manifestation of sound philosophical clear-mindedness. And that, this being done, we could, in any field, reason without qualms as if we had to do with a reality "in the intuitive sense", that is, "per se." Moreover, the fact that these philosophers made great use of the linguistic procedure consisting in deleting the epithet "empirical" in the labeling of the notion thus introduced psychologically helped a great deal in rendering it acceptable. For the scientists then finally felt entitled to set the distinction between "per se" and "empirical" within brackets, not only when calculating and elaborating theories, but also in their intuitive way of accounting for intersubjective agreement between their colleagues and themselves. It was without even noticing this intellectual move—so justified they thought it was—that they built up deductions of the type "all of us see a glass on the table, hence there really is a glass and it is really on the table."

Should they—should we—give up drawing inferences of such a type? Obviously not! But still, if, like our pointer above, the glass were replaced by a quantum system lying in a superposition state, the inference would, as we saw, be basically flawed. In other words, the inference is justified only when it bears on objects that look macroscopic to us. Hence its outcome is meaningful, only when the word "really" is taken in a weakened acceptation, referring to this very "empirical reality" notion of which we found out that it greatly depends on us. Otherwise said, the domain of validity of our inferring from intersubjective agreement to the existence of what we call "reality" is limited and very much depends on human aptitudes. This confirms the fact that, in view of physics, the—observationally based, hence philosophically sound—"empirical reality" notion cannot be universally substituted to the one—considered unacceptably metaphysical by many—of "reality, period." Since our intuitive notion of reality is the hidden string that moves our everyday thinking (remember the teapotlike explanation of intersubjective agreement), such a limitation raises problems. The point is that, since our common way of thinking proves not to be universally valid, the "feeling of reality" on which it rests obviously cannot be regarded as being in any way unalterable. The examples we analyzed therefore demonstrate the vacuity of the hope that, once the philosophical jump from reality-per-se to empirical reality has been accepted, we could apply to this new and meta-physics-free object all the notions and modes of thought that used to fit objectivist realism.

5-4 Conceptual Glimpses; Carnap, Quine, Primas; Relative Ontologies

The foregoing considerations are grounded on physical data that could be described here but in a succinct way but are quantitative and most precise nevertheless, and constitute a remarkably consistent whole. Clearly they lead to considerable conceptual changes, to the study of which the second part of the present book is devoted. But, in order to form a well-balanced understanding of the content of this and the foregoing chapters we should, already at this stage, examine how and to what extent the notions of strong and weak objectivity fit with some purely philosophical twentieth-century investigations.

Of course, in this perspective we first think of Wittgenstein, of the role he attributed to language, and of his well-known axioms: "The World is the set of the facts, not of the things" and "We build up tables of facts." On the other hand, however, it must be granted that, in the eyes of a present-day physicist, steeped in relativity and (even more) quantum theories, Wittgenstein's terminology entails some obscurity, concerning, for example, the very word "fact." Indeed, when reading the *Tractatus* (Wittgenstein 1921) we may sometimes fall under the impression that its author aimed at preserving some kind of a subtle ambiguity he himself introduced between a realist and a mind-centered acceptation of one and the same word. Admittedly it is conceivable that such an ambiguity should, in some sense, constitute the end result of this philosopher's quest. But even if this is the case, the said impression imparts to a physicist reader the feeling that these texts still leave much to be clarified. This is why, rather than to Wittgenstein's theses, we shall here be interested in a certain notion due to Carnap (1950)—though closely connected to some views of Quine (1943)—the advantage of which is just that it is quite clear and unambiguous. The notion in question is that of "linguistic framework," or of "(relative) ontology" in the last-named author's terminology.

As a positivist, hence concerned, above all, about the referential meaning of words, Carnap considered the "reality-per-se" notion to be meaningless. For him, therefore, the above-defined strong objectivity notion would have been meaningless as well. According to him, the meaningful notions are exclusively those that relate to experience. Hence, for him as for Wittgenstein, the high importance of language, a standpoint that finally led him to the "linguistic framework" notion just alluded to. For

Carnap, the linguistic framework was not what it used to be for Kant, namely an unalterable mould made up of fixed, unchangeable "categories." According to him, what linguistic framework we use results from a choice. The one that serves in ordinary life is the one of the "world of things" but, he pointed out, philosophers may legitimately make use of others, such as the one of sense data. And it is only within a given linguistic framework that we may ask—and answer by means of appropriate investigation procedure—the question whether such and such object or attribute thereof exists. Clearly, the "empirically true" and the "epistemologically true" statements defined in section 4-2-4 correspond, in Carnap's terminology, to those that are true in, respectively, the "world of things" and the "sense data" frameworks. As for Quine, at the place (loc. cit.) where he mentioned "the ontology to which one's use of a language commits him" he expressed a view obviously very similar to Carnap's (albeit in terms that the latter thought misleading[15]). Hence we may presume that, for him, our "empirically true" and "epistemologically true" statements would have fitted in two different but equally legitimate "relative ontologies."

Such a use of the word "ontology" calls to mind the expression "regional ontology" considered above in relation with the complementarity approach. Unfortunately, in the eyes of a quantum physicist what is most striking there is that, finally, neither one of the two expressions labels a genuine ontology. As regards the first one this is immediately clear as Carnap himself most sensibly stressed by pointing out that it just designates the use of a certain form of language. Concerning the second one it is perhaps not so immediately clear as long as we only think of the use Husserl made of similar views. But if we consider its use following Lurçat in connection with Bohrian complementarity (here section 4-2-1), we find that the role this approach imparts to the choice of the experimental arrangement unquestionably implies that some reference to human action is involved in it. In both cases, such ultimate referring to human aptitudes is obviously incompatible with the notion the word "ontology" was originally meant to express, namely, the (justifiable or unjustifiable) one of a description of "Reality as it really is." For a thinker such as Quine, who chose simply not to use the word "ontology" in this—traditional—sense, the reemployment he suggested of this word was void of semantic inconvenience. But for us, who did not take that decision, it has the drawback that we shall be forced, whenever we use the said word, to specify in which one of its two presently accepted senses we take it. A convention that has started to spread is to write "Ontology" with an initial capital when used in its traditionally accepted meaning. We shall try to abide by it.

---

[15] See Carnap (1958), I, section 10.

For all that, and independently of this semantic problem, we cannot but be impressed by the final converging of two quite independent exploratory enterprises. On the one side, the one hitherto described in this book, grounded on information stemming from physics (nonlocality, weak objectivity etc.). And on the other side the one carried on by a number of distinguished philosophers (including, along with Quine, such thinkers as Putnam, van Fraassen, etc.) who, without grounding their views on scientific considerations, also infer from their analyses that our discursive knowledge bears on nothing akin to a "reality-per-se."

For the upholders of the view that "the World ('as it really is' being implied) is intelligible" this convergence may come as a surprise. Note, however, that it is but a partial one. Between the two "end results" two differences, in fact, remain. One of them lies in the fact that the philosopher tends to speak of a genuinely free choice between several linguistic frameworks (at least, Carnap does). On the contrary, the physicist, though he/she does not dismiss the idea of a choice (remember the two approaches mentioned is section 4-1-1), has to stress that serious constraints are there at work. For indeed, if she chooses to express herself, at any scale, within the "world of things" linguistic framework she will have to take nonlocality most seriously into account. Now, whichever model is chosen, this nonlocality has disconcerting implications. For example, in the Broglie-Bohm model (taken in its simplest and less unattractive version) one such so-called "thing" is the pilot wave; in other words a field but—here the surprise lies!—a nonseparable one, a function, not of three space coordinates as all classical fields are, but of an enormous number of coordinates (those of all the particles in the Universe). In other words we have here to do with an entity assumed real and that, nevertheless, has nothing in common with what we normally call a "thing," even when we go so far as to give this name to such entities as electric or magnetic fields. As we see, such constraints severely limit the "free choice" the philosopher believes in. Any serious philosophical discussion of philosophical relativism should take them into account.

As for the second one of the two mentioned differences, it bears on the way of appreciating the very nature of the "arrival point." It is a consequence of the great differences there are, not only in the starting points but also in the modes of reasoning. In the conceptions of most representatives of the philosophical trend under study, the idea that we cannot get any (strongly) objective knowledge of whatever could be called "reality-per-se" is implicitly identified with the view that the very notion of such a reality is meaningless. In such circles, because of an age-old philosophical tradition that goes back to the neo-Kantians, such an identification is more or less taken for granted and the objections that, rightly or wrongly, the "outside people" have against it are neither discussed nor even thought of. Consequently, it is there considered quite natural that words from the classical vocabulary, such as "reality," "ontology," and so on, that it would somehow be a pity to drop, be redefined along these lines. To some extent this justifies Quine's uncouth use of the word "ontology" to mean something relative to us. There, "ontology" just refers to the fact that the requirements of everyday life force people to continually make use of the language of things, so that they can derive but advantages from considering things as beings-per-se, even though, in the thinker's eyes, this is but an illusion. As for the scientists, most of them entertain quite a robust conviction that anyhow some "outside stuff" has to be there; and in fact they can back up their view with substantial arguments on which we had some glimpses already. On the whole they therefore generally share the opinion that language is not all that is, that everything does not boil down to words.

Not all physicists, far from it, take the trouble of going thoroughly into the resulting philosophical problem. But some of them do. In particular, this is the case with the physico-chemist Hans Primas. (1981). On the one hand, in a manner that may remind us of Quine's approach, Primas stressed the fact that, notwithstanding the holistic character of quantum physics, we keep on speaking of molecules and other partial systems, attributing them definite forms positions etc. Moreover, he noted that, for us, it is impossible not to make use of such an objectivizing language. This led him to confer (still somewhat parallel with Quine) the label "ontic" to such partial systems and their dynamical properties. But on the other hand—and this is the gist of his analysis—he also stressed that the said systems and dynamical properties are merely intersubjective. They depend—he wrote—on the abstractions we make but "whoever makes the same abstractions ends up with the same phenomena." In some later writings (see, in particular, Primas [1994]) he developed this point further by imparting to these "ontic aspects of Nature" the name *exophysical ontologizations* or equivalently *contextual ontologies*. In his views, this notion is opposed, on the one hand to that of simple *statistical interpretations* (bearing on observational predictions) and on the other hand to that of *endophysics* (a—conjectural!—description of reality-per-se). As for the, mentioned, "abstractions," they amount to purposely refraining from taking some quantum correlation effects into account. More precisely, they consist in treating some ensembles of systems lying in state $A+B$—with the notations of section 4-2-2—as if they were mixtures of systems in state A and systems in state B.[16] The propositions expressing such em-

---

[16] This abstracting procedure corresponds to what I called "applying the empirical reality axiom" (d'Espagnat, 1989).

pirical ontologizations are nothing else than the "empirically true" (or "false") propositions defined in section 4-2-4. In other words in the, there specified, vocabulary they are those that have a *strongly empirical* truth-value.

In sections 4-2-2 and 4-2-4, when discussing quantum measurement theory, we met with such "empirically true" propositions. They were those bearing on pointer positions. It is easily understood that there should exist more general ones. In fact, according to Primas all of our assertions bearing on well-specified macroscopic objects are of such a nature. For indeed it is impossible that they should be elements of a description made in the realm of endophysics—that they should bear on reality-per-se—since the latter is a nonseparable Big Whole. In other words, all such assertions depend on abstractions we make. On this basis[17] Primas put forward an attractive idea. He suggested that since *we* carry out such abstraction procedures, their outcome might well, after all, depend on our way of performing them. More precisely, his claim was that, depending on what we decide to abstract from, it is one or the other of two or several mutually incompatible exophysical ontologizations that gets validated. The examples he gave are startling. One of them consists in the pair of concepts "molecular structure" versus "temperature." Starting from indivisible quantum endophysical reality, we arrive at an "exophysical description" in terms of molecules by performing one given abstraction. And we arrive at an "exophysical description" in terms of temperatures by performing another abstraction, incompatible with the first one. So that neither one of these two descriptions is more "basic" than the other one.

All throughout our study we shall, in this or other forms, meet with what constitutes the gist of these ideas, namely, the fact that unquestionably some reality exists on its own, but that it shows itself to us in forms depending a lot on ourselves and the abstractions we perform.

---

[17] Supplemented—this is worth noting—by a theoretical generalization of quantum mechanics that Primas himself conceived of and called "algebraic quantum mechanics."

CHAPTER 6

UNIVERSAL LAWS AND
THE "REALITY" QUESTION

§ۂ

6-1 The "Theoretical Framework" Notion

WHAT IS IT THAT makes some developments in physics manifestly much more significant than others? On pondering this question I, for one,[1] came to consider that in pure physics there are, in fact, two kinds of theories; those that deserve the name *theoretical framework*, and those that we may call *theories in the ordinary sense of the term*.

A glance at the history of science immediately clarifies the nature and motivation of this distinction, for, in it, we find an elementary, clear-cut instance of a case in which it applies. It consists in the two most famous Newton discoveries, Newtonian mechanics and universal gravity. For indeed the first one is a near-to-perfect example of a "theoretical framework." By this I mean that the three "Newton laws" constituting the said mechanics are statements considered fully general—applying to all conceivable situations—but, correlatively, not including the detailed information needed for quantitatively predicting any phenomenon whatsoever. In particular, they do not stipulate the "law of forces." In contrast, the gravitation law is an example of what—for lack of a better name—I just termed "theories in the ordinary sense of the term." Such a theory should be specific enough to enable the physicist to determine at least the main lines of a set of phenomena. Newton's gravitation theory does this by specifying the precise nature of the "law of forces." Throughout the long period of time during which these two Newtonian discoveries were considered strictly true, forces other than gravity were discovered, obeying other laws and governing other phenomena (electricity, magnetism, and so forth). It was then necessary to admit that gravitation theory is not general enough to deal with such cases. But all phenomena, including those concerning which the gravitation law does not apply, were still thought of as being governed by Newtonian mechanics. As fitting into the *theoretical framework* it constituted. Acceleration of a moving body was

---

[1] (d'Espagnat 1989), pages 192, 193.

still considered to be proportional to the applied force, whatever the law governing that force might be. And the proportionality coefficient (the mass) was still taken to be a constant of the body.

Once the difference between *theoretical frameworks* and *theories in the ordinary sense* is clearly defined—say, on the basis of this example—it remains to be seen whether or not such a distinction may still be of use at present. The question amounts to inquiring whether within our present-day immensely complex sciences it is still possible to discern a genuine theoretical framework, that is, a set of laws endowed with complete universality. This is in fact a rather intricate issue, that we shall try to somewhat clarify in some of the following sections. Here let us just briefly note in what way the situation has evolved in this respect.

About one century ago, classical physics could reasonably be regarded as constituting the—conceptual and partly methodological—foundation of the other sciences. It could therefore be deemed capable of becoming, in the future, the basis of a complete explanation of the whole set of all phenomena. And it must be owned that still today a certain way of popularizing science shows the mark of—or, should we say, reveals some yearning for?—such a way of looking at things. In fact, however, this conception had to be given up since classical physics led to erroneous predictions within the very field that formed the core of its subject, namely, the innermost structure of matter. In contrast, while quantum mechanics is nowadays brought into play in a great variety of fields, in none of them did it ever yield observational predictions that turned out to be at variance with experimental data. From this it follows that if, today, some theoretical framework exists at all it can only be quantum mechanics or, more precisely, the basic general laws of the latter, that is, the observational predictive rules that constitute its hard core and may be called the "quantum framework." Same as, formerly, Newtonian mechanics, the set of the said rules in no way specifies the "laws of forces" (instead of mentioning "laws of forces" present-day physicists rather speak of "Hamitonians" or "Lagrangians," but on the issue at hand the corresponding shades of meaning do not matter). These "laws of forces" are not the same depending on whether atoms, quantum electromagnetic field, crystals, quarks, etc., are invloved. Within each one of these domains they were found out by investigations specific to the domain in question. And they constitute the cores of *theories in the ordinary sense,* which, in the just mentioned cases, bear, respectively, the names *nonrelativist quantum physics, quantum electrodynamics, solid-state physics, elementary particle physics, and so on.* But in whatever ways the said theories represent—or, better to say, generalize—the force concept, they all presuppose, without exception, the above mentioned "quantum framework."

Is this a sufficient motivation for believing in the universality of, at least, some "laws"? As is well known, the universality concept nowadays has, generally speaking, quite a bad press. Renowned thinkers pertinently pointed out that, when carried over from the hard-science realm to the immensely more complex one of the soft sciences, the concept in question gave rise to simplistic extrapolations which, in the end, induced disasters. This explains why, among philosophers, a tendency developed at subjecting the idea of universality to an intensive criticism, which soon extended to all of the domains in which the idea appeared, including even the hard sciences one. In the next few sections we shall examine various aspects of the thus arisen problem.

6-2 Antiuniversalism and "Realism about Entities"

It seems clear that, at the start, practically all of the people who take up a scientific career are at least partly motivated by the, precise or hazy, idea that universality is a valid and inspiring notion. On the other hand—to repeat—that same idea is nowadays sharply criticized on the basis of the fact that, under some political systems, it generated simplistic modes of administration that proved utterly unfitted to the complexity of the real world. Now to tell the truth, even though this observation is pertinent, still, from a strictly logical point of view, in the issue at hand endorsing its validity qua an objection is not easy; and everybody will surely grant that the universality notion cannot be rejected just on its basis. For it would amount to answering a question bearing on an objective fact ("are the scientific laws universal?") on the basis of the good or bad consequences the fact in question entails for human beings; much as if we claimed that we could determine whether or not there is water on Mars on the basis of the advantages and inconveniences that the presence of water may have for future astronauts.

But of course other objections deserve attention. For example, in one of his books Ian Hacking mentioned a *Scientific American* advertising leaflet announcing articles on "How to strike a karate blow with bare hands," "The enzyme clockworks," "The evolution of disc-galaxies," and "The divination bones of the Shang and Chow dynasties." How, Hacking wondered, could a complete theory of these four subjects be at all conceivable? Not to mention a complete "theory of everything" including these four subjects (Hacking 1983).

Clearly, this "commonsense" remark should prompt us to seriously question the universality of scientific laws. But it must be clear as well that, except perhaps for a mind of a very naive complexion or extremely bent on efficiency, such a questioning will not automatically lead to rejection. It is of course true that, in practice, the four subjects in hand belong to four different fields and that, to investigate them, we must possess the relevant pieces of knowledge. But, in principle at least, could not the said pieces of knowledge be derived from other, less disparate and deeper ones? Does not the electric field ensuring the stability of the atoms composing the muscles of the karate addict also ensure that of the atoms in clock-works, the enzyme ones included? As well as the stability of those consti-tuting the "divination bones" of the Shang and Chow dynasties? Should we question its role in the evolution of the disc-galaxies? And, finally, shall we deny that it, and the atoms, and their combinations obey laws that nowadays are quite well known? The Hacking argument cannot therefore be thought of as invalidating, all by itself, the universality idea. Strictly speaking it does not even refute its most extreme variant (not to be upheld here), according to which a totally agile mind exactly informed of the laws of the Universe would be able to faithfully reproduce the de-tailed structures of the muscles of karate addicts, the enzymes, the divina-tion bones and the rest. I call this variant "extreme" because others, issu-ing smaller claims, exist. Remember, for example, the one due to Hans Primas we met with near the end of section 5-4. This approach, we noted, reconciles in a natural way the notion of a high-level universality with the presence within every scientific branch (physics, chemistry, biology, neurology, etc.) of a subset of laws peculiar to the said branch. In this respect it is, I think, quite attractive, and may well yield a valuable indica-tion concerning the nature of science. This, I think, is true even though Hacking's sally appropriately stops us from being overconfident in such matters, and should be kept in mind for this reason!

On the other hand, never did universality, either in its "strong" or its "weak" version, rally the whole set of the traditional realists (the "naive realists" in the philosophers' language). For example, those amongst them that were once called "vitalists" claimed that living matter obeys, not the laws of physics—more precisely, of inanimate matter—but quite different ones. Of course they conflicted with materialists, who held the opposite view and were, in most cases, universalists. Today, among the "tradi-tional" realist philosophers it is appropriate—following Hacking (loc. cit.)—to distinguish the *realists about theories* and the *realists about enti-ties*. The first named ones consider that our theories—that is, those that are correct—describe the true order of the World as it is. Most of them adhere to universalism since, for them, not to do so would imply drawing boundaries between the domains of validity of various theories thought of as being radically foreign to one another, and this—we noted—is ticklish.

As for the realists about entities, they claim that the objects are what make up the furniture of the World. In their eyes both the macroscopic *and the microscopic* individual objects do really exist as individuals. Now, it seems that a great many epistemologists choose to reason merely on the basis of ideas that seem convincing to them, and consider that actual physical data yielded by contemporary physics pertain to a branch of knowledge having no bearing on their own. In their eyes, realism about entities is a standpoint quite as defensible as any other one. And when not associated to realism about theories it constitutes a view that moves away from universalism, just because of the existential primacy that it imparts to each *individual* object, for in science it is only concerning laws, principles and so on that some universality may meaningfully be thought of. If it is claimed that the laws lie nothing remains to which the universality idea might be related.

6-2-1 *The Criterion "Real since Movable"*

The—at least apparent—convincing power of realism about entities comes from the fact that, to check whether an object is real there is a criterion that, at first sight, seems unfailing, namely, "can the said object be manipulated?" I may question the reality of ghosts, flying saucers, and even that of the quasar that I see (or believe I see) in my telescope, but—some thinkers claim—I cannot doubt the existence of the stone I can lift up and throw away. Nor, they say, can I doubt the one of the electrons a beam of which I can direct at will. But on the other hand, whoever has a smattering of modern physics must count as objections to "realism about entities" such facts as, for instance, the one that—as we saw in chapter 1—according to quantum theory of fields a particle is *not* a reality in itself. It would be more adequately characterized as an element of an inter-subjectively most efficient theoretical representation.

The point is in fact quite a basic one for the claim that an electron is a small permanent object endowed with its own individuality has consequences strictly at variance with what is observed in atomic, molecular, and solid-state physics. Whereas, on the contrary, giving up this claim, as quantum mechanics requests, made it possible to formulate, on the basis of the latter, a great number of observational predictions all of which were proved correct. The turn of the balance is thus clearly *against* realism about entities. And this is all the more true as the, above mentioned, allegedly "unfailing" reality criterion is, in fact, nothing of the sort. To show how unreliable it actually is it suffices to consider again the "broken stick experiment" and note once more that nothing is easier than to move at will the "break" along the stick, since we do this whenever we gently raise or lower the said stick. If the criterion were as reliable as at first sight it seems to be, we should have to infer from this that the break does really exist, whereas, quite on the contrary, we feel completely sure that it does not! It is therefore clear that the criterion is not a crucial one and that the mere fact we can act on an "entity" and see that it moves as we expect is by no means a watertight proof that the entity in question is "real." This is significant for it sometimes happens that this so-styled criterion is implicitly made use of and considered conclusive. We already noted how a recent discovery, that of the tunnel-effect microscope, made it possible to manipulate individual atoms on crystal lattice surfaces and how—as a tacit use of the criterion—this was considered to be an "obvious proof" that atoms do exist as localized individual objects. Remember, however, what we then observed. It was, first, that, in view of all the "conceptual peculiarities" pervading atomic and subatomic physics it is necessary to be careful in interpreting what is seen. From the experiment in question we cannot derive the conclusion that atoms are localized objects even outside this experimental framework. But it was mainly the analogy with the "break of the stick"—which is not real although we manipulate it— that proved revealing. For it shows that, even within the said framework, the "obvious fact" claimed to be conclusive in fact is not. To be able to move something that we see—atom or "break"—does not prove this "something" is more than just an appearance.

6-2-2 Halfway between Objectivist Realism and Logical Positivism

More generally, let us note that the problem under study raises *by itself* questions of modes of reasoning, for we may think of two methods for treating it that, in effect, correspond to two different ways of stating it.

The first one consists in just naively asking ourselves: "Is the stick broken or is it not?" Within this approach it is implicitly considered that the questioning bears, not on our collective human representation of the stick but on the stick itself, as it "really is." In other words, it is taken for granted that in virtue of the actual configuration of the stick, the break either does or does not exist. If we take up this view, we shall stress the fact that merely observing the half-immersed stick is quite a coarse and incomplete experiment, and that, to form an opinion, we must therefore supplement it by other, more refined ones such as fingering the object on its whole length. The upholder of this approach will presumably grant that no complete certainty (with respect to either the stick or the atoms in crystals) will ever be reached this way, but he will insist on the point that we can come up really near to it. He will stress that if we finger the stick as said, if we redo the initial observation with a light whose frequency is such that the refractive index of the liquid is, for it, very nearly equal to one and so on, we shall, in the end, presumably convince ourselves that the stick is not broken. Maybe some parallel move will show that atoms are very likely to exist.

The other approach implies evincing a greater conceptual agnosticism. Most philosophical books—at least all those philosophers write for a general public—begin with a chapter showing that regarding the physical world appearances are deceitful. They demonstrate that the table on which I am writing and the one classical physics claims is the true one have practically nothing in common. The first one, they stress, is a solid, compact, colored, robust object. The second one is made up of nuclei, electrons and, most predominantly, vacuum (and let us not consider here *quantum* physics, which would take us even farther!). Some of these books also mention the fact that neutrinos are the more numerous constituents of the Universe, that practically no obstacle can ever stop them, and that if we perceived them rather than photons everything would look dramatically different. Obviously, such well-known and highly pertinent remarks are still reinforced to a high degree by the above made observation that things we think we clearly "see," such as trajectories in bubble chambers, sometimes are just illusory.

More generally, note that while very many thinkers quite rightly stress the importance of facts, quite a number of them just as rightly point out that, in contemporary physics the "fact" notion takes up a full-fledged meaning only in the realm of the macroscopic. Now, this limitation has a considerable philosophical bearing. Indeed, such a view was foreign to old-time physics. It was there commonly assumed that microscopic facts necessarily underlie macroscopic ones and that the former are just as independent of our "human way of perceiving" as are the latter. Today, we grasp much more clearly how conjectural this idea was (it would force us to turn to ontologically interpretable models such as the Broglie-Bohm one, with the corresponding difficulties). As a rule, we therefore tend to speak of "facts" and take the notion seriously, only when dealing with macroscopic states of affair. But, to repeat, the borderline between the two domains is hazy. And, what is more, it is but weakly objective, for it is quite clear that it crucially depends on our abilities and lack thereof.

All this cannot but throw additional doubts on the implicit starting point of the above "first approach", that is, the view—called "near realism" in chapter 1—that we directly dispose of the "good" concepts, those enabling us to ask meaningful, answerable questions about reality-per-se. The second approach essentially consists in rejecting this idea, that is, in claiming that our discursive knowledge primarily consists in a synthetic ordering of our collective human experience.

Note though that "primarily" does not mean "exclusively." In other words, no logical necessity compels us to take up in their entirety the views of the early-twentieth-century positivists, those of the Vienna Circle in particular. One of the principles these thinkers set forward was that, to be at all meaningful, any scientific statement must be such that, in the last analysis, its meaning can be entirely explained in terms referring to observation (such as: "the pointer *is seen as lying* here or there"). Of course, he who sets such a great principle forward must comply with it. The positivists were therefore consistent when they claimed that the whole knowledge we may have *exclusively* consists in a synthetic ordering of human experience. But here there is no question of adopting positivism or any similar view as a starting point. In this book we let ourselves be guided, as much as possible, by physics. And we make a point not to forget that physics was built up—and is still today mainly developed—by people most of whom were or are intuitively bent toward realism. It was physics, not at all an a priori philosophical stand, that progressively made us aware of the very serious deficiencies of all received forms of realism. Hence no principle bounds us except that of remaining in conformity with all observational data. And, to repeat, there is therefore no question that we should part with realism, or with any nonoperationally defined notion, just for a reason of principle. Only, we have to take into account the most powerful synthesis of communicable experience that present-day physics is, and we must therefore part with the views that either are flatly at variance with it or match it only artificially and in parts.

Among such views are, as we saw, near realism and its closest neighbor, realism about entities. Consequently, any approach that focuses on the individual reality of objects of any size remains to us unacceptable. It can be taken into consideration only formally, that is, if and when it is freed from any explicit or implicit reference to reality-per-se and conceived of as merely being the description of a part of our communicable experience. This, of course, is the part—called "empirical reality" in section 4-2-3 above—that can be reported on in an objectivist *language*. The Carnapian notion (met with in section 5-4) of a "linguistic framework of the world of things" helps making this more precise. As Carnap explained, once this framework has been chosen, usual questions concerning existence or attributes may be asked and are to be answered by empirical investigations. For example, in order to know whether or not a half-immersed stick is broken, we may perform the various operations described above and legitimately infer from them (in most cases) the definite answer "no, it is not."

Can we do the same concerning atoms on crystal lattice surfaces? Yes, in a way, but with the above mentioned reservations concerning possible extensions to other contexts. And in fact, in view of the quantum properties of atoms the said reservations must be taken with utmost seriousness. Moreover, in both cases we must of course remember that such inferences are by no means statements bearing on "what exists." Conceptually, Carnap's very definition of what he called the "world of things" refers to us. Incidentally, this we most clearly realize if, within the framework of the broken stick experiment, we imagine a world inhabited by intelligent robot computers deprived of most of the moving abilities and especially palpation possibilities that we possess. Let us assume they still have the ability of moving the stick up and down. Having observed that they thereby have the power of moving the "break" at will along the stick, these computers would apply the criterion "movable at will, hence real" and, in line with Carnap's conception of "reality of the world of things," would therefore claim such breaks are real. Within Carnap's conception they would be entirely right, just as we, who have different abilities, are absolutely right in asserting the opposite.

In the foregoing we met with a number of reasons that finally induced us to discard realism about entities, that is the conception that most naturally accommodates antiuniversalism. We are thereby led to take interest in the views most radically opposed to the conception in question, that is, realism about theories and the universality thesis. Under the name *Pythagorism*—alias *Einsteinism*—the first named one in now to be investigated.

6-3 "Pythagorism" (Einsteinism")

In *Enigmes et controverses*, Jean Largeault (1980) noted that, with scientists, pure positivism is nowhere present and claimed that, in fact, it is just an epistemologists' lucubration. He wrote, "the theoretical physicist is at once deeply committed to empiricism and on the look for symmetries. He tempers instrumentalism by means of Pythagorism." We shall comment on instrumentalism but before that we must emphasize how right Largeault was in stressing the importance present-day physicists attribute to symmetries. To claim that this concept (and perhaps even more its "offspring," the notion of symmetry break) dominated the whole of twentieth-century theoretical physics would hardly constitute an overstatement. This fact is patently obvious in high-energy particle physics but it is also true in other fields. In many domains it was the quest for symmetries and the discovery of small violations of the latter that were at the origin of the most important discoveries. The name "Pythagorism" is appropriate for qualifying investigations of this type since it was in the harmony between proportions, in the symmetries within graphical representations of numbers and in other similar considerations that Pythagoreans of the sixth century B.C. expected to find the essence of things. Its sole inconvenience is that it applies to too large a spectrum of views, ranging from a questionable mysticism of numbers to the mathematical realism many present-day pure mathematicians firmly believe in. In view of this we might, in spite of the disadvantage of neologisms in general, feel here tempted to introduce one. We then would call "Einsteinism" the conception of the relationship between mathematics and physics that it seems Einstein had in mind. This variant of Pythagorism is favored by those among present-day physicists who, in view of the failure of Cartesian mechanism, turn to mathematics in the hope of finding in them (at last!) the "true" concepts, those that should make possible a faithful description of "what really is."

It should be observed that such a Pythagorism à la Einstein (which most epistemologists, up to now, serenely ignored) radically parts with both objectivist realism and positivism. It parts with the former in that, within it, well-balanced relationships between mathematically defined entities are what *constitutes* the very essence of physics; so that what supremely exists is not a set of simple or complex objects but, precisely, such entities and relationships. And, by the same token, it also obviously differs from positivism, since, in it, the basic elements of our knowledge are not "protocol sentences" either (I mean, they are not observational statements of the type: "the pointer is seen lying in such and such a graduation interval"). In fact, it is not wrong to consider that, in both its "pure mathematical realism" and its "Einsteinian" versions, Pythagorism, contrary to positivism is not far from being an Ontology. Indeed, in his mature age Einstein seems to have considered that a theory such as general relativity does yield a genuine access to concepts (curved space, geodesics, etc.) adequately describing the basic structures of what is.

Admittedly, however, this statement should be watered down to some extent. In such essentially philosophical matters responsible physicists are wary of abrupt judgements and, as Michel Paty appropriately pointed out (Paty 1993), there are significant Einstein texts that introduced shades of thought in his mature-years realism. According to them, it is only to a whole array of concepts, not to an isolated one, that a truth value may be imparted. What therefore matters, in other words, the way to give physics its true meaning, is to check the coherence of a whole indissociable set of ideas, concepts and data. And Einstein added that this is true, even concerning the concepts of physical reality, outside world and real state of a system. Indeed he asserted that a priori it is no more legitimate to consider them absolutely necessary for our thinking than to discard them straightaway. That only verification makes it possible to settle the matter (Einstein 1953b).

Within such an approach a Pythagorean ontology is not laid down as an a priori framework. Rather, it is expected to come out as the result of a successful consistency quest. And indeed, it surely was because Einstein considered the quest in question to have been (at least to a great extent) successful that a realist standpoint seemed to him amply justified. But anyhow the significant point is that he finally judged it was[2] (and that many physicists took up the same standpoint). Hence we are justified after all in *defining* Einsteinism as being an ontological interpretation (with or without a capital O; see section 5-4) of all or some of the mathematical entities contemporary physics introduced. Note in this connection that some most important such entities do not correspond to any "a priori mode of our sensibility" as Kant used to say, and would therefore have been rejected by the latter. This may be counted as an advantage Einsteinism presents over some at least of the approaches grounded on moderate or radical idealism (more on this in chapter 13).

However, the said Einsteinism nowadays encounters serious difficulties within the very realm of physics. Already in chapter 2 we could observe that, while, in the physicists' minds, the Feynman formalism readily gives rise to ontological pictures, these, when all is said and done, constitute but a pseudo-ontology. In the same vein, in section 3-2-3 we incidentally noticed that opting in favor of instrumentalism—thus giving up any ontology whatsoever, Einsteinism included—removes at one stroke some supraluminal influences that, within realism, seem incompatible with relativity theory. And, finally, in chapters 4 and 5 we made ourselves acquainted with the considerable difficulties generated by any attempt at ontologically interpreting the mathematical symbols constituting the framework of quantum mechanics.

Some physicists are not taken aback by such a bundle of momentous difficulties. They meet with the Platonist mathematicians in asserting the existence of a world of pure mathematical beings, lying outside the human mind and explored by the latter. And their interest for theoretical physics comes from the fact that they see in it a possible—admittedly indirect but nevertheless practicable—way of having access to that world judged more truly real that this one is. This is a beautiful and thought-provoking intuition. We may well feel inclined to believe it is true. Concerning its "physical" part—linking such a world with the experimental data, which involves a belief in the pertinence of the latter—it nevertheless seems quite clear that hope in a positive, veritable success should be given up. In view of all the above reviewed data it seems unquestionable, to repeat, that

---

[2] As shown—for example—by a quotation of his part of which was already given in section 1-2 under the heading "Realism (Einsteinian)" and that, in full, reads as follows: "There is something like the 'real state' of a physical system, which [this state] objectively exists independently of any observation or measurement and can in principle be described through the means of expression of physics [which adequate means of expression, that is, which basic concepts are to be used for this purpose, this, in my opinion, is not yet known (material points? fields? modes of determination still to be discovered?)]" (Einstein 1953a). The original text is in German.

the mathematical formalism of the "framework theory" we call quantum mechanics may not be considered to yield an access to mind-independent reality. And it follows from this that the same must be true concerning all the "theories in the usual sense"—supersymmetry, supercord theories, and so on—grounded on the said framework theory.

That much being said, let us note that, even though the structures of independent reality do not seem truly describable by means of mathematical physics, Pythagorism—i.e., the search for symmetries, etc.—still obviously constitutes a most reliable guide for research. And this strongly suggests that the great mathematical laws that go to make up its field of study might well reflect, after all, "something" of reality.

6-4 Remarks Concerning Two "Macrorealisms"

All physicists are aware of the fact that in its most general form (the one people spontaneously have in mind) realism about entities is not scientifically justifiable (even the Broglie-Bohm model appreciably parts from it). Some, however, strive to defend the view that it is true but merely concerns the macroscopic realm. In such an approach, often called "macrorealism," it may be considered that universality, at least when taken in its very strict sense, is given up. For, in it the macroworld—the one of things on the human scale, measurement apparatuses, and so on—is taken to be compatible with the realism of the entities and in fact ruled by classical mechanics which, it is claimed, describes it as it really is; while on the contrary, it is considered that our knowledge of the microworld, the one of atoms, electrons etc. is definitely not of such a nature. Concerning the latter, macrorealism admits, in fact, of several variants, two of which, at least, are worth mentioning.

Within the first one (see, e.g., W. M. de Muynck [2002]), quantum physics, far from describing the microscopic world "as it is" is seen as merely yielding rules predicting what we shall observe on our instruments (which, of course, are macroscopic), when we perform such and such measurements. The outcomes of the latter must therefore be viewed as being merely indirectly related to the microscopic world and yielding no reliable analysis of it. The upholders of this approach often call themselves "empiricists" and they do have some good motivations for that. For indeed, in some respects their conception may remind us of the older—pre-Berkeleyan and pre-Kantian—versions of empiricism, which did not question the intrinsic reality of the observed world (then limited to the macro realm) while insisting that our whole knowledge about it comes from the senses. Viewed from another angle the conception in question may also be called "realism of signification" and (under this name) it is to be examined in section 9-7. But in my opinion the fact that the notion of a macroscopic object is ill defined constitutes anyhow a serious objection to macrorealism proper.

In the other variant of what was called "macrorealism" above, also the microscopic world is assumed known—and even, in a sense, known "as it is"—but merely through the medium of logics differing from the usual one. This variant itself splits into several approaches. Some of them, involving logics of included middle, remained essentially qualitative but others, such as the one involving so-called "partial logics" (Omnès 1988) were developed quantitatively and in detail and are indeed quite instructive. On the other hand, analyses of the way such partial logics operate seem to show that, all things considered, the theories in question are not truly compatible with a strict realist standpoint and, finally, should not be labeled "realist" (either "micro" or "macro").[3]

6-5 Quantum Mechanics as a Universal Theoretical Framework

We noted that in our times the only possible candidate for the role of a universal theoretical framework is quantum mechanics or, more precisely, the *quantum framework* composed of the great observational predictive rules lying at its core. But we also noted that, in view of the (often justified) criticisms nowadays addressed to the very notion of universality, the very idea of a theoretical framework might be questioned. Here let us therefore consider the arguments for and against the view that the idea holds good and that the quantum framework does indeed fulfill the role in question.

Of course one condition that, to this end, any theory must comply with is that of generality. In this respect quantum mechanics is faultless. Conceived of, at the start, as a theory of atoms and molecules, it was progressively found to apply to all branches of advanced physics. In favor of its universality this, unquestionably, is a powerful argument. But still, it is too vague and general to carry conviction all by itself, and the question therefore arises whether it can be backed up by other ones.

The answer is yes. But to show this some care must be exerted since, along such lines, we might think of an argument that, at first sight, looks quite all right but that, in fact, cannot be kept. For future reference let it here be called *the atomicity argument*. The atomicity argument is grounded on the observation that, in the material world, everything seems to be composed of atoms, that atoms are themselves composed of particles

---

[3] The interested reader may find in *Veiled Reality* (chapter 12) a more detailed analysis of this aspect of the question.

and fields, and that quantum theory is just precisely the theory that accounts for the behavior of particles and fields. Hence, so the argument goes, this theory is necessarily universal, at least in the sense that its laws apply to everything.

Admittedly the argument looks attractive. In fact, however, it is disputable for reasons that follow from some of the results the foregoing chapters reported on. One of these reasons is nonlocality. For indeed, the atomicity argument looks convincing just because we implicitly ground it on the Cartesian principle of divisibility by thought, that is, on the notion of a mind-independent external reality composed of very many small interacting but distinct parts. In the form—nonseparability—that it takes up within quantum mechanics, nonlocality disproves the principle in question, so that the atomicity argument cannot be based on it. If we nevertheless strove to keep it we should have to reformulate it in terms of empirical reality. We should then try to take advantage of the facts that, for a large number of predictive purposes, the things we see may be thought of as composed of atoms and that, for other predictive purposes, atoms can be treated quantum mechanically. But it does not seem that any convincing argument showing the quantum framework is universal could be derived from merely the conjunction of these two facts. Another reason that invalidates the atomicity argument is of course to be directly found in the circumstance that "orthodox" quantum mechanics is only weakly objective. In other words, it lies in the fact, noted in section 4-1-1, that the Born rule cannot be viewed as yielding the probability that such and such an event *takes place*, but only as yielding the probability of such and such an *observation result*. Otherwise said, quantum physics provides us not with descriptions of what is, but merely with observational predictions. Again, the atomicity argument is thereby deprived from its substance. More precisely, it can no more even be stated.

Finally, it may be recalled that, as explained in section 3-3-3 (Remark 3) an essentially instrumentalist interpretation of physics makes it possible to reconcile relativity theory with the outcome of Aspect-like experiments. And that, in fact it seems there is no other way to get at this, conceptually necessary, result. We have there an additional reason of regarding basic physics to be essentially a source of observational previsions. Concerning the atomicity argument this is obviously in line with the content of the foregoing paragraph and corroborates its conclusion.

All this therefore shows the atomicity argument is not strictly self-consistent.[4] However, it so happens that the very reasons why it is not yield

---

[4] Which does not mean that any variant of it should necessarily be rejected for it is unquestionable that quantum physics is both general and fundamental.

elements for another, quite different and most powerful, argument in favor of universality.

In order to properly understand the nature of this argument, let us first derive from what has been recalled above the obvious lesson that (as already repeatedly noted) quantum mechanics is an essentially predictive, rather than descriptive, theory. What, in it, is truly robust is in no way its ontology, which, on the contrary, is either shaky or nonexistent. It is the set of its observational predictive rules. Consequently the whole of microscopic physics—governed by quantum mechanics as it is—essentially reduces to observational predictions. But then, with respect to macroscopic physics is it not true that the most reasonable the conceptually least risky—standpoint we may take up is to, similarly, set its ontological interpretation somewhat aside and focalize on its observational predictions? After all, the validity of the observational predictions of microphysics is certain and so is the one of the observational predictions of macrophysics; and if we compare two certain things we are on safer grounds than we would be if one of them were uncertain. Under these conditions, however, the question whether or not the quantum framework is universal takes up quite a clear, unambiguous form. It boils down to that of knowing whether or not the predictive rules of macroscopic physics may be accounted for by means of those of quantum physics. Now, we shall see later (sections 8-2-1 and 8-2-2) that, in view of the outcomes or recent investigations, we now know the answer to be positive. Finally it therefore appears that universality may reasonably be considered established.

Of course, this universality substantiates the, above hinted at, analogy between the "quantum framework" and the three great laws of Newtonian mechanics (also, in their time, considered universal). And this analogy, in turn, helps us to find out, within the quantum realm, which questions are the most fundamental ones. Quite obviously they are those concerning, not the "theories in the usual sense" but the theoretical framework. This is the reason why, in this book, such topics as the classification of quarks, the unification of the four (or five) fundamental forces, the harmonization between general relativity and quantum mechanics, etc., will not be treated. All such subjects, however important, belong to the realm of the "theories in the usual sense." And—through quantum field theory or otherwise—all said theories are finally grounded on the basic quantum predictive rules, that is, on the quantum framework defined above.

6-6 Antirealism

Various authors make use of the term "antirealism" for labeling standpoints that are opposed to strict realism but differ from one another in other respects. For instance, some texts style "antirealist" a view limiting realism to the macroscopic objects, much as is the case concerning the type of macro-realism described above as being a variant of empiricism. A different definition—and one to which the qualifier *anti*-realism seems to fit in a closer way—was put forward by M. Dummett. Observing that problems concerning the concepts of reality, existence and so on are obviously in close relationship with those regarding the truth of *statements*, Dummett (1978) considered that it is by referring to the latter that realism and antirealism are best characterized. Along these lines, he first pointed out that statements about which realists and nonrealists may dispute divide into several classes (mathematical, physical, and so on). Within the subject matter we are here dealing with only two of these classes may possibly be of interest to us, the one of statements bearing on general laws—call it class *L*—and the one of statements bearing on contingent facts—call it class *F*. (In fact, we shall be led to focus essentially on the latter). Dummett then defined realism relative to a given class (which he calls "the disputed class") as being "the belief that statements of the disputed class possess an objective truth value, independently of our means of knowing it: they are true or false in virtue of a reality existing independently of us." Correlatively he called antirealism the view that "the meanings of these statements are tied directly to what we count as evidence for them, in such a way that a statement of the disputed class, if true at all, can be true only in virtue of something of which we could know. . . ."

According to the above definition, realism may be associated with a large variety of statements, including some that do not fit with the "realism of the accidents" (alias "objectivist realism") standpoint. Observe for example that, in the said definition no condition that things should correspond to sets of real numbers is laid down. Correlatively, statements may easily be found that, unquestionably, are meaningful according to a Dummettian realist whereas a Dummettian antirealist should declare them meaningless. Following Dummett, let us take as an example the statement "Mr. X (who, we assume is now dead) was a brave man." Let it be called "Statement *S*." Still following Dummett let us assume, for simplicity's sake, that nobody acts out of character and that character never changes. And let us moreover suppose, as Dummett also did, that Mr. X led, in fact, a sheltered life. That he never was in a situation of danger. The question then is: is statement *S* meaningful? Does it have a truth value (right or wrong)? A Dummettian realist will answer "yes." He will observe that, by assumption, Mr. X had his own character, unknown to us but which existed all the same and involved either courage or cowardice. And he will claim that the (contingent) fact that Mr. X never had any occasion to show it in public is therefore quite irrelevant to the question raised. On the contrary, according to Dummett himself, a Dummettian antirealist of strict obedience will answer "no." He will observe that the "character" of a human being is nothing that other people can have a direct knowledge of, that all the evidence that can be gathered with regard to it must therefore come from the person's actions, and that it is likely that, however much we knew about the behavior of Mr. X, we never would know enough to determine how he would have behaved in a situation of danger. Under these conditions it seems it can safely be claimed that this is a case in which we can have no evidence—not even in principle—either in favor of statement S or in favor of its negation. The antirealist must then judge that statement S is meaningless. And whoever thinks that, to be at all meaningful, statements should relate to some possible, communicable experience must consider that the antirealist has a point here.

Dummett's approach has its limits in that it considers statements to be wholes, while, of course, they are made of words, standing for concepts. As long as the concepts in question are known and familiar—as that of courage is—this is not a source of difficulties. But there can be no question of stating contemporary physics just in terms of good, old, familiar concepts. New ones have to be introduced, which is usually done by combining mathematics and observational data. It is therefore clear that, at least in the realm of physics, what the words "realism" and "antirealism" actually mean concerning the concepts themselves has to be defined.

This is a problem that we shall have to consider (see, in particular, the notion of an "operational definition," section 7-2-2). In short, let us note that, as it here arises, it is linked with the question—well known to the philosophers—of whether or not the meaning of a word should strictly be limited to the set of what it refers to in our experience. A consequent antirealist will normally tend to say "yes it should." He will assert that the meaning of a concept—its content in terms of intellectual representation—can by no means extend beyond the set of the factual data this concept was coined to describe. In this he will, in most cases, be entirely right. This general rule of prudence is what enables us to avoid the intellectual pitfalls into which we always run a risk of falling whenever we make use of a concept outside its well-ascertained validity domain. Fundamentally however it should be acknowledged that such a limitation is inspired from the old *tabula rasa* axiom put forward by the English empiricists of the old school ("nothing is in the mind that has not passed through the senses"), and that, venerable as this axiom is, it nevertheless is nothing but a postulate. When all is said and done, nothing proves that no notion whatsoever, not even such extremely general ones as those of *existence* and *the infinite*, falls outside the realm of this rule. Otherwise said, there is no way to *prove* that such notions are not *necessary ideas*. For the reasons explained, in particular, in sections 5-1-3 and 5-2-3 I, for one, consider indeed the notion of *existence* to be of such a nature.

Let it be noted however that this opinion of mine does not run counter to all forms of antirealism. For example, Lena Soler (2000) pointed out that an antirealist may quite consistently either reject or accept metaphysical realism. The only thing he or she may not do, she stressed, is to conceive of the link between theory and such a referent as having the nature of a correspondence. And she even added: "Such an interdiction in no way prevents the antirealist from admitting that the extra-linguistic referent exerts some constraints on the elaboration of the scientific theories, that is, has an influence on the content of the latter," for example by barring "some conceptually possible models." Since, after all, metaphysical realism, taken in a broad sense, hardly differs from open realism it is clear that the positions adopted in this book are compatible with antirealism interpreted along these lines.

CHAPTER 7

ANTIREALISM AND PHYSICS;
THE EINSTEIN-PODOLSKY-ROSEN PROBLEM;
METHODOLOGICAL OPERATIONALISM

§§

7-1 "Value of a Quantum Physical Quantity"
in the Antirealist Framework

FROM ALL WE HAVE seen it follows that contemporary physics fits smoothly neither with objectivist realism—the realism of realist and materialist thinkers—nor with Pythagorism, taken as an Ontology. Obviously this is the main reason why, in practice, antirealism is so extensively used in physics.

By asserting this we do not mean that most present-day physicists explicitly adhere to antirealism. Many of them have not even heard about it! Only, without having theorized it, and without necessarily agreeing with any "deep philosophy" that may be attached to it, they spontaneously apply the methodological rules it advocates, just because in physics they happen to be the most reasonable ones. Indeed it might be claimed that, from Galileo's time or so, most responsible physicists were keen on using but statements obeying the meaningfulness condition that antirealism nowadays formally expresses. They practiced antirealism without being aware they did. On the other hand, it is also true that, due to peculiarities of quantum systems, applying antirealism to quantum physics requires some care. Distinctions should be made that in turn imply that, in this field, antirealist statements should be formulated in a special way. This is the point that will now hold our attention.

To examine it we have to make use of Dummett's above-stated definition of antirealism, which, in substance, we reproduce here for convenience.

Dummett's definition. Antirealism concerning statements of a certain class is the view that the meanings of the statements of this class are directly tied to what we count as evidence for them. More precisely, a statement of this class, if true at all, can be true only in virtue of some-

*thing of which we could know and which we should count as evidence
for its truth.*

Let us remember that the class we are interested in is the one of statements bearing on contingent facts (which rules out factual statements
such as "protons bear an electric charge"). More specifically still, it essentially is the class of propositions attributing a value to a physical quantity
(such as "the speed of such and such an object is x mph"). Such propositions practically always deal with situations that take place at well-specified places and times. For them, the foregoing definition simplifies to "For
such a proposition to have a truth value (be true or false) it is necessary
that, in the situation in question, the physical quantity be measurable
(either directly or indirectly.)"

We may wonder whether this condition is also a sufficient one. This
question is significant. For indeed people who take science to be a synthesis of experience—that is, who incline towards antirealism—even though
they stress that not any conceivable physical quantity necessarily has a
meaning, still generally consider that, at least, any quantity that can be
measured ipso facto has one. Unfortunately, in atomic physics we have
here to face an ambiguity for what do the words "can be measured" exactly mean? Imagine for example that we are interested in the position,
at a given time, of the electron of, say, a hydrogen atom. In theory we do
dispose of instruments enabling us to measure as accurately as we like the
position of an electron. Still, we cannot infer from this that the proposition "at such and such a time the position coordinates of the hydrogen
electron lie within such and such a range of values" has a truth value (is
true or false). This is due to the fact that the measurement in question, if
it were actually performed, would merely inform us about the values these
coordinates have *after* it has been made. For indeed this measurement—
depending on the theory or model we believe in—either creates or modifies the said values. It does not just simply register them.

7-1-1 The Einstein-Podolsky-Rosen Problem

The fact that the above-stated necessary condition is not a sufficient one
does not imply, of course, that no sufficient condition can be found. It
merely shows that such a condition should be stricter than mere measur-
ability. And the nature of the encountered difficulty suggests that requir-
ing the measurement to be a non-disturbing one might be of help. Let
us therefore tentatively consider the following proposition, to be called
Proposition P.

Proposition P. *For the statement "On system S the physical quantity A has value A" to have a truth-value it suffices that it should, in principle, be possible to measure A without disturbing S.*

Note that here the phrase "it should be possible" explicitly signifies: "We should be able to imagine, at least in theory, means though which the measurement might actually be carried out in case we decided to perform it."

Quite unquestionably Proposition *P* is in the spirit of the antirealist approach. Are we sure, though, that it is true? This is the question. Admittedly, asking it raises a preliminary one, namely: "are there at all measurements of which we may be sure they are nondisturbing?." To this question Einstein and two of his pupils, Podolsky and Rosen (Einstein, Podolsky, and Rosen 1935) gave a simple answer. They pointed out that in some cases—at least, according to our usual modes of thinking—such measurements do exist. They are indirect ones and bear on systems strictly correlated with other systems. For instance, let us think once more of the pairs of correlated darts we considered in chapter 3. Some time after the initial "burst," the two darts composing a given pair are quite far from one another, and it is therefore intuitively clear that what is done on one of them in no way disturbs the other one. But, on the other hand, they are both oriented along the same direction. It is therefore obvious that we can (indirectly) measure the orientation of, say, the right-hand-side (rhs) one just by (directly) measuring that of the left-hand-side (lhs) one. And it follows from the remark just made that we do not disturb the rhs one by doing so.

Clearly, we expect the same to be true when, by thought, we replace the darts by photons and their orientations by the polarization vectors of the latter (or when we replace the darts by protons and their orientations by vectors called "spins," attached to such particles). Now, it is indeed possible to measure (with outcome + or −) the sign S_n of the component of the polarization vector of a given photon along a direction *n* chosen at will. Moreover it is known that if this is done on each one of the two photons of one and the same "Aspect pair"—choosing the same direction *n* for both—the two outcomes always are the same. We may then consider making—just as in the dart case—a nondisturbing (indirect) measurement of the sign S_n attached to, say, the rhs photon, just by measuring the same sign S_n on the lhs one. According to Proposition *P*, the mere fact that we *can* do this—that we have such a possibility at our disposal—should suffice to guarantee that the statement "the sign S_n attached to the rhs photon is +" has a truth value (is true or is false).

Unfortunately, making such an inference immediately sets us at variance with a basic feature of the quantum mechanical framework! For indeed, since the foregoing reasoning is grounded, not on some *actually made* measurement bearing on the lhs photon but just on the fact that we enjoy the possibility of doing it, it can be reiterated, *concerning the same pair*, with direction n changed into any direction m we like. Hence it unavoidably entails that, on the rhs photon, not only the statement specifying the value of S_n but also those specifying the values of the S_m corresponding to all possible directions m in space have a truth value. Now, concerning polarization (and spin as well) the quantum formalism implies that, at a given time, only one of these quantities may have a definite value. At best (within the weak completeness hypothesis, see section 2-8) the other ones could be "hidden variables" but nothing more. And, while the antirealist does not (or at least should not) dogmatically deny that hidden variables "may exist," he, at least, must reject the view that we could somehow measure them. In substance, this is what is often called the Einstein, Podolsky, and Rosen (EPR) "paradox." It shows that, relative to the "microscopic" or "quantum" systems the conditions stated in Proposition P are not yet restrictive enough. If, concerning such systems, we are on the look for a definition of truth—concerning value statements—in line with antirealism, we must ground it in some other condition. For example, we may tentatively think of considering the one expressed by Proposition E below, in which the notion of possibility is replaced by that of actuality (the indicative tense replacing the conditional).

Proposition E.   For the statement "On system S the physical quantity A has value a" to have a truth value it suffices that, in the situation under study, data be in fact available making it possible to predict what value would be obtained if A were measured.

For example, take A to be a "constant of the motion." Then, if A has already been measured on a system S, with outcome, a, say, we know the value we would find if, on S, we measured A again, since it is just a (in our quantum jargon we then say that S lies in an "eigenstate" of A). Consequently, according to Proposition E, after the first measurement has taken place the statement "A has value a" is true. Of course, this first measurement may also have been an indirect one, à la Einstein, Podolsky, and Rosen. But the point is that the above mentioned difficulty is here removed. For indeed, we never can have *simultaneously* at our disposal two pieces of information concerning the rhs photon, one enabling us to predict the outcome we would get if, on the said rhs photon, we measured S_n and another one enabling us to predict the same concerning S_m. For example, if, on the lhs photon, we measure S_n we disturb the state of this photon, thereby destroying the two-photon correlation. So that if, on this same photon, we later measure S_m the outcome of this measurement will provide us with no information whatsoever concerning the rhs photon.

To sum up, all this shows that (at least concerning microscopic phenomena) Dummett's antirealist condition for a statement to be meaningful is necessary but not sufficient and that it is only by considering more specific conditions, such as Proposition *E,* that we may hope to get at a sufficient condition.

Admittedly it could be argued that it is only "strong completeness" (section 2-8) that turns E.P.R. incompleteness into a genuine "paradox," and that therefore switching to weak completeness should make Proposition *P* a sufficient condition, after all. However, just assume for a moment that we kept to Proposition *P,* hence accepted the consequence we found it has, namely the fact that the signs of the components of the polarization vector of one photon along all directions in space simultaneously have definite values. We then could not escape a consequence of this (see appendix 1, section B), namely, that the Bell Inequalities hold true. However, as we know, these inequalities are at variance with experimental results. Unquestionably therefore, this assumption is not tenable.

7-1-2 *Prediction of Observation versus Factual Knowledge*

The foregoing analysis showed that within antirealism a sufficient condition for the statement "On system *S* the physical quantity *A* has value *a*" to have a truth value may indeed be found. But it also showed that in the case of microscopic systems the only candidate (*E*) we could reasonably think of differs from the one (*P*) suggested by Dummett's definition of antirealism and is, in fact, more restrictive. Indeed, Dummett's condition of meaningfulness is quite general and, to a large extent, applies irrespective of the factual situation under study. On the contrary, Proposition *E* most precisely refers to the knowledge we have under the circumstances in hand.

Another significant point is that Dummett-like antirealism focuses on the notion of the meaning of statements and decides which ones are meaningful by means of a criterion based on our abilities at gaining *knowledge*. As we just saw, Proposition *E* takes us even farther away from realism since what it attributes special importance to is our ability at *predicting observation outcomes*. Unquestionably it thereby leads us in the direction of operationalism. At this point it is therefore appropriate that we should reflect a moment on the kind of knowledge operationalism provides us with.

7-2 Operationalism (Alias "Instrumentalism")

This section is aimed at showing that, contrary to a widely held opinion, operationalism, when appropriately understood, is a conception that confirms the value of science and reinforces the certainty of its statements.

7-2-1 Reliability of Science in the Fields of Synthesis and Predictability

As we could observe at several places, a very important contribution from physics to the theory of knowledge consisted in that it proved the need for new concepts. Here, of course "new concepts" does not simply mean "names for new objects." As depicted in popular imagery, that is, as corpuscles, quarks for example, are not representatives of a new concept, even though their bearing fractional electric charges is something "new." At a given moment in the history of science, a concept is genuinely new only if, to be understood, it requires performing a real thinking effort. This was the case, in Newton's time, concerning the one of forces at a distance; it also was, later, concerning that of fields; and, still later, concerning those of curved space, wave function, etc. Proceeding, as a rule, from a combination of experience and theoretical reflection (involving some set of new equations) the new concepts used, for a long time, to be interpreted along the lines of realism. And it should be observed that this is still the way they are introduced in teaching. At school, it is explained that an experiment such as Oersted's shows some interaction to take place between an electrical network and a magnet. A theory, Ampère's in this case, makes it possible to quantitatively describe this interaction by means of equations that involve a certain field. The latter is then conceived of as being some self-contained physical reality. And finally knowledge of the said reality makes it possible to predict some new effects, to be tested by future experiments.

Notwithstanding a few reservations expressed at quite an early stage—in Ampère's time or before—for a long time this conceptual scheme seemed, on the whole, quite justified. Indeed the bulk of its commentators, epistemologists included, regarded it as describing the very nature of science. Admittedly Mach, his followers and, later, the logical positivists were exceptions. But today most thinkers who write about science look upon positivism as being outdated. And reading their books reveals that the notions they put in the forefront have, in fact, a role very similar to the one that objectivist realism imparted to its own basic notions (more on this in section 11-2).

Moreover, as we know, most of the thinkers just alluded to are, in general terms, critical about science. Not, of course, without some reasons. The one that consists in judging its results to be, on the whole, more harmful than beneficial was already mentioned above. As we noted, even if true (which, incidentally, is questionable!) it would in no way show the scientific findings to be wrong. The real and sound reason these thinkers have of questioning the validity of science is an altogether different one.

It is that science itself quite often challenged its own foundations or at least seemed to do so. Already in Poincaré's time the fact that the scientific community, after having adopted Fresnel's theory of light, gave it up and turned to Maxwell's created serious misgivings that Poincaré himself (1902) strove to alleviate. Since then, however, such cases multiplied. We already encountered one of the most surprising ones, the shift from Newton's gravitation theory to general relativity. The basic component of the former unquestionably was the gravitational force. But then general relativity theory asserted that such a force simply does not exist, and replaced it by something radically different, namely space-time curvature. This really was a "conceptual revolution", and others followed. As already noted, the end result was that many epistemologists—and sociologists after them—started claiming that science is not cumulative and scientific theories are finally nothing but cultural products, mirroring definite social contexts.

This is not yet the proper place for us to engage into a detailed discussion of this conception. Let us therefore limit ourselves to stressing once more a point the foregoing considerations set particularly in the forefront, namely, the fact that this whole criticism of science is implicitly grounded—be it indirectly—in too naive a realism. For indeed it rests on a *descriptive* interpretation of the scientific laws. And it must be granted that within such an interpretation the criticism in question *is* relevant. What credit does a science deserve that, after having invented the gravitation force and made it the central pillar of its world view now claims this force does not exist? Which, after having considered light to be composed of corpuscles, asserts that it is wavelike, then, again, that it is made of particles, and finally (?) that it is neither this nor that but somehow both at the same time? To such criticisms we might, admittedly, retort that they caricature the real process, which actually consisted in a progressive maturation of our knowledge concerning highly complex data, in no way reducible to familiar notions. But still, we now know that all theories aimed at describing have something in common with civilizations as Paul Valéry saw them: they are mortal. We are no longer unaware that a theory grounded in some given concepts has every chance of, at some time or other, being superseded by another theory, grounded in different concepts. If we really view science as being aimed at *describing* facts and objects, knowing that much may legitimately render us skeptical concerning it.

But does science reduce to that? Nay, should it be conceived of as being *primarily* just that? Great minds did not think so. Concerning the preference given to Maxwell's over Fresnel's theory, Poincaré wrote: "Does this imply Fresnel's work was fruitless? No, for Fresnel's purpose was not to find out whether there really is an aether, whether or not it is composed of atoms, and whether or not these atoms move in such and such a way.

It was to predict the optical phenomena. And Fresnel's theory still predicts them." And he went on stressing that Fresnel's equations remain valid and that the predictions that can be inferred from them therefore remain entirely correct. This adequately illustrates a nowadays quite insufficiently recognized fact, the remarkable reliability of science as regards prediction of observations.

In this connection it is sometimes pointed out that the relationship between Fresnel's and Maxwell's theories is somewhat special, and that, as a rule, the equations of the former theory do not exactly coincide with those of the latter. That they merely approximate them within some restricted domain of the variables. But Poincaré's remark remains quite significant also in such cases. It may be expressed by stating that the richness of science primarily consists, not in the—changing—descriptions it yields of "matter" (or, more generally, "reality") but in its ability to provide us with a rational, and hence inspiring, synthesis of observed phenomena, which means, in particular, an increase in our ability at predicting them. And it must be stressed that, on this latter point, the forward move of science was, up to now, quite faultless and continuous. As happens also in other fields, censors of science generally ignore this latter fact. To their readers they tacitly suggest that, since it is obvious, there would be no point in mentioning it. But it is clear that ignoring a crucial point under the pretext that it is obvious is no genuinely rational reasoning.

Moreover, the stress we lay here on the steadily increasing predictive ability of science is not to be confused with any trivial utilitarianism. True, considering the richness and often quite spectacular nature of scientific applications it seems unavoidable that laymen should increasingly take science to be directed towards basically practical aims. Such an opinion is nonetheless erroneous, as all scientists very well know. Primarily, science is knowledge. But, just like several other fundamental notions— think of that of *cause*, for example—that of knowledge is much subtler than it looks. It may be reduced to the simple one of "description of things" only when all the terms that such a description requires have been imparted fully definite and generally acknowledged meanings. When great advances in physics take place, this condition is, in general, not fulfilled. And—as we know—one of the most efficient means the scientist has of overcoming the thus created difficulty is to focus his enterprise on the purpose of producing a synthetic account of communicable human experience. Now, such an account obviously requires using statements of the type "if we do this, we observe that," that is,—and here we are again!—statements of the predictive kind. Nothing therefore prevents us from considering such statements to be expressions of quite genuine pieces of knowledge.

7-2-2 Of Criticisms of Operationalism

Above, analyzing various ideas progressively made us impart some genuine "cognitive weight" to statements and laws whose nature is but predictive. And of course, as soon as we take quantum theory specifically into account the quite basic role this theory attributes to the observational prediction rules still reinforces this standpoint. Consequently, we must take an interest in the philosophical approach called *operationalism*, and in the class of theories labeled *instrumentalist*. In short, operationalism or instrumentalism (we shall not try to distinguish between the two) is the view according to which the "hard core" of a theory—here we think particularly of theories of mathematical physics—is its ability to be used as an instrument for predicting observations.

Operationalism was severely criticized by thinkers of various schools. Before analyzing these criticisms—and in order that the meaning of the comments to be made of them be adequately understood—let us first remember what we noted above, when discussing open realism (section 5-1-3), concerning the notion of explanation. We noted there that, even though the great physical laws do depend very much on us, still the idea that this dependence is exclusive meets with serious difficulties, so that we can hardly avoid considering that they also depend on "something else." This argument calls of course for more extensive analyses, to be undertaken in the second part of this work. But—as we showed—its validity depends in no way on conjectures concerning the accessibility of the "something else" in question. In view of the following discussion the conjecture of interest (which is logically tenable though unusual, we come back to this point later) is that this "something else" while not being scientifically knowable still has structures that indirectly have some effects on our laws. So that such an elusive "independent" or "ultimate" reality may be thought of as "explaining the existence" of the said laws. Now, of course, when, within the framework of an operationalist theory, we are led to make use of the word "reality" it is not *that* reality we refer to. It is—it can only be—the notion, introduced in section 4-2-3, of an *empirical* reality, composed of the totality of repeatable and communicable human experience. So, let us firmly keep in mind the fact that the word "reality" is not univocal, that it refers—or may refer—to either one of these two concepts. The distinction is crucial.

That much being said let us examine the main objections to operationalism that have been made.

One of them concerns the very notion of explanation. Mario Bunge, for example (1990), noted that, according to operationalism, hypotheses and theories are nothing but concentrates of experiences and that consequently no genuinely explanatory role is imparted to them. But, he pointed out, the main role of a theory is, precisely, to explain. Which, he claimed, theories only do to the extent that they yield at least a rough sketch of what reality really is.

The point is worth considering. When discussing the "no-miracle" argument (sections 5-1-2 and 5-1-3) we ourselves noted, in the same vein, that, short from granting that miracles occur all the time, we have no other "way of thinking" at our disposal than to consider that "something" exists, in which our great laws are rooted. However, at these places and elsewhere we also noted that (for reasons explained in chapters 3 and 4 in particular) it is extremely risky, if not strictly impossible, to identify this "something" with any *scientifically and experimentally accessible* reality. The weak point of Bunge's criticism analyzed here entirely consists in that according to him such an identifying is requested. In fact, if we give it up we can easily reconcile the operationalist's discourse with a demand for an explanation. Along with the operationalist we shall consider theories to merely be syntheses of experience. We shall assert that they primarily predict communicable observations. We shall add that, within vast but limited fields, they may nevertheless be formulated as if they bore on things, that is, in terms of an objectivist language that syntactically is descriptive. And finally we shall, to repeat, decide to call *empirical reality* what such a description refers to. We shall unreservedly grant that, conceived of in such a manner, empirical reality is much too anthropocentric to be of any use with respect to our desire of explaining physical *laws*. But that much we shall not demand. And we shall not, not at all on the supposed ground that we should deny any explanation is needed but because we found out that the explanation in question can only lie in the "something" mentioned above.[1] Admittedly, realists will object that such an explanation is purely ad hoc. But we saw (section 5-1-2, "First Question") that, finally, this same objection may be (and indeed was) raised against the realist approach. And since what the opponents to operationalism aim at proving is that realism is better, it is clear that a criticism that may be addressed to realism as well as to our version of operationalism is not decisive against the latter.

Moreover, it must be noted that another objection expressed by the same author consists in just an "act of faith" in favor of the realism of the accidents. Of course, Bunge granted that, in physics, validation of ideas rests on experiments of various types. He claimed however (*loc. cit.*)

---

[1] In other words, Mario Bunge's objection is pertinent only sconcerning an operationalism or an instrumentalism raised to the level of an a priori philosophical standpoint bearing on the theory of knowledge. A purely methodological operationalism, such as the one here upheld, is immune to it.

that experiments do not constitute the *object* of this science. He forcefully asserted that any physical idea explicitly refers to a real object and that the said idea becomes void of any value if this object turns out not to be real. Taken literally however, such a claim is, in view of present-day physics, hard to endorse, for indeed physical laws involve a number of ideas that cannot properly be said to refer to objects. Such, for example, is the notion of probability, of which it can only be claimed, either that it refers to an ensemble of imaginary objects or (subjective interpretation) that its object is of the nature of a thought, hence not subjected to physics. And looking at the situation in more detail merely reinforces this observation.

Let us, for example, consider two ideas at the same time: that of "particle" and that of "wave function" (alias "state vector"). If we consider wave functions not to be real this, according to Bunge, implies that the very notion of wave function is void of any value. However, in quantum physics it is quite impossible to do without this notion or equivalent ones, so that, if we decide to follow Bunge, we must consider wave functions to be real. Similarly, in the same science it is impossible to do without the particle notion so that, still following Bunge, we must consider particles to be real. In other words, if we take Bunge's assertion literally we must consider that both wave functions and particles are real, either simultaneously or alternatively. If simultaneously, we must accept the interpretation of quantum mechanics yielded by the Broglie-Bohm model, although it encounters serious difficulties as we already know (see also chapter 9). Moreover also this model, when examined closely, turns out to be hardly compatible with Bunge's claim. To see why, consider the momentum p of an electron such as may be measured by means of the Compton effect (a detailed description of the corresponding method is to be found in Belinfante [1973]). For an upholder of the realism of the accidents this entity, measurable and controllable as it is, is obviously real. And yet, according to the Broglie-Bohm model it is not the "real" momentum of the electron! By definition, the momentum of a particle is, as we know, the product of its velocity and its mass, and, still according to the model, *this* momentum—call it π—is in general different from p. For example, according to the model in a hydrogen atom lying in its ground state the electron is at rest (the Coulomb force is exactly balanced by the one due to the "quantum potential") so that its π is zero, whereas, according to the predictive rules common to standard quantum mechanics and the model, its p is nonzero. In such an instance the idea that a physical notion is significant only when it refers to some real object is confusing rather than illuminative since, far from clarifying the picture, it merely raises pointless problems.

The alternative possibility would of course be to consider that what is real is sometimes the particle and sometimes the wave, and assume that the latter changes to the former through "collapse." Indeed, models are available that more or less materialize such a view but, as we know, conceptually and quantitatively they also encounter serious difficulties.[2] And as for the notion that only wave functions should be real, it has to face hard problems related to nonseparability and the measurement riddle (see sections 8-1-1 and 8-4 for details about the latter). From all this it follows that, when all is said and done, Bunge's view here under study appears unduly dogmatic. It can definitely not be considered a scientifically acceptable objection to an operationalist approach.

Other objections are of a partly psychological nature. On the ground that operationalism limits the aims of science to synthesizing communicable human experience its opponents often claim that taking it up prevents invention and speculation from playing their parts in research. This view has only a very small amount of truth in it. True, the theorist working in a field governed by—say—the quantum laws will not strive to invent naive mechanistic theories since he or she knows beforehand that such constructs, grounded on concepts of "near realism," would automatically be doomed to failure. But the said theorist—operationalist, might we say, of necessity—still has many opportunities to imagine and speculate. Two possibilities along these lines are available and were indeed made use of—often simultaneously—by the most creative twentieth-century physicists. One of them consists in letting oneself be guided—or, better to say, inspired—by naive notions and pictures while being fully aware they are naive. One example is the astounding notion Dirac put forward of a "sea" of negative energy electrons; and another, closely related, one is his, equally quaint, idea that electrons may pop out from it, leaving "holes" that "are" positrons. Let us remember that the two basic notions of antimatter and pair creation just come from this.

On the other hand, it must be realized that the fruitful concepts of "sea" and "holes" were inspired by a most precise but also—until the relevant experimental checks were made—quite conjectural mathematical formalism. This brings to light the other manner the theorist has of putting his or her imaginative powers into action. It consists in reflecting on the mathematical formulas, in wondering how they might be made more general and/or more beautiful (the two go often hand in hand), in discovering possible symmetries hidden in their bosom, and in striving to extend them. All this being, of course, subject to a posteriori experimental monitoring. In short, we recognize here the Pythagorean standpoint and observe once more that, after all, it is not conditioned by realism. Even though the theorists who make use of these methods incline in most cases (sometimes unconsciously) towards some kind of Einsteinian realism, such a realism

---

[2] For more about this see, e.g., *Veiled Reality*, chapters 10 and 13.

hardly serves them in their work and is not therefore what guides them in practice.

Lastly, besides some other minor and refutable objections the details of which would be tiresome, operationalism has to face a basic philosophical criticism, which however, as we shall see, comes from inappropriately extending its scope. What is here at stake is the notion of *operational definition*. The criticism in question is based on the—unquestionable—facts, first that the operationalist greatly stresses the need of defining the concepts brought into play and, second, that he or she makes use, to that end, of operational definitions. That, otherwise said, he/she considers a physical quantity is reliably defined only when the nature of the operations by means of which we could measure it is precisely stated. It is on these two points that his/her opponents attack.

First of all, they point out that, generally speaking, it is quite impossible to define all concepts. As Mario Bunge (*loc. cit.*) appropriately noted, when a concept can be defined, the definition involves other concepts, which necessarily implies that some concepts remain undefined.

This fact is clear, and it should make us better realize that we must not consider operationalism to be an a priori and self-sufficient philosophical thesis. As Giuliano Toraldo di Francia (1981) appropriately pointed out, it should primarily be taken as part of a *method* aimed at gaining wider and better ascertained knowledge. Now, from a method what we expect is, first of all, efficiency. And in this respect the objection does not stand up. After all, we all know that dictionaries suffer from the shortcoming in question. They cannot define all the words they use and are termed circular for that reason. Nevertheless, dictionaries are useful. To put it otherwise: from operationalism it should not be demanded that it define every concepts. It must be granted that, at the start, we already dispose of some notions that seem clear and self-evident. Are they inborn? Did evolution and natural selection coin them? Never mind the answer. What matters here is that we make use of them for building up other concepts. However it should also be noted that—contrary to what both Descartes and Kant believed—such primary concepts are not strictly sure. And correlatively they are not absolute. Even just with respect to what, practically, "can be done with them," their domain of validity is but finite. This, however, does not preclude using them for describing, say, the structure and running of a measuring instrument for, still from a practical point of view, such a description lies within the domain of validity of the concepts it makes use of. And this is just what makes it possible to define the other concepts by means of operational definitions (more about which is to follow). It can even happen that the latter procedure leads to redefining a concept long believed to be primary. Einstein's 1905 paper on relativity theory is probably the best example. By analyzing the way in which time

can be measured Einstein showed that time—hitherto considered an absolute in the highest sense of the word—is, to some extent, relative to the
observer. As Toraldo di Francia (*loc. cit.*) puts it, "no brooding on a purely
aprioristic time concept could have led to so much."

The other "line of attack" of operationalism opponents has even more
directly to do with the operational definition notion. Such a defining procedure basically refers to measurements, and two basic criticisms are accordingly addressed to it. One of them is that, as a rule, several different
procedures for measuring a physical quantity are available. Whence the
question: How, in such an approach, are we to justify our intuitive view
of having to do with one and the same quantity? The other one is that a
measurement yields but a number, while the numerical value of a physical
quantity attached to an object is merely *one aspect* of the latter.

We already answered, at least in part, these objections. They could be
raised against a philosophical system claiming that the mind of the physicist is, at the start, a *tabula rasa,* and builds up its concepts exclusively by
referring to measurements, that is, to numbers. But the version of operationalism that we are here considering is nothing of the sort. In it, as we
saw, it is considered that, prior to any operational definition, mind already
disposes of some concepts that it acquired otherwise. When, for example,
Bunge points out that the electric field is a function, that measurements
merely yield sets of points on a diagram and that they therefore never will
provide us with the field notion if we do not possess it already, we can
but endorse his remark. True, we have ideas coming from sources other
than numbers. But, this being granted, we still have to find out in which
cases these ideas are likely to scientifically describe physical quantities.
And at this stage we must face the fact that even if experience corroborates
such and such ideas we have, this can only be within some domain. Suppose, for example, that, applying to observed facts our—preacquired—
notion of "function" we were able to define two fields. Quite naturally,
we shall then be inclined to think that at any point in space-time both of
them possess a precise value. May we keep this view within an elaborated
theory? To answer, the operationalist puts a rule forward. According to
him the answer should, by definition, be "yes" if, at least in principle,
nothing prevents us from (exactly) measuring, at that place, both fields
together. It should be "no" in the other case. More generally, operationalism considers that, within the realm of physical quantities (that is concerning empirical reality), what is not measurable has no meaning.

Summarizing, operational definitions are to be viewed, not at all as
helps in a quest for not yet thought of physical quantities but as kinds of
handrails. As barriers that should prevent us from clinging to deceptively
clear-looking ideas or inadvertently over-extending the domains of validity of notions proved valid within some well-defined range. We just noted

how efficient this approach was concerning relativity theory. But it is mainly within quantum physics that its role was determinant, as we found out in connection with Bohr's conception and as we shall observe on other occasions.

7-3 On "Meaning" and "Prediction"

In classical physics the meaning of a concept is most often linked with the idea that the entity the concept designates is physically real. In it, as already noted, a chain of reasoning combining theoretical hypotheses and experimental data induces, as a rule, the emergence of a concept, for example the one of electric field, assumed to designate a physical reality. It is then claimed that "there exists" an electric field, endowed with such and such properties, obeying such and such equations, and some consequences are drawn, which are checked by experiments. This imparts to the concept a meaning intimately linked with the reality notion. However, the foregoing discussion showed that such a conception of "meaning" raises questions. True, it was observed that meaning does not emerge from just a combination of operations. But it was also noted that—contrary to some realists' claim—there are concepts that their relationship with experimental data makes unquestionably meaningful and that, nevertheless, are not unambiguously and certainly linked to some specified physical reality. In quantum mechanics let us, for example, consider once more the wave function of a system. This notion is extremely useful, and hence meaningful. But still, for reasons linked, among other things, to nonseparability, taking it to constitute the whole reality of the system—or even just a part of it—would be, as we know, a very questionable move. When we have to do with such a notion it is natural therefore that, to make its meaning explicit, we should refer to the notion of *prediction*.

Already in classical physics concepts had, of course, a predictive role. Obviously the one, for instance, of electric field served for predicting observable effects. But such a function was there considered to be but a derived, secondary one, a mere consequence of the fact that the field is real. In quantum physics the situation is not the same, essentially due to the fact that, in it, the wave function concept is made use of not only for analyzing stationary phenomena but also for predicting observation outcomes. Admittedly, if it only had the first role we could repeat concerning it what was just stated concerning the electric field.[3] However it also has the second one, in which the involved quantum axioms are but weakly objective. As

---

[3] It may be considered that between the hydrogen atom wave function and the function "electric field within a given cavity" there is no outstanding conceptual difference.

we saw, considered in what might be termed their "initial purity," that is prior to the various—and laborious—attempts at interpreting them we hinted at, these axioms (and especially the Born rule) yield information, not on what *is* but just on what will be observed. More precisely: on the outcomes that we shall read on the measuring instruments. Moreover, they also state that, correlatively, the wave function should then be changed in a manner that takes the measurement outcome into account. In these circumstances, attributing physical reality to this function becomes a source of conceptual difficulties and it is quite definitely more natural to link its meaning to the notion of observational prediction.

We saw that if there is such a thing as a universal theoretical framework this cannot be classical mechanics—either classical or relativistic—, which, in some domains, is disproved by experiment. We noted on the other hand that it may well be quantum mechanics, which, admittedly, is but weakly objective but the observational predictions of which never proved wrong. We also saw, in section 7-2-1 in particular, that the attacks on science that took extension in the last decades were essentially based on an interpretation of the latter that was, if not explicitly realist, at least quite "descriptivist." We observed that within this realm they do show some degree of pertinence. But we noted that not much of them remains valid when we cease judging science on the basis of such a criterion and choose to replace it by one grounded on the power of synthesizing and predicting phenomena. Lastly, we found that the objections to (method-ological) operationalism are, all things considered, unconvincing. In fact, to an appreciable extent, they seem to be ascribable to a yearning after objectivist realism that, admittedly, is understandable but that is more and more clearly found to match the facts very poorly or not at all. All this indicates that, when all is said and done, theories that focus on predicting observations are more robust and powerful that those aiming at descrip-tion. It would therefore be preposterous to deny them quite a strong signi-fication, even though, for powerful reasons we saw, we must consider that, in isolation from any notional context reaching farther, they are not endowed with the "ultimate" explanatory power.

To sum up, the aspirations of the opponents to methodological opera-tionalism in physics are understandable and worthy of respect. True, it would, in a sense, be intellectually gratifying if all that is known were describable in objectivist realism terms. If, as Einstein seems once to have both hoped and expected, the "World-in-itself" were fully intelligible or at least were to become so, asymptotically in time. Unfortunately, "grati-fying" does not mean "true." Indeed we can but observe that actual phys-ics is clearly very much at variance with this ideal.

CHAPTER 8

MEASUREMENT AND DECOHERENCE, UNIVERSALITY REVISITED

§§

8-1 Introduction

HOW DO I KNOW that there is a stone on the path? Obviously by having a look (which is tantamount to making a measurement).[1] Hence, if I were extremely cautious not to make unwarranted statements I should not bluntly say that there is a stone on the path. I should say: "We know that if we had a look at the path to check whether or not we have the impression of seeing a stone there, we should actually get the impression in question." As long as we remain within the realm of pure thinking, this remark does not amount to taking an option for or against objectivist realism. It is just a matter of cautiousness, that is, of taking care not to make unjustifiable claims. It may be that objectivist realism is true. But, since it is not provable, maybe it is not. So, we keep on the safe side by not implicitly postulating it.

In ordinary life making use of such long, intricate sentences is quite impossible. For all practical purposes we are therefore fully justified—even if we are not diehard realists—in using the shorter, so called "realistic," sentences, that describe objects as "really being" here or there. In the quantum mechanical realm the situation, however, is different. As we already know, this is a domain in which too "realistic" sentences, implicitly postulating that all the quantities of interest always *have* values, would lead us astray. True, there are circumstances in which we know (for sure) beforehand the result we would get if we measured a given physical quantity on some given physical system. For example, if, after having been accelerated in such and such a way, a particle is left free, we know exactly (to within trivial measurement errors) the value we would get if we measured its momentum. (And in the quantum formalism the particle is consequently attributed a plane-wave wave function corresponding to this value.) In such cases, according to Proposition *E* in sec-

---

[1] For its main part, section 8-1 reproduces with but inessential changes the beginning of an article of mine published in *Physics Letters A* (d'Espagnat 2001).

tion 7-1-1, it is possible—and convenient!—to make use of a realist *language* and just simply say that the considered quantity (here, the momentum) *has* the value in question. But then, the formalism forbids us to even *think* that, on the physical system in hand, such and such other physical quantities also have definite values that we do not know and a measurement would reveal. It can be proved that within the said formalism such an apparently harmless conjecture would make us utter inconsistent or erroneous predictions. Concerning them we can therefore do no better than to keep to statements of a purely predictive and probabilistic nature.

When, assuming the quantum formalism is universal, we apply it to the macroscopic realm we must expect to meet with similar limitations. So, within the example put forward is section 4-2-2, if we know that the electron initially was in state *a*, then, even before looking at the pointer, we know for sure that when we look at it we shall find it in state *A*. And in this case, just as in the one of the stone on the path, we may choose to just simply say that, after the electron has interacted with the instrument the pointer of the latter *is* in state *A*. Of course, the same holds true if symbols *b* and *B* are substituted for symbols *a* and *A*, respectively. But when we know that initially the electron was in state *c* = *a* + *b* (section 4-2-2) we cannot express ourselves in such a way. We cannot any more predict with certainty what impression we shall have when we look at the pointer. Hence we cannot any more make use of the semantic convention (linked with Proposition *E* of section 7-1-1) that we could safely apply concerning the stone on the path. A conceptual problem therefore arises, bearing on the question whether or not we should still assert that the pointer lies at some definite place.

More precisely, the problem consists in the fact that the above described general approach may be applied in two different ways, depending on the amount of a priori knowledge we consider we have, and that these two ways lead to different conclusions.

One of these "ways of arguing"—call it "Option 1"—consists in clinging to the realist philosophy, which claims that the reason why we see macroscopic objects to have definite forms and places is that they really are endowed with these qualities, quite independently of us, that is, of our sensorial and intellectual equipment. We see them at definite places because they *are* at definite places, period. Hence we do not merely know that whenever we have a look at, say, a pointer we shall see it lying within some definite scale interval. Before looking we know (which is more) that this pointer *is*, in itself, within such a definite (yet unknown to us) interval. In other words, within Option 1 we have to consider that, whether or not we looked at it, a pointer can never be lying within two scale intervals at the same time and, more generally, that a macroscopic object can never be simultaneously within two macroscopically distinct states. We infer from this that the quantum mechanical symbol meant to describe the physical state of the pointer cannot, at any time, be one of those describing a quantum superposition of such macroscopically distinct states.[2] If this symbol is a state-vector, this vector can therefore not be of type $A + B$ according to the conventions and notations of section 4-2-2.

Yet, as we saw in this section, if the symbol in question is a state vector (which, by far, is the most natural idea) by virtue of the equation that governs its evolution (the Schrödinger one) it is, in some cases, automatically of the said type! This is the great difficulty that all the quantum measurement theories came up against, for indeed, practically all of them were developed within the Option 1 conceptual framework. Their authors therefore strove to solve the problem by describing the instruments with the help of symbols other than state-vectors and by some other stratagems. But, as I showed some decades ago (d'Espagnat 1976) (a simpler and, in a sense, more general proof is to be found in Bassi and Ghirardi [2000]), all things considered such an enterprise is, in fact, unrealizable.

The other "way of arguing"—call it "Option 2"—consists in keeping close to a standpoint taken by a number of ancient Greek philosophers (Plato foremost), adopted as a "starting point" (though finally dropped) by Descartes, and forcefully argued for by Kant.[3] To an appreciable extent some of the comments made above on the facts and data reviewed in the foregoing chapters already made us acquainted with it. Basically, it is the view that, quite generally, the testimonies of our senses are deceitful and should not be taken at face value. More precisely, it consists in claiming (contrary to the opinion of Galileo, Descartes, and Locke) that, when all is said and done, the distinction between "primary" and "secondary" qualities is in fact irrelevant. The qualities Locke called "primary" (shape, position, motion, etc.) should be considered to be human dependent precisely in the same sense as are those he called "secondary" (color, taste, smell, etc.). According to this trend of thought (taken to be the most reasonable one by, perhaps, the majority of contemporary philosophers), the fact that we perceive some "things," namely, the macroscopic objects, as lying at distinct places is due, partly at least, to the structure of our sensory and intellectual equipment. To consider that it informs us on a property of the "things" themselves is an unwarranted extrapolation. In other words,

---

[2] Within Option 1 this consequence could be avoided by granting that state vectors yield but an incomplete description of physical systems. However, taking up such a view would be tantamount to shifting to models of the Broglie-Bohm type, which, as we know, also meet with difficulties.

[3] As is, in particular, most adequately explained in Putnam (1981).

within the framework of Option 2 only the set of the observational predictive rules is considered to be certain.[4]

As we see, when we take up Option 2 the most formidable of the difficulties the authors of quantum measurement theories had to cope with simply vanishes. This explains why, lately, quite interesting new schemes in this field—based on decoherence (see below)—could be developed. But of course this is not to say that, all by itself, Option 2, which is but qualitative and general, removes at one stroke all difficulties. Two of them arise here with special acuteness. One has to do with the obvious fact that our existence does not seem to boil down to sequences of (possibly probabilistic) predictions grounded on past impressions and bearing on future ones. Let us say, as a minimum, that we find it most convenient to think in terms of elements of an empirical reality, that is, to picture the things— especially the macroscopic ones—as truly having properties even before we look at them. In particular, we find it necessary to be able to think of the latter as having at all times well defined spatial properties. The other difficulty has more specifically to do with the *behavior* of these macroscopic things. We do have a feeling of truly grasping the said *behavior*, and of understanding it essentially on the basis of laws (the laws of classical physics) that refine views we, apparently, naturally have. Now, a priori the quantum mechanical formalism seems to stand totally opposite to these two intuitions. Concerning the first one it disconcerts us by giving rise to puzzles of the $A + B$ type. And regarding the second one it takes us aback because of the fact that its laws considerably differ from the classical ones, that they are probabilistic instead of being deterministic and so on. In sections to come we shall examine these problems. But before this we have to enter into some details concerning what, in quantum mechanics, is called a "measurement" although, in some cases, the operation referred to under this name is not at all a "measurement" in the usual sense of the word.

8-1-1 Defining "Measurement"

In chapter 5 we noted that, contrary to appearances, in science objectivist realism does not constitute the necessary implicit background of any physical argument aimed at explaining. We pointed out that the set of the quantum observational predictive rules offers a valid substitute to it, while being of a wider scope since it extends to the atomic and subatomic realm. And finally we observed above that dropping a presupposition of

---

[4] For further discussion on this conception of the role and nature of decoherence see d'Espagnat (2001).

objectivist realism (namely, Option 1) removes what was the main difficulty in quantum measurement theory. Clearly, all this paves the way to a better understanding of the latter.

The dropping in question entails some focalizing on the quantum predictive rules. The latter are applied to the information at hand, which is condensed as we know (section 4-1-1) within some mathematical symbols (wave functions or similar signs). In textbooks these symbols are in general implicitly assumed to describe physical states in which the quantum objects under study are thought to "be" quite independently of what the physicist knows about them. But we already found out that such a realistic interpretation runs into difficulties of various sorts. Hence—discarding it—we shall go on considering that the said symbols represent but some knowledge the physicist possibly has.[5] Making use of a terminology introduced in section 4-2-4, we shall state that they stand for *epistemological realities*. Here—as there—the qualifier "epistemological" is meant to signify an opposition to ontology even stronger than the one the word "empirical" implies. On the other hand, in order to avoid neologisms we shall go on using the word *state* for designating "realities" of this kind. But in order to make it clear that it does not refer to states in the usual sense we shall write it within quotation marks.

In particle physics what shall we then call a "measurement" (of some physical quantity)? Within the quantum mechanical framework it is appropriate to define it to be an operation (i) that makes us become aware of some datum and (ii) of which it is known that, if it is reiterated (immediately and on the same system), the same datum will reappear.[6]

This definition is purposely grounded on the notion of observation. But this should not make us forget that, for the experimentalist who performs the "measurement" in question, the latter is first of all a physical process in the course of which some micro-object (an electron, say) interacts with a measuring instrument assembled so that condition (ii) be satisfied. For shortness sake, let us call the physical process in question just simply "the process," and save the word "measurement" for designating the combination of the latter and the, above mentioned, "becoming aware" event.

Of course, in view of the universality assumption the measuring instruments, just as any other systems, must obey quantum mechanics. In chapter 4 and here, again, in section 8-1 we had a glimpse on the questions

---

[5] And, as I, for one, pointed out (d'Espagnat 1995, p. 371), a knowledge that essentially consists in the possibility of making *predictions* relative to future observations. Consequently, the conception upheld here, even though it focuses on human knowledge, has nothing to do with the rather widespread view that the wave function yields *partial information* concerning the contingent states of systems assumed to exist per se.

[6] Possibly, after some preassigned correction is made, which depends on the nature of the instrument. For maximal simplicity we disregard such rather special cases in what follows.

this raises. To study them more in detail let us first drop the simplifying hypothesis made in section 4-2-2, according to which the electron gets absorbed by the instrument. When it is not, which is the case in most experiments, after the process is over the electron and the instrument constitute together a composite quantum system. Now, if the measurement was performed on an electron initially lying in "state" a, according to the quantum formalism the said composite system lies in a state labeled aA concerning which we have two pieces of knowledge that are, for us, of the nature of being *certain*. The first one is that when we observe the pointer (if we do) we shall have the impression of seeing it in state A (that is, lying within the instrument scale interval here conventionally labeled A). The second one is that if, behind the instrument in question (call it I_1), we were to put another one, I_2, identical to I_1 and operating on the emerging electron, and if we looked at I_2, we would have the impression of also seeing its pointer in state A. Since these two pieces of knowledge are both certain, we may here make use of the convenient language convention that, by means of the stone example (section 8-1), we showed, in such cases, to be harmless. More precisely, we may then say that after I_1 has interacted with the electron its pointer *is* in state A. And similarly— by, here, extending in a natural way the said convention and the use of the verb "to be"—we may say that the electron then *is* in "state" a. Clearly, the use we make here of I_2 has concerning the electron a role paralleling the one played, concerning the pointer, by our impression of seeing it in state A. Note, however, a difference. For, obviously, I_2 itself has no "impression of seeing" whatsoever. It is we who, by looking at I_2, have an impression. And what we then have an impression of seeing is not the electron itself. It is just the I_2 pointer. In other words, nothing and nobody ever has an impression of seeing the electron. This is why, above, in line with what we just decided, we wrote the word "state" within quotation marks at the place where it referred to the electron.[7] We had to do so for, clearly, after the electron has emerged from I_1 the symbol a that we attach to it has, in our approach, merely a predictive role (the one of specifying the outcome of some possible future observation).

Clearly, all of what we just said concerning a and A should be repeated word for word concerning b and B. Here, as in section 4-2-2, the case that raises interesting problems is the one in which the electron initially lies in a "state" such as $c = a + b$. And what must be noted in this respect

---

[7] Of course, it comes as no surprise that, in a theory that imparts such a central role to observational predictions the notion of "becoming aware" should not be reducible to any other one. And we may note that, quite strictly speaking, within our conventions the word *state* should appear without quotation marks only in expressions "to have the impression of seeing X in the state Y," in which no attribution of an ontological nature is understood.

is that, according to the quantum formalism (whose predictive validity is, as we know, confirmed by its impressive successes in a large number of different fields), the corresponding process sets the composite system in the final "state" $aA + bB$. At the experimental level what is then observed is quite similar to what we noted previously (section 4-2-2) concerning the slightly simpler case in which the electron gets absorbed. With regard to the pointer, what was asserted there may indeed be repeated literally. The full-fledged theory states, and facts confirm, that if, on a whole ensemble of "electron-instrument" composite systems, once all the, individual, "processes" are over, we look at the instrument pointers we shall see some of them in state A and the other ones in state B. In other words, if we have to do with one such composite system only, there is a certain probability p that we should have the impression of seeing the pointer in state A and another one p' that we should have the impression of seeing it in state B ($p = p' = 1/2$ in our example). With regard to the electron we must, for describing the factual data, resort as above to the fiction of a second instrument I_2 placed behind I_1. The formalism then predicts that we shall have the impression of seeing the I_2 pointer in state A on some elements of the thus constituted ensemble and in state B on the other ones. And furthermore it predicts that this will take place in strict correlation with what we shall have observed concerning the I_1 pointer. If the I_1 pointer has been seen in position A, the I_2 pointer will also be seen in position A, and similarly concerning positions B.

At this stage the conceptual difficulties due to the superposition effect (section 4-2-2 and above) are by no means alleviated. They stop us from considering that, once the "process" is over, "there is a probability p that the pointer should *really be* in state A and the electron in 'state' a and a probability p' that they should *really be* in state B and 'state' b, respectively." If, nevertheless, we are absolutely bent on attributing a "state" to the electron alone, then by putting together the various pieces of knowledge gathered above we may, tentatively and formally, express ourselves as follows. We may still say that just after the first "process" (the electron-I_1 interaction), the electron "is," with probability p, in "state" a or, in other words, that the "process" has a probability p of setting in "state" a any electron initially lying in "state" c. However, when expressing ourselves this way we make use of a language that, formally, is descriptive, and we must keep in mind that it is just formally so. That, as long as the superposition troubles have not been cured or at least somewhat alleviated, the "process" cannot be considered to really be one that sets the electron in a "state." And that, even when (and if) they *are* alleviated (see section 8-2), in view of what we noted in the introduction we may merely expect it to set the electron in what we called an epistemological state, just synthesizing the observational predictions we can, in practice, make about it.

REMARK 1. THE NOTION OF "ENTANGLEMENT"

A "state" such as $aA + bB$ above is said to be an "entangled" one (or, equivalently, in such an expression the component "states" a, A, b, and B are said to be entangled with one another). The notion of *entanglement*, which we occasionally met already (sections 3-1 and 5-2-5) and the importance of which Schrödinger forcefully stressed, has indeed a crucial role in quantum physics. It is a specifically quantum one, with no equivalent in classical physics.

REMARK 2

Note that, when the electron initially was in "state" c, the above-described procedure does not let us know what this "state" was. To the extent that the observed quantity—the pointer position—does show something concerning the electron, this "something" only has to do with the *final* "state," a or b, of the latter. Nevertheless, since the *procedure* used here is identical to the one made use of when initially the electron is known to be either in the a or in the b state, it is still, by extension, called a "measurement." We see therefore that what is called a "quantum measurement" may actually differ very much from what we use to call a "measurement" since, in ordinary life, a measurement is supposed to inform us on the value the quantity had just *before* it was measured. Some physicists, either because they failed to realize the unavoidable character of this consequence of the formalism or because they thought they could somehow overcome it, tried to restore in full generality the notion of a measurement informing us on the *initial* "state" of the measured system. However, their efforts in this direction never were really convincing.

REMARK 3

Incidentally, let us note that some physicists, considering that, apart from having some peculiarities, measuring instruments are just ordinary systems, choose to say that when an electron interacts with any macroscopic system this, under some fairly general conditions, constitutes a "measurement" made by the system on the electron. Correspondingly, they hold the involved process to imply that the electron undergoes a transition from one "state" to another "state." The content of the next section will show that indeed such a way of expressing oneself turns out to be acceptable in some respects. That, in particular, it is so in the sense that, when extended to more macroscopic systems, it synthesizes in a simple, intuitive way the main data that impart weight to the empirical reality notion. But it will also make it clear that (as the above already indicates) it carries with it quite serious risks of conceptual misunderstandings.

As we already know, many physicists are adepts of forms of realism—mathematical or otherwise—that bar any kind of weak objectivity as well as any notion of mere *empirical* reality. They think therefore that the wave functions (or equivalently the state-vectors since these two expressions have the same referent) truly refer to realities existing per se. As we noted, this induces them to consider that, upon measurement, the wave function, viewed, to repeat, as a reality, suddenly gets reduced: "collapses." As von Neumann stressed, such collapses, if they do occur, obey an evolution law that is discontinuous and hence radically differs from the continuous one we considered up to now (which satisfies a differential equation, namely, the Schrödinger one). One of the questions this conception raises is of course the one of knowing at what stage the collapse (or the "cut," or the "jump," as is also said) actually takes place. Within the above sketched measurement theory it is natural to consider that it happens at the time when the measured quantum system interacts with the instrument. But this implies that the collapse hypothesis is somehow at variance with the universality assumption. For if the instruments are genuine quantum systems, if, as any other ones, they obey the quantum mechanical laws, it is not understandable that they, and they alone, should have the "power" of reducing the wave functions. Moreover, as we shall see (section 8-2) such an idea leads to awkward consistency difficulties.

To try and remove these deficiencies von Neumann suggested (in line with his "von Neumann chain" idea; see section 4-3) that the cut be considered movable. Assume once again, for example, that we have to do with a composite system made up of a microsystem S (a particle, an atom, etc.), an instrument I, a second instrument I' that performs measurements on S and I, a third instrument I'' that performs measurements on S, I, and I', etc. In order to predict what we shall observe on S alone we may place the cut either between S and I or between I and I' or between I'' and I''. It may be shown, as we know, that whatever method is taken, the predictions will be the same, while, on the contrary, in order to predict all of what could, in principle, be observed on both S and I, viewed as making up a composite "grand system," we may only place the cut either between I and I' or between I' and I'', and so on.

With regard to observational predictions this "movable cut" procedure is fully suitable. And in fact it stands in close parallelism with the ways calculations are quite often done in practice. (Note its similarity with the coexistence of the two "methods" described in section 3-1). Let it be observed, however, that, ironically, if it is taken up it must be at the price of giving up the aim at a realistic description that, above, was seen as a motivation for introducing the very notion of a cut. For indeed, if human-independent reality truly obeys two different evolution laws it must be the case that the circumstances in which either of them operates be definable independently of the actions *we* intend to perform. Regarding ideas this—I think—reveals the tension induced in the mind of many physicists by the coexistence of strong realist convictions on the one hand and, on the other hand, working methods that, due to the very nature of the investigated facts, can be but operational. In the case in hand this tension leads, as we just saw, to some sort of a logical "shaky balance." For the idea of a fixed cut, which realism requests whereas localization arbitrariness makes it unbelievable, gets, for this reason, changed into the one of a movable cut, entirely in line with the physicist's methodology but irreconcilable with physical realism.

8-2 Decoherence

As we know, after the interaction process discussed above is over, observations made on the electron alone will be consistent with the view that it lies either in "state" *a* or in "state" *b*. But on the other hand—as we also know—we may not take this "at face value." We may not assume that the electron then is in a state in the ordinary sense of the word. At best we may assume it to lie in an "epistemological state." There are at least two reasons for this. One of them is that, contrary to what takes place in classical physics (where what is observed always matches the equations of the relevant theory), here the assumption in question is at variance with what the general formalism indicates, namely, that the final "state" is the entangled one $aA+bB$. Such a mismatch may be, by itself, considered a difficulty by the physicists who both take up a strict realist standpoint (ignoring the epistemological reality notion) *and* consider—in line with Galileo and Einstein—that the book of Nature is written in a mathematical language, that the formalism is a true picture of what is. True, other "realist" physicists do not take up such a square standpoint. They grant that some features of the mathematical formalism may well correspond to nothing real, and merely demand that scientific descriptions be logically consistent with respect to facts and predictions. Rightly or wrongly, these physicists may feel not very much worried by the mismatch we just noted. However, even they have to face another source of puzzlement, already hinted at in sections 4-2-2 and 8-1 and the nature of which is now to be analyzed in detail. It is linked with the, already noted, fact that we find it extremely convenient to think of things—the macroscopic ones at least—as truly possessing properties, even before any sentient being looks at them.

Concerning this problem let us first remember that for testing experimentally the validity of a theory involving probabilities there is no other way than to resort to the law of large numbers. This implies performing—at least "in thought"—a whole statistical ensemble of experiments of the same type. To this end, let us consider an ensemble E of composite "electron-instrument" systems, generated by imagining that a large number N of "processes" of the type considered in section 8-1-1 took place, in each one of which one electron interacted with one instrument. The quantum formalism implies, to repeat, that after all the interaction processes are over the "state" of each one of the composite systems is represented by the symbol $aA+bB$, which is called a "quantum superposition" of "states" aA and bB. It is impossible to impart to this symbol a descriptive—not to mention an ontological—meaning, and it is well known that, soon after quantum mechanics was discovered, Schrödinger stressed the acuteness of the problem by replacing—though in thought only!—the pointer by a cat and the states A and B of the same by the states "alive" and "dead," respectively. To consider $aA+bB$ to be a real physical state, in the ontological sense, would be tantamount to granting that a cat can be alive and dead at the same time. We shall therefore carefully avoid such a point of view. For us, such "states" should merely be—to repeat once more!—symbolic representations of our *information* concerning the "preparation" of the involved systems, that is, the events or manipulations we know they were subjected to. What quantum mechanics actually yields, when the "state" in question is known, are the probabilities we have of getting such and such results upon either directly observing the pointer or indirectly observing the electron (by means of other instruments, see above), or both.

Yet, it remains true that if, in E, we look at the pointers we have the impression of seeing some of them in state A and the others in state B and that it is, therefore, very tempting to think of them as really *being* in the said states. We already have a qualitative idea of how we could manage to recover this possibility, at least "in words" and in practice. It consists in making use of the distinction between reality-per-se and mere empirical reality and in imparting to the epithet "really" a sense parallel to the one of the latter expression. But along this line serious difficulties are, alas, in store for, as we saw, it so happens that the formalism also yields information concerning other, conceivably observable, physical quantities, and that this information has disquieting features.

To clearly apprehend what these difficulties consist of let us simplify the picture by imagining, just for a moment, that the pointers under study are microscopic systems, such as atoms.[8] Each element of the final ensem-

---

[8] It should be clear that, here as elsewhere, it is the assumption that quantum theory is universal that is of interest to us. Within this framework, the only basic difference between a macroscopic system such as a pointer and a microscopic one such as an atom is that, as pointed out below, the macroscopic one strongly interacts with its environment. In the later

ble E is then a composite "electron-atom" system. Now, on such rather simple systems it is possible to perform, by means of appropriate instruments, measurements of a more complicated nature than just becoming aware of the *positions* of the atoms. And the formalism informs us of the relative amounts of the various possible results that we should get if we actually did such measurements. Are these relative amounts those we should find if, still using the quantum formalism, we calculated them, not on the ensemble E whose N elements all are in "state" $aA+bB$ (such an ensemble is called a *pure case*) but on an ensemble E' called *mixture* or, more appropriately, *proper mixture*[9] made up of composite systems, about $N/2$ of which would (in this case) lie in "state" aA and $N/2$ in "state" bB?[10] The formalism yields the answer, which, as already mentioned, turns out to be negative. So that the situation in which we find ourselves is rather strange. We know that if, on the various atoms in the ensemble, we directly performed the "trivial" measurements corresponding, in the case of pointers, to measuring their position, the outcomes on E and E' would be the same, namely $N/2$ atoms in state A and $N/2$ in state B. In other words, as long as we forget about all possible measurements except these "trivial" ones we may consider that the mixture E' is a good representation of E. (For future reference let such a representation of E as a mixture be called *Representation R*.) But if we think that the quantum predictions concerning everything that *could* be observed are all correct (which, to repeat, is our working hypothesis) we are *not* entitled to take up that view. For the above-stated reason we cannot consider E to be a mixture of composite systems in "state" aA and composite systems in "state" bB. Replacing, then, our atoms by cats, we find ourselves facing again the Schrödinger cat paradox; however, this time, no longer because of just the structure of a mathematical symbol but in virtue of verifiable observational predictions. (Below we shall see that in the cat case these predictions are not really testable in practice, but in the atom case they

---

stage (described in the next paragraph) of the analogy suggested here this difference will vanish since the atom will there be assumed to appreciably interact with something else (in effect, a large molecule).

[9] In quantum mechanics the word "mixture" is made use of in two senses which, when an interpretation of the formalism is looked for, turn out to be conceptually different (d'Espagnat 1965, 1966). The epithets *proper* and *improper* serve to mark this distinction. It is essentially the *proper mixture* concept that is relevant in the present discussion.

[10] Do proper mixtures "really exist?" The answer is not straightforward for it depends on the sense imparted to the verb "to exist" (see, e.g., *Veiled Reality*, section 7-3, Remark 5). But, in fact, the very relevance of the issue may be questioned. What counts is that, because of our realist turn of mind, when we think of an ensemble whose elements seem not to be identical to one another the notion to which the expression "proper mixture" refers is the one that naturally comes to mind. And correspondingly what is actually significant concerning some abstract quantum ensembles is that they are *not* proper mixtures.

indeed are.) From now on this conceptual difficulty will be referred to as the Schrödinger-cat problem.

To study this difficulty let us proceed step by step. And let the first one consist in imagining that each one of the considered atoms nonvanishingly interacts with another system conceived of as being relatively complex; let us say, a large molecule endowed with a great many degrees of freedom. Let us assume that all conceivable measurements involving the electron and the atom can indeed be done. But let us imagine that those also involving the molecules will not be done (either because they are too complicated, or because we are too lazy, or for some other reasons). And, in fact, let us take a step further and assume that these measurements are *so* complicated that nobody will *ever* perform them. With a few further but relatively minor specifications, the formalism then shows that the difficulty in question vanishes. For, as a consequence of its equations, it shows that under such conditions the outcomes of all the measurements bearing on the electron-atom composite systems are, at least to within an extremely good approximation, compatible with Representation R. In fact, what it shows is that the only measurements that could make the difficulty reappear are those—which the foregoing assumption discards—of highly complex (and correspondingly hard to measure[11]) physical quantities simultaneously involving the electron, the atom, *and the molecule.* Hence, while the (imaginary) "superphysicist" able to perform all the said measurements could not consistently take ensemble E to *be* a (proper) mixture of systems in "state" aA and systems in "state" bB, deprived as she is of such aptitudes the "just human" physicist can. The first named one cannot, because she knows that if she happened to perform such a measurement its outcome would prove her wrong. The second one can because she knows that whatever measurement she makes (amongst those she can make, of course), its outcome will not falsify what she asserts. In other words, to the "just human" physicist ensemble E *appears* as if it were a "real" mixture in the strong (i.e., realist) sense of the epithet. (For this reason it has indeed been called a mixture but only an "improper" one). Basically, this is the essence of the *decoherence* phenomenon (so named by reference to the "coherence" that quantum states of the aA+bB type are said to possess).

If now—this is our "second step"—we get back to the initial problem, the one relative to pointers (or cats), we observe that such a decoherence does indeed take place in them, for it never is the case that pointers—

---

[11] In some limiting cases involving systems with an infinite number of degrees of freedom it can be shown (Hepp 1972) that these quantities are themselves infinitely complex, and hence genuinely nonmeasurable, and may therefore be considered to be void of physical meaning.

nor, quite generally, macroscopic systems—are absolutely isolated. They interact with other systems same as our "atoms" with our "molecules." Already in year 1914, Emile Borel made the surprising observation that, just because of gravitational effects, shifting a small piece of rock as distant as Sirius by a few centimeters would radically modify the behavior of the gas molecules contained in a vessel here on Earth. From the point of view of quantum physics what is crucial in such matters is that the energy levels of macroscopic objects are extremely close to one another, so that even an inordinately small disturbance may induce level shifts in such systems. It was found (Zeh 1970; Joos and Zeh 1985] that, in this respect, even a speck of dust lying far away in interstellar space is not totally isolated due to its interaction with cosmic background radiation. In other words, any macroscopic system interacts in a non-negligible way with its environment.[12] Now, the complexity of the latter is fabulous. It makes it strictly impossible that we should be able to perform, on the electron-pointer-environment composite system, measurements analogous to the—already most complex—ones that we should have to make if we aimed at experimentally disproving Representation *R* on the electron-atom-molecule triplet.[13] In other terms, from an observational (or operational) point of view practically everything takes place as if the above-defined ensemble *E* were a proper mixture. Correlatively, everything takes place as if both the ensemble of the pointers and the one of the electrons were (correlated) proper mixtures, the first one of pointers in state *A* and pointers in state *B*, the other one of electrons in "state" *a* and electrons in "state" *b*. In still other terms, whatever measurements (not involving the environment) we might imagine to perform, subsequently to the electron-instrument interaction process, on the electron-pointer pairs, everything will look as if the instrument had set the electrons either in "state" *a* or in "state" *b*, thus turning the initial electron ensemble into a mixture.

Incidentally it may be noted that in fact the main role in this process is played, not by the instrument (which merely sets the composite system in the *aA+bB* state) but by the environment (or the molecules in the foregoing allegory). This may explain why some authors choose to say that the

[12] As well as with what is sometimes called its internal environment, that is, the myriad degrees of freedom of the atoms, etc., constituting it.

[13] Detailed calculations made, in particular by Caves (1993) and Omnès (1994b) vividly illustrate this point. They show that the complexity of the instruments that would be required for circumventing decoherence is indeed absolutely prohibitive. In other words, these authors' claim that "one cannot circumvent decoherence," taken literally, is strictly true. On the other hand, note that the pronoun "one" stands for "human beings in general," and that therefore the said statement is typically of an antirealist kind. It crucially involves a reference to the evidence human beings are able to gather. For an antirealist it settles the matter at hand. But not so for a realist.

environment "continuously measures" the physical system (here the pointer) on which it acts. There is only one danger inherent in this "way of speaking." It is that it should be understood along the lines of objectivist realism, that is, as implying that the environment continuously imposes collapses to an "independent reality" composed of distinct objects. As we saw, such a conception is at variance with the general implications of the formalism.[14]

As we observed, it is also such a decoherence effect (in a broad sense) that leads to the observational predictions described in section 4-1-1 above concerning the thought experiment of Young slits with gas. In the conventional Young slit experiment (without gas) observation of interference fringes is what gives evidence of the fact that the particle ensemble is not a "proper mixture" of particles having passed the upper slit and particles having passed the lower one. But, as we already noted, when the particles are assumed to appreciably interact with systems that escape measurement—here the molecules of the gas—the situation becomes similar to the one we just considered. In it, the gas molecules precisely have the role the "molecules" play in our analogy. The ensemble of the observed partial systems—here the particles—then constitutes an improper mixture that, normally, can be distinguished from the corresponding proper mixture by no observation bearing on it alone.

This analogy is interesting in that it sheds light on the "philosophical message" of decoherence. The point is that in the "Young slit with gas" thought experiment two possible attitudes of mind are at our disposal concerning the two blobs we see on the screen. As theorists, we may think of the way in which the particles got there, that is, the evolution law they were subjected to. In that case we shall consider that, just as in the no-gas case, each particle, in the quality of a wave, went through both slits. And, in line with Plato (and in the spirit of Option 2), we shall then say that our senses betray us by letting us believe that each one went, particlewise, through but one slit. But we may also take a standpoint (somewhat artificial in that case but fairly natural in comparable instances) consisting in deliberately ignoring how the particles were produced and taking what we see for granted. This makes us consider that beyond the diaphragm and before reaching the screen the particles are not all of them in the same state. That they are already divided in two bunches, an "upper" and a "lower" one, corresponding to the two slits. True, it would be inconsistent on our part to totally ignore the "production" side of the story, so that it is quite out of the question for us to believe that the

[14] It is to be regretted that even most reliable commentators often express themselves in such a way that such an erroneous interpretation of decoherence is the one that seems to be implied by their discourse.

representation that we thus construct has anything to do with the "real stuff." But still, we may—not illegitimately—consider it describes *empirical* reality.

At least in their broad outlines such views are easily transposed to the measurement problem. There are differences, of course, some of which are technical. But the main one is of a qualitative—nay, even of a psychological—nature. It simply consists in the fact that our tendency at mistaking "empirical" for "independent" reality is much greater in the measurement case than in the particle one. Nevertheless, the structural similarity between the two phenomena makes it quite clear that it is decoherence that makes empirical reality arise. It is decoherence that makes us apprehend an object world. The next section will back up this conclusion even more.

8-2-1 Decoherence and Appearance of a Classical World

As long as the expression "a new theory" is understood to mean a theory grounded on new laws, as was the case concerning Newtonian mechanics, relativity, or quantum mechanics, decorerence cannot be considered a new theory. For it merely consists in applying the basic quantum mechanical axioms to the fact that macroscopic systems always appreciably interact with their environment (including their "internal" one). But still, as pioneering calculations showed (Joos and Zeh 1985) decoherence effects prove to have a major role concerning our apprehension of the world, and in particular in accounting for the fact that macroscopic objects are always seen localized (situated at definite places). Within the here taken up quantum universality hypothesis this localization, as we know, had to be justified. Indeed, in view of the basic features of quantum mechanics and the general consequences they have—such as wave function entanglement—the, perceived, locality of the macroscopic objects is a feature that not only is not obvious but even cannot be imparted to them as elements of human-independent reality. Presumably it must be considered a universal appearance. But at least decoherence theory—and from a philosophical point of view this is its major interest—opens a way towards justifying this appearance.

To understand how this comes about, note first that, just because of the entanglement phenomenon, once a physical system S initially taken to be in some definite initial "state" has interacted with some other physical system T, in general it is not any more in any well-defined quantum state. Its "state" is entangled with the one of T, same as, above, the "state" c of the electron got entangled with that of the pointer. In other words, even if, initially, S could be described by a wave function, after the interaction took place this is no more the case. Only the composite "S plus T" system can be so described. But still, it turns out that in quantum mechanics, when we consider not just one such S but a whole ensemble F of replicas of S, it is possible to write down a mathematical representation of F, called *reduced state*, that yields at least some information on the S.[15] Let us now assume the systems S to be macroscopic (think, say of tiny sand grains or dust specks). Even if small on the human scale such systems, as we know, appreciably interact with their environment, which thus plays the role of our system T. Hence they cannot be attributed a wave function. But still, the ensemble F they constitute can be attributed a "reduced state," and indeed the information the latter yields is quite instructive and enlightening. It shows for example that if a bunch of systems S belonging to F is passed through a diaphragm with two slits, as in the Young experiment, no fringes should appear on the screen. And it shows more generally (which is the aspect of decoherence we are presently interested in), that, except in special cases, macroscopic systems are not expected to exhibit quantum effects. This appropriately reconciles theory with ordinary experience. Note however that this matching has a "conceptual price." In fact, due to some mathematical properties of matrices the "reduced state" in question adequately represent not just one but an infinite number of possible proper mixtures of pure quantum states. And it so happens that most of the latter are made up of pure quantum states that are in no way localized, which would, in a way, sort of jeopardize any tentative interpretation of them in terms of empirical reality. Luckily, one of them at least is composed of pure quantum states that *are* explicitly localized at various places.[16] On the assumption that the very structure of human understanding is what forces us to perceive the world under the figure of locality, it goes without saying that this physical ensemble is the one that, for us, should constitute the "correct" one. To put it differently, decoherence theory does indeed prove that the quantum predictions on the one hand and, on the other hand, the particular "a priori form of human sensibility" (in Kant's language) that locality is, are compatible with one another.

Further, let it be noted that the interest of decoherence is not limited to just this. For example, it helps accounting for the friction phenomenon. It also was made use of in connection with chirality, a phenomenon consisting in that molecules, with the exception of the smaller ones, appear as a rule in two forms that, same as our two hands, cannot be superposed by rotation. In other words, it proves to be a most useful notion in several fields of physics. In the last decade it was made the subject of extremely

---

[15] This is called "writing down a *partial trace*."

[16] Erich Joos (1987) drew attention to this point as part of a comparison he made of the relationships between decoherence theory and an ontologically interpretable model. Later I independently hit on the same observation (*Veiled Reality*, section 12-3) in connection with a specific study of the conceptual problem at hand.

interesting experiments (Brune et al. 1996). A generalized "measurement" procedure such as the ones considered in sections 4-2-2 and 8-1-1 above sets a macroscopic (more precisely, mesoscopic) system in a "superposed" "state" *A* + *B*. Immediately after this sudden event has happened, the system in question cannot, of course, be any more in a state of equilibrium with respect to its environment. And the experimental device is so contrived as to reveal the quantum properties the system then has (thereby yielding another evidence that quantum mechanics also apply to nonmicroscopic systems). But because of the system-environment interaction the system is not expected to remain in that state. And indeed it is possible to measure the—extremely short but nonzero—time it takes for decoherence to "set in" or, in other words, the time before the system, in virtue of the system-environment interaction, recovers the classical features that, due to that same interaction, it normally exhibits. By corroborating decoherence theory, the outcome of this experiment clearly constitutes quite a strong argument in favor of quantum universality. For indeed it is difficult not to regard it as a convincing indication that between quantum systems and those we perceive as being classical there is no basic difference. That it is just because of their coupling with their environment that the latter are perceived as such.

In section 5-4 we already noted that (in the case of macroscopic systems), once the "world of things" linguistic framework has been chosen the usual questions concerning the existence and attributes of various things can legitimately be asked. In a way, decoherence theory makes this approach considerably more precise in that it explains in detail *why* the realist, classical language (the "world of things" one) is useful concerning all the everyday-life objects (all of which indeed are macroscopic).

8-2-2 Classical Laws Seen as Emerging from the Quantum Ones

When we strive to understand, on the basis of the quantum rules, why the world appears classical we soon realize that we may not rest content with, say, just observing that the Schrödinger equation nicely accounts for all atomic and molecular energy levels. We have to inquire further. For indeed we all feel that classical mechanics gives us a clear understanding of the apparent *behavior* of macroscopic objects, that is, their evolution in time, their interactions and so on. We therefore have to wonder whether or not, within our new approach, centered on quantum physics, we may preserve the feeling in question. A necessary condition for this to be possible is that the quantum laws should account for the said behavior at least as well as the classical ones do. Our problem is to inquire if this condition is fulfilled.

This is a far-reaching program. And since this field of study is a very active one at the moment, to try and give some kind of an overall view of it would be risky. However, a few guiding ideas are already clear. The main one of them has to do with the fact that any macroscopic object—a stone, a table, etc.—is endowed with immensely many degrees of freedom that may be viewed as those of all the atoms composing it. These parameters may be distributed in two sets: on the one hand the physical quantities—length, width, positions, velocities, etc.—that are more or less directly accessible to us, on the other hand all the remaining ones. If we assume, as a first approximation, that these two sets do not interact with one another we may focus our attention on the first one, the elements of which are termed "collective variables." In short, the problem then may be stated as follows. Since we take as our starting point the idea that the quantum formalism constitutes the basic theoretical framework, the variables in question have to be described within the said formalism. They must therefore be represented by "operators"[17] the time evolution of which should obey the laws of the latter. But, on the other hand, in classical physics they are represented by numbers and their evolution obey classical mechanics. Our problem is to check that these two modes of descriptions are compatible. In other words, it is to start from the quantum formalism, which yields observational predictions, and show that the classical way of describing the said variables and the evolution thereof finally yields observational predictions not differing from those derived by means of quantum theory.

Now, it turns out that to really prove that much—on a level of strictness high enough to justify calling the outcome a proof—is not an easy task, far from it. In fact, it is only comparatively recently that such an enterprise was undertaken and that convincing developments (that we already alluded to) started appearing. But they were successful. And indeed—due, in particular to Roland Omnès' work (1994b)—it may presently be claimed that we do dispose of a genuine proof. Hence we have now no reason to believe that there are, on the one side the microscopic systems, obeying one definite set of physical laws and on the other side the macroscopic systems, obeying quite another set of laws bearing no relation to the former. Quite on the contrary we have to consider that it is because they follow from the quantum rules (under some definite but practically quite often satisfied conditions) that the classical mechanics computation rules yield the (successful) observational predictions that we know of.[18]

---

[17] In the "Heisenberg representation" (section 4-1-3).

[18] The "Ehrenfest theorems," which have been known for a long time, did already give some indication in that sense, but much too vague to be conclusive. In particular, they did not explain why classical mechanics is determinist.

In other words (and to be somewhat more precise), let us consider a macroscopic system concerning which some amount of macroscopic information has been gathered. We might represent the latter by means of a set of mathematical data of, very roughly speaking, the type of those called D in chapter 4, derive from it, for any later time t, a corresponding symbol D_t, and finally derive from D_t what impressions we should have were we to observe the system at time t. But, of course, we may just as well represent the information in question by means of a set of numbers representing the values at time zero of the quantities (positions, velocities, etc.) we are familiar with, and use the classical mechanics computation rules to calculate their values at time t. It follows from the above-mentioned investigations that the results of the two methods will coincide. In other words, the observational predictions yielded by the values in question will be the same as the ones the first method provides us with. Since the second method is considerably simpler and more familiar to us than the first, it is of course the one that is made use of in practice. And this habit we have of using it, together with the fact that it is realistically interpretable (and always successful in practice), results in that we quite naturally reify by thought the various quantities the values of which it yields. Of course, there are some well-identified cases—when, for example, the system is composed of a particle accelerator and a bubble chamber—in which the interaction between the collective variables and the other ones may not be neglected and concerning which the above does not hold. But such cases are well specified and, on the whole, exceptional. In other words, in standard situations quantum theory itself shows that the observational predictions from classical mechanics will prove correct. And, the latter being deterministic, it is understandable that the notion of a strict cause-effect link should quite naturally have arisen. We know that such and such impression we have implies that if, later, we decided to observe this or that we would get such and such result. This suggests to us the converse. From the fact that we now have such and such an impression we infer that if, in the past, we had looked at this or that we would have had such and such sensation. We hypostatize this view, that is, we raise it to the level of a representation of the past conceived of as being the cause of the present, and so on. In other words, our elaboration of empirical reality is at work with respect not only to the present but also to the past and the future. (But, to repeat, let us not interpret such an elaboration as corroborating radical idealism. We build up empirical reality in somewhat the same way as the gardener and the dragonfly both build up their own vision of the garden. And since these two visions differ from one another, certainly one of them at least is not a trustworthy picture of reality. However, granting this does not imply negating the reality of the garden.)

REMARK ON COUNTERFACTUALITY

In view of the tight connection we noted between counterfactuality and realism, the above-described "reifying by thought" process (the building up of empirical reality) truly accounts for our macroscopic experience only if it makes it possible to "recover" counterfactuality, as defined in sections 1-2 and 3-3-5. We must therefore wonder whether or not this is the case. The answer is not quite obvious for, if we consider again the Aspect-type experiments and the attempts at interpreting them within the framework of realism, we have to remember that, in view of the Bell theorem, the outcome of the said experiments *violates* counterfactuality. Is the situation the same when, in conformity with the above expressed views, we apply quantum mechanics to correlation-at-a-distance experiments structurally similar to the Aspect-like ones but bearing, this time, on macroscopic object pairs, such as the dart pairs of section 3-1?

At first sight, we might fear it is, but fortunately it is not. One way of ascertaining this is to study the question within the framework of the Broglie-Bohm model. (Making use of the well-known fact that within the nonrelativistic realm this model yields the same observable predictions as standard quantum mechanics.) In such a deterministic, hidden variable model, if the outcome of any measurement made on one of the particles composing a pair were predetermined by (hidden) variables pertaining exclusively to that particle counterfactuality would obviously be satisfied. Now, as we know, this is not the case. The "Bell calculation" (section 5-2-4) shows, as we saw, that in the "microscopic" realm what determines the physical quantity "on the right" is not (or not only) the value of the seemingly relevant local parameter at that place. It essentially is what happens to the, very distant, particle "on the left." And it is easily seen that this is what corresponds to the just mentioned counterfactuality violation. Fortunately, however, if we ask how the considered model accounts for the correlation between *macroscopic* objects such as our darts, we soon find out that in fact the just described mechanism has no role in it. The orientations of the darts are macroscopic variables, as such they are but sorts of mean values in which microscopic effects such as the one under study may easily be washed away. And in fact this is what takes place, for these orientations should be numbered among the collective variables and we know (see above) that the latter must obey classical physics, which, of course, satisfies counterfactuality. (More precisely, we know that all the observational predictions concerning them necessarily coincide with those we would derive from classical physics, and the latter obey counterfactuality.)

8-3 Decoherence and State Robustness

In section 3-2-4 it was already noted that the quantum "state" of an individual microscopic system is not knowable in general. Assume for example that what we know concerning a particle is just that it is described by some wave function (alias, "state vector"), without knowing what the latter actually is. Then it may be shown that it is in general impossible for us to increase our information, by means of measurements made on this particle, in such a way as to find out what this wave function really is. True, by measuring what is called a "complete set of compatible observables" we may get to know the "state" in which the particle lies *after* all these measurements are made. But since the latter usually disturb the particle "state" this does not let us know what this "state" was *before*. More generally, it is impossible, concerning any individual microscopic (hence nondecoherent) quantum system, to conceive of a strategy enabling a noninformed observer to find out, by measurements, in what "state" this system is and be sure that *he* did not—Procust-like!—put it there himself by the very manipulations implied by the said strategy. In contrast, as Zurek (1998) could show, a strategy yielding such a security exists concerning systems that are macroscopic enough to undergo decoherence, provided that the system be characterized, not by its (nonexisting) quantum state but by its so-called "reduced state." In section 8-2-1 we already encountered "reduced states" apropos of ensembles of macroscopic systems. As we know, they are mathematical symbols ("matrices") yielding descriptions of the latter. These descriptions take their preparation, their evolution but also—globally—their whole interaction with the environment into account. Concerning them, two important points may be shown. First, there are physical quantities Q whose values[19] remain unchanged when measured on (ensembles described by) such reduced states. And second, (in contrast with the microscopic systems case) it is possible to know which ones these quantities Q actually are, even when the initial state of the ensemble is unknown. Hence, if we focus our attention on one of the systems composing one such ensemble we realize that the observer knows in advance which physical quantities he may measure on it while being quite sure that he will not alter their values by doing so. Obviously, if he does actually perform such measurements his feeling of genuinely learning something about the system "as it really is" will be much more intense than if he had to consider that the measurements he performs are Procust-like ones. Whoever is interested in concepts and ideas

---

[19] In the sense, "values that would be observed if. . . ."

must hold this to be an important application of the decoherence notion. Indeed, thanks to it we grasp more clearly the nature of the "impression of reality" that macroscopic systems give us; for, undoubtedly, this impression is linked with our idea that the properties we see them to have were theirs already before we had a look at them.

8-4 The Everett-Zurek Semirealist Approach

Within the general course of our investigation the present section should be considered to be, in a way, a parenthesis. The point is this. Already in chapter 4 we came to recognize that what is really robust within quantum mechanics is the set of its observational predictive laws. And as we went on the difficulty of imparting to this theory an interpretation on line with physical realism became more and more apparent. In this first part of the book we shall continue our study of quantum mechanics essentially along these lines. But still we cannot forget that, on the whole, physicists legitimately go on worrying concerning realism and the corresponding problems. Hence we must, at some stages, have a look on the attempts made at salvaging at least some elements of the latter. Before tackling—in chapter 9—those of them that are aimed at restoring a full-fledged realism let us here, in line with the content of the foregoing section, have a quick look on an approach that succeeds in combining semirealist views with faithfulness to the letter of quantum formalism.

We know that, generally speaking, the said quantum formalism greatly relies on the statistical ensemble (or, for short, "ensemble") notion. And the foregoing clearly showed that decoherence theory is no exception to that rule, for indeed, as we saw, its basic element is a mathematical representation of an ensemble of systems interacting with their environment. But on the other hand, there is no denying that our experience is that of *individual* macroscopic systems, and that what we are expecting from our theories is a set of, descriptive or predictive, statements concerning such individual systems. Unfortunately as soon as, within the quantum formalism, we aim at going (in the direction of some form of realism) somewhat beyond the realm of just the observational predictive rules, we discover that shifting our thoughts from ensembles to individuals is not easy. In particular, as we know, it is then quite specially worrying that, while a symbol such as $aA + bB$ involve at the same time both A *and* B, we always observe either A *or* B. Such a replacement of conjunction *and* by conjunction *or*—sometimes called the "and-or problem"—is nowhere suggested by the equations. Some therefore consider it to constitute an enigma, and it must be granted that, concerning it, decoherence theory did not, up to now, lead to fully satisfactory results. The reason for this may be considered to be that the problem in hand is of a conceptual, rather than of a mathematical, nature. In way of an introduction to the inquiries presently made in this field let us however mention an interesting development due to Zurek (*loc. cit.*) that explicitly relies on decoherence.

The development in question is grounded on a former one, due to Everett (1957). Briefly, Everett's approach—called by him "relative state theory"—was explicitly aimed at realism. In it no distinction between the concepts of state and "state" was therefore made. But still, it claimed that when a measurement is performed on a system lying in a state such as c—the notations being those already used—no collapse takes place. Otherwise said, the mathematical symbol that represents the final state of affairs was, in it, considered to be $aA + bB$. However, this quantum state was conceived of as being somehow split into two. In other words, two "branches" were thought to simultaneously exist, in one of which, aA, the pointer is in state A and the electron in state a while in the other one, bB, the pointer is in state B and the electron in state b. (The observer himself is somehow assumed split in the same way.) Everett's claim is far from being conceptually clear, but it is suggestive and it was much commented on within the relatively closed theoretical physics world.

This book is not a proper place for engaging into a detailed discussion relative to this theory (see, for instance, *Veiled Reality*, chapter 12). True, its ground notion of a splitting reality ("branches," etc.) seems quite fantastic at first sight. But still, there are good reasons why many physicists took and still take it in earnest. For, as we see, the theory amounts to taking the mathematical formalism entirely at face value, without putting up against it any resistance grounded on received views or concepts. Now, it can be proven that when this much is accepted it is quite impossible to demonstrate that there exists but one reality. Every observer, on his or her branch of the universe, sees the latter as being unique, and the very formalism prevents her from becoming aware of the other branches.

The truth is that, right from the beginning, Everett's theory was a kind of a conceptual Chinese puzzle for the physicists. One point, however, is sure. Its author never considered that such branching effects happen in virtue of somebody becoming aware of something. In fact he viewed them as taking place, not only upon measurements but on a great many other occasions as well. Consequently, the theory implied that Universe branches should proliferate at a tempo past all imagining. How then could these branches be conceived of? Some thought of them as being full-fledged worlds, that is, they considered that, on every occasion when a branching occurs the Universe (observers included) actually gets multiplied. Others pointed out that such an idea of universes getting physically multiplied does not appear in Everett's original paper, and tried to impart to the content of the latter less disconcerting interpretations.

Unfortunately, none of them may be claimed to be both fully consistent and conceptually entirely satisfactory. Nevertheless, many physicists keep on being attracted by the theory because of its mathematical clarity. They point out that trusting formalism rather than commonsense may lead to quite major discoveries, as the existence of the two Einsteinian relativity theories demonstrates quite convincingly.

Such a motivation incites them not to linger over the just mentioned conceptual riddle but, rather, investigate the question of the concordance between the theory and basic features of human experience such as the impression we have of the existence of real enduring objects. It is, I guess, in this spirit that Zurek (1998) proposed to throw a light on Everett's approach by means of decoherence and conversely. For this he built upon the important remark of his mentioned above (section 8-3). Imagine a person who, within some branch of the Universe, is on the point of making some trivial observation on a macroscopic system. Zurek essentially pointed out that, in virtue of decoherence and of the remark in question, this person may think of the state in which this system *is* (and that her observation will reveal) in the same spirit as she would if classical physics were true. More precisely, she can proceed without having to fear that the state that her observation is on the point of revealing to her is but the one in which her very observation act will set the system. And—here, I would say, is the crucial element in his approach—it is this reliability that Zurek finally *decided* to adopt as his very *definition* of what reality *is*. For characterizing the, obviously somewhat esoteric, sense that the verb "to be" is here endowed with, he suggested the expression *relatively objective existence*. Of course, he is quite aware that in so doing he constructed a kind of an oxymoron ("look here, the plurality of branches is still there!" a diehard realist would, no doubt, exclaim). But, he answers, it is a deliberate one.

Considering the exceptional and universally acknowledged difficulty of the problem it tackles, Zurek's theory is unquestionably attractive. On the other hand, there is no denying that, in view of the importance of the conceptual points it leaves in the dark, it must make any strict upholder of realism unsatisfied.

8-5 Universality Revisited

It must be granted that, within the physicists' world, the judgements concerning the decoherence notion—its bearing, its meaning, its interest— are contrasted. Roughly, the persons who express their views on the subject may be considered divided in three groups. First, there are the objectivist realists, who, observing that decoherence merely accounts for "what appears to us," take note that it falls short of their expectations. Next—it must be said—there are some objectivist realists who mix up appearance and reality and erroneously believe decoherence solves, in realist terms, the measurement problem.[20] And finally there is a third group, whose members take the realists' expectations to be over-ambitious and consider that science is *not* aimed at describing reality per se. These physicists are neither objectivist realists nor even supporters of physical realism. And since decoherence opens the way to a scientific explanation of the *phenomena* they have all reasons to consider it a major advance in physics

The latter standpoint is to be explained in detail in chapter 10. But before that it is appropriate that we should examine to what extent and in what way decoherence contributes to the analysis of a question already touched upon above, to wit, what is the likelihood that quantum mechanics be universal.

To restate the main lines of the problem, let us first remember that the general quantum mechanical laws apply in practically all advanced parts of physics and that whenever they led to precise observational predictions the latter agreed with the observed facts. In favor of universality this of course is a simple argument that carries weight. In section 6-5 we noted that, at first sight, we might be tempted to supplement it by an even simpler one, just consisting in observing that quantum mechanics is the theory of atomic particles, that all existing bodies are composed of the latter and that therefore quantum mechanics *must* be the "theory of everything." This we called the "atomicity argument." But we also took notice of a serious weakness of the latter, due to its ontological nature. As we saw, a condition for it to make sense obviously is that the theory in question be descriptive, and standard quantum mechanics fails to fulfill it. We noted, of course, that, contrary to what is the case with respect to classical probabilities, those that quantum mechanics involves do not just proceed from our ignorance. But we saw moreover that the Born rule is self-consistent only if the probability it yields is interpreted, not as a trivial "probability of *presence*" but as the probability that the particle—or whatever—*be found* at the place where a position test is made. Similarly, we noted that there is a conceptual gap between the phenomena, which we perceive as localized in space, and a notion of mind-independent reality that, in view of the Bell theorem, can only be a nonlocal one. All this shows—to

---

[20] There might be a purely linguistic reason for this. It so happens that in English (nowadays, the universal language in which science is couched) the word "appearance" has two meanings. In one of them—the one in which it is taken here—it refers to what we perceive, in (possible) contradistinction with what *really is*. In the other one it has (as in the sentence "mammals made their appearance during the tertiary era") the sense of "coming into existence." It is conceivable that some of the people in the second group just mixed up these two meanings of the word.

repeat—that what quantum physics yields in a clear and indisputable way is but a set of observational predictive rules. Besides, as we noted, macroscopic physics might also well boil down to a set of observational rules, in which designative terms and sentences involving them just stand as convenient shorthand expressions referring to frequently occurring sequences of operations and observations. Within such a conceptual framework the atomicity argument is quite obviously meaningless. But, as we also noted, correlatively the very statement of the problem that we are investigating takes up a new form. The question "is quantum mechanics a universal theory?" becomes "do the predictive rules of classical mechanics and of the rest of physics follow from those of quantum mechanics or do they not?"

So, do they? Well, from all that we have seen we may conclude that, with an appropriate shade of meaning, the answer is "yes." And this, in spite of the fact that, at first sight, appearances favor a negative one (since the phenomena are localized while quantum theory is holistic, etc.). Indeed, after quantum mechanics was discovered the idea that the classical approach is somehow more basic than the quantum one was still, for a long time, held to be true. Clearly it underlay Bohr's assertion that our everyday language—suitably refined through the adjunction of technical words—is, ultimately, the basic one. It also underlay Landau's often quoted remark according to which quantum physics has to rely on classical physics for its very formulation. In fact, what turned the scales in this respect was the advent of decoherence theory. For, as we saw in sections 8-2-1 and 8-2-2, this theory is a great step forward towards understanding the classical appearance of a quantum world. And what is even more remarkable in this respect is the way in which experiment confirmed the validity of the theory. We already explained (section 8-2-1) the reasons why the work of, in particular, the Haroche group (Brune et al. 1996) constitutes a truly convincing factual argument in favor of the universality of quantum mechanics.[21]

The above mentioned "shade of meaning" in no way lessens the importance of the developments just alluded to but, still, should be noted. It consists in the fact that, contrary to what might be guessed at first sight, the developments in question do not at all imply that such sciences as chemistry, biology, etc. could, in principle, be strictly deduced from quantum mechanics and nothing else. Decoherence theory does not imply that

[21] Provided, however, that the words "quantum mechanics" be understood in a rather wide sense, not setting aside the quantum mechanics "with additional terms" to be mentioned in section 9-8. Obviously the argument considered here does not refute these theories, even though, as long as no ontological requisite is set, introducing the said terms may, in view of decoherence, appear as being but a fruitless complication.

an angel knowing nothing but quantum mechanics and an appropriate set of initial conditions would be able to calculate what he or she would see were he/she to come down within our world. The reasons why this is so were explained in section 5-4. In order to proceed from quantum mechanics to, even just simply, chemistry, it is necessary to make—by hand, so to speak—some abstractions. In the formulas, some terms that the quantum formalism definitely shows are there must purposely be dropped. It seems that it is only by doing so that it becomes possible to infer from quantum mechanics facts such as the shapes of large molecules and many others (Primas 1981, 1994). In other words, quantum mechanics achieves the feat of reconciling the universal character of the laws with a nonreductionist world view, in which our human way of conceiving things has a crucial role in determining what we finally perceive.

CHAPTER 9

VARIOUS REALIST ATTEMPTS

❦

9-1 Introduction

T HIS CHAPTER CONSTITUTES one more digression. True, it is in a way a pity that we should, at times, have to take such sideways steps. We are engaged in a search for a world-view fitting in with our present-day physical knowledge and, admittedly, when mind is focused on a quest of such a nature and feels its enterprise is advancing, the idea of having to look sideways or backward is a priori distasteful. Here, however, such a move is necessary for we must be careful. For reasons circumstantially explained above, the view that started taking shape in the foregoing chapters is definitely at variance with the conventional realist one. But it is a fact that, notwithstanding the existence and weight of the reasons in question, various ideas aiming at restoring realism were put forward, which were—and still are—considered interesting by quite a substantial minority of theorists. We cannot spare going into a critical analysis of them, be it only for verifying the validity of our own approach.

Of course, a systematic review of these attempts here would be out of place, all the more so as it would amount to just reproducing analyses of the main ones already put forward elsewhere (*Veiled Reality*, chapters, 11, 12, and 13 in particular). In what follows we shall therefore limit ourselves to examining those of them that may be considered most interesting. But before we objectively engage into this, let me try and make clear the reasons why I, personally, a priori sympathize with the adepts of realism, and part from them (though not radically!) merely a posteriori grounds. I mean at the end—not the beginning!—of my quest.

9-2 On Our Intellectual Craving for Realism

Even though physical realism is not provable (for millennia philosophers have known it to be but a metaphysical option), its upholders admittedly have arguments that do raise echoes in our minds. One of them is just simply the apparently absurd character of the opposite standpoint. They note, for example, that nobody really succeeds in believing that the Moon does not exist when not looked at. Some physicists, and not the lesser ones (in his mature years Einstein was, as it seems, quite avowedly one of them) are fully convinced by such commonsense arguments. Similarly, John Bell considered them to be decisive even after he discovered his theorem. So that, even then, he went on believing in the existence of a reality that, in principle, can be known—in "physical realism," in fact—and was content, in view of the said discovery, to hold this reality to be nonlocal. This standpoint is also the one of Sokal and Bricmont (1997) even though these authors are, obviously, fully aware of the classic philosophical objections the realism in question has to cope with. As we see, this whole set of problems is disconcerting, so that it may prove useful to reconsider the whole question right from the start. Let us therefore think of a physicist—call him Jack—who, without a priori adopting one particular philosophical system tries to have a fresh look at the question. Let me imagine him building up the two stages of the following short reasoning.

First Stage

Jack starts from the view (the "fact") that there is an external reality. Of course he does not assume it to coincide with direct external appearances. But he thinks that we may legitimately propose to discover its true features by performing appropriate experiments. Having in mind such discoveries as those of electricity, atoms, Hertzian waves and genetic code, he notes that, unquestionably, up to now, such a program was successful. And he considers that, with all due respect paid to the manes of Berkeley, Kant, Hegel, and others, the sole reasonable manner of explaining such a success is to assert that, by proceeding this way, the scientists at least partly lifted the veil of appearances. That, in the domains they investigated, they did describe elements of reality more or less as they really are. This argument convinces Jack that the raison d'être of scientists is to firmly go on along these lines, without letting themselves be taken aback by difficulties. And that the idea of resting content with mere synthetic descriptions of phenomenal human experience would constitute for the scientific community a clear-cut forsaking of its basic role. In other words, he claims that there is no common measure between a theory that may be taken to describe reality-per-se and all the theories however brilliant they may be that merely yield syntheses of communicable experience. He judges that even in the case in which the former one seems strange, or remains fruitless for quite a long time, still it is it, and it only, that constitutes actual *knowledge*.

SECOND STAGE

Jack of course is informed of the objections his conception may raise. One of them is quite old (it goes back to Hume) and quite general. It consists in stressing that our experience only reveals our sensations: How could we know whether or not the latter are in any way in correspondence with reality? In line with Sokal and Bricmont (*loc. cit.*), Jack dismisses the argument on the ground of its excessive generality. True, he says, Hume's skepticism is irrefutable. But, as a matter of principle, it applies to everything, including even our most trivial pieces of knowledge, those that our daily life is sprinkled with. Now, he claims, of the veracity of the latter we do not doubt. Why then should we doubt of the veracity of the ones science provides us with?

An up-to-dated version of Hume's remark is to observe that once experiments have been performed it is necessary to state what they let us know concerning reality, For this, words are needed. And, here again, nothing guarantees that the words we have at our disposal—I mean, the concepts they designate—have a one-to-one correspondence with some "bricks" of reality. To this, Jack answers anew in the same spirit. He notes that we do have at our disposal quite a large number of fundamental, simple concepts—those of object, position, motion, etc.—that work beautifully, help us in managing our life, and enable us to describe a myriad of things with the feeling that we understand them. And he claims that the sole sensible explanation of all this—barring unreasonably sophisticated ones—is that these concepts do essentially correspond to some elements, or features, of reality. So, he reflects, we do have at our disposal a whole set of concepts of which it is most likely—let us say, practically sure—that they correspond to reality. From all this he infers that if a general theory matching experimental data and expressed in terms of the said concepts happens to be available it is the one that must be considered true, rather than any one of those linked more or less to idealism or Kantian criticism. Observing that the Broglie-Bohm theory precisely fulfills the just stated condition, he entertains quite a strong bias towards it.

Taking then a look at the present-day scientific literature he regretfully observes that most of the physicists who claim to be sharing his views—who are physical realists just as he is—actually are inconsistent in that they rest content with the formulations of standard quantum mechanics. These people, he notes, remain blind to the obvious fact that some of the basic statements of the latter are merely weakly objective. And, of course, he deplores having to realize that, consequently, his own aiming at strict conceptual consistency sets him somewhat apart from the rest of his working community. However, we shall assume that he is not of a gregari-

ous temperament and that this observation is not therefore of such a nature as to make him change his mind and join the flock.

The following analysis of the Broglie-Bohm theory will enable us to
appreciate to what extent the latter meets the expectations Jack entertains
concerning it.

9-3 The Broglie-Bohm Approach

In the foregoing chapters we repeatedly had to turn towards the model
in question. This illustrates the extent to which it constitutes a useful
"theoretical laboratory," that is to say, a conceptually clear framework,
making it possible to apprehend the nature as well as the unfamiliar aspects of novel notions, such as nonseparability, that contemporary physics
imposes upon us. But, each time, the necessity of not breaking the course
of the analysis we were then engaged in prevented us from going beyond
just a short mention or a mere sketch of arguments. This is now the place
to—without of course going into long technical details—fill up to some
extent this information gap. Essentially, what must be understood is the
reason why a small but significant number of most competent physicists—
like the "Jack" of our story—go on considering this model to be attractive; as well as the nature of the difficulties that induce a majority to hold
the opposite standpoint.

9-3-1 Advantages

There are several of them.

1. As we know, at least within the nonrelativitic framework the model
exactly reproduces the quantum predictions while being ontologically interpretable. It deals, or claims to deal, not, primarily, with predictions of
observations about things but with the very things themselves and the
evolutions they undergo (in other words, it deals not with "states" but
with states!). Admittedly, in chapter 5 we reached the conclusion that,
contrary to our prime intuition, this is not a condition that a physical
theory *must* fulfil in order to be a sensible one. Far from being provable,
physical realism is merely grounded on plausible conceptions, themselves
weakened—what is more!—by the fact that they ultimately emerge, at
least in part, just from human ways of acting and thinking. But still, it
remains true that, however human they may be, such conceptions are
most strongly rooted in our minds. Even philosophers who, intellectually,
feel quite sure of their relative nature fail to be inwardly convinced of it.
What then of the man in the street, or the physicist! Hence it is only

natural that if a theory is ontologically interpretable this should be considered by many as a powerful argument in its favor.

2. A second advantage, closely related to the first one, is that, compared with standard quantum mechanics, the model gives us the impression of being much more genuinely an *explanation* of what takes place. Further in this book—in chapter 15 in particular—we shall study the notion of explanation in more detail and realize how complex it actually is. But it is clear that a theory such as the Broglie-Bohm model, that claims to describe the constituent parts of objects and account for the way they evolve in time, and that, for this purpose, makes use (or seems to make use) merely of familiar concepts (particles and forces at a distance), is bound to give (at least at first sight) an intense impression of being genuinely explanatory. For example it gives to its supporters the theoretical possibility of explaining the existence and forms of the dinosaurs bones we now see in museums by means of a discourse truly bearing on the past, stating what really took place, independently of us, 65 million years ago. Contrary to what most physicists believe this in not the type of discourse that standard quantum mechanics enables us to literally justify. And not even in Feynman's form, for the fabricated ontology that Feyman's approach introduces is, as we saw, but a pseudo-ontology (more concerning this in section 9-6). Now, even though this "feeling of having explained" proves, when all is said, to be of but a questionable pertinence, still it is endowed with an unquestionable attractive power, so that it should also be counted among the advantages the model has.

3. Finally, an advantage of this model which, in the mind of any physicist with a realist bent, must have great weight is that it does not suffer—or only suffers in an attenuated way—from the "and-or" difficulty mentioned in section 8-4. This is just because the final state wave functions corresponding to the different possible states of the pointer do not overlap and, in configuration space, the system "representative point"[1] must lie in the "support" of one of them (the region in which it is nonzero).[2] The other ones are said to be "void". The fact that there exists at least one

---

[1] As we already know, the *configuration space* is an abstract $3N$-dimensional space, N being the number of particles of the system. In this space the set of the $3N$ coordinates of the latter defines a point, called the *representative point* of the system.

[2] This advantage, however, is not as clear-cut as it may seem because of the fact that, in the model in hand, the reality of a physical system is represented, not by the representative point alone but by it together with the wave function. In the case of a generalized measurement process such as those considered in sections 4-2-2 and 8-1-1 the reality of the final composite system is thus represented by the conjunction of a representative point that corresponds to the pointer lying in one definite interval on the scale and a wave function that is nonzero within several of the various scale intervals. It is only because we perceive only the so-called "hidden" variables that the system appears to us to be localized. In principle we could observe the corresponding nonlocality by indirectly measuring on ensembles of

realist interpretation of the formalism in which this problem does not, or does "hardly," appear indicates that the occurrence of the "and-or" problem does not constitute a definite refutation of realism. In a sense, we may thereby feel encouraged to consider that even if, as it seems, independent reality is neither known nor knowable, still, strictly speaking, there is nothing in contemporary physics that actually forbids us to take seriously the idea of its existence.

9-3-2 Inconveniences

They are many.

1. The appearance in the model of "void" branches of the wave function is sometimes held to be a conceptual difficulty. In fact however since, nowadays, it has become clear that no theory totally in line with our elementary intuitions will ever be able to account for all of the basic phenomena, this peculiarity of the model may hardly be considered a major defect.

2. Its unfruitfulness is more significant. In fact, during the eighty years or so that elapsed since it was invented the model never led to any specific experimentally verified observational prediction. True, the verifiable phenomena that it predicts are indeed verified, but they were already predicted, and in mathematically much simpler ways, by conventional quantum mechanics. And the latter has on it the considerable advantage of predicting many other ones that the model alone could not have led us to suspect.

3. In the same vein, we may find it somewhat disappointing that a model that, at the start, was essentially grounded on the idea of only using familiar concepts based on locality—corpuscles, trajectories and so on—eventually turned out to be nonlocal. If we feel that the explanatory power we at first sight credit the model with comes from its basic use of clear, common-sense notions, the final appearance in it of a nonlocality extraneous to everything we are familiar with may make us take the said power to be a lure. Indeed, this turn taken by the model may even result in it being considered hardly consistent by those who judge the reliability of a theory by the matching of its guiding ideas and its results. Let it be granted nevertheless that this objection is not such as to shake the trust put in the model by the persons who, notwithstanding the arguments mentioned here in chapter 5, go on considering physical realism to be an indisputable truth.

4. To the foregoing difficulty another one is linked, which bears on the particle-creation process. As already noted, in order to account for such processes within the Broglie-Bohm model it is necessary to drastically

---

such systems complex quantities also involving the environment but, here as in section 8-2, these quantities cannot be measured in practice.

change the very nature of the entities taken within the latter to be the ultimate constituents of reality. More precisely, it is necessary to replace the particles we call bosons (i.e., photons, pions, etc.) by abstract quantities chosen to be the Fourier components of some given fields (or, in equivalent versions, these fields themselves). Now, independently from worrying questions of mathematical strictness, the fact that, as in Planck's pre-Einsteinian theory, photons are here reduced to the status of mere appearances is seen by some as a partial renunciation of the guiding idea of the model. And this renunciation, which facts dictate, unavoidably brings some discredit on the theory. Concerning the other particle family—the fermions—the situation is no better. Technical reasons make it impossible to treat them in the same way as bosons and it seems fair to say that the problem concerning them is still open, although quite interesting suggestions were made.[3]

But the most worrying difficulties of the model are still to come.

5. One of them consists in that there are several such models, all of which, by construction, correctly reproduce the quantum mechanical observable predictions (at least concerning mean values) and between which it therefore seems extremely difficult to experimentally discriminate. Clearly this is a field in which the devil of the underdetermination of theories by experiment is quite especially noxious.

6. Another one is that, while the model focuses on the use of quite familiar concepts, taken from everyday experience, still, in some cases it—perforce—imparts to them meanings that make them refer to quantities differing very much from those that are commonly observed and designated by the same names. In section 7-2-2 we already had an example of this, concerning momentum. The momentum that, according to the model is the "true" one (the mass—velocity product) is altogether different from the measurable momentum. Similarly, it may be shown that, as a consequence of the model, there must be cases in which the detectors are "fooled." They respond just as if a particle had reached them, while, actually, no particle did (Brown et al. 1995). In a similar vein, let it also be noted that, according to the model, a "free" particle—lying at an arbitrarily large distance of any known center of forces—does not, in general, travel along a straight line. For it is submitted to a "quantum potential" due to all the other particles in the Universe and this potential, surpris-

---

[3] One of them essentially consists in making use of the old "Dirac sea" notion mentioned in section 1-1 (which, historically, came before full-fledged quantum field theory). Another one amounts to considering also the fermions to be "appearances," at least in the sense that they do not continuously exist. A singularity gets formed at some site, then disappears; another one gets formed at an immediately adjacent site, then disappears, and so on. The whole yields an impression of forming a trajectory. This second idea is akin to that of implicit order (see next footnote).

ingly enough, does not decrease with the distance. Hence, as Michel Bitbol notes, "the descriptive structures that [such models] graft on the predictive quantum formalism are . . . such that they entail their own experimental inaccessibility" (Bitbol 1998).

7. Another serious difficulty of the model—somewhat linked to the former one—has to do with contextuality, in particular in the form in which, within the framework of the Aspect-type experiments, "Bell's calculation" (section 5-2-4 above) reveals it. The equations this calculation leads to were purposefully kept by Bell in general form. But it is instructive to consider in some detail the way they work in the simplest possible case, the one in which the left-hand-side and right-hand-side registering polarizers-analyzers are oriented along one and the same direction. As we already know, within such a layout the theory predicts—and experiments confirm—that (as in the dart case) a strict correlation is to be found. What the calculation actually yields (see appendix 3) is a detailed description of the "mechanism" by means of which this correlation gets established. And, quite surprisingly, it shows that in the model in hand (which, as we know, is determinist) an essential asymmetry exists between what determines the outcome of the first measurement and what determines the outcome of the second one. For it proves that the initial values of the parameters relative to the measurement that takes place first determine the outcome, not only of *that* measurement *but also of the other one*; and this, irrespective of the distance between the places where these two measurements are made.

This fact has rather peculiar consequences. For example, within the framework of an Aspect-like experiment (section 3-1) let us consider again a situation in which the two registering devices are oriented along the same direction. And, first, let us think of pairs on which we imagine that only one measurement (either the left-hand-side one or the right-hand-side one) is made. Is it conceivable that the hidden variables of some of these pairs should be such that if only the left-hand-side measurement were made its outcome would be + and if only the right-hand-side measurement were made its outcome would be −? In view of the (observed) fact that, whenever both measurements are made, a strict positive correlation is registered, we would naively answer "no." It turns out, however, that we would be wrong. According to the model some of the involved pairs may quite well be of such a type. For indeed imagine we have to do with a pair of this kind and consider a case in which, on it, the left-hand-side measurement is performed first. Its outcome, of course, will be +. But the above-mentioned "detailed calculation" then shows that, in fact, the outcome of the right-hand-side measurement, if we finally decide to make it, will also be +, in accordance with the (quantum-mechanically predicted) positive correlation that is observed. If, on the contrary, the measurement that is performed first is the right-hand-side one, its outcome will, of course, be – and the "detailed calculation" shows that the outcome on the left-hand-side will also be –. Hence, it follows from the model that (however large the involved distances are) the individual outcomes concerning one given pair may well depend on the time order of the two involved measurements.

This being the case, it is clear that the piece of knowledge provided by the measurements in question is not one that, together with others of the same nature, might be considered to constitute a genuine knowledge of some (contingent) data concerning the particle world "as it really is." Here again, the model proves to be seriously disconnected from any conceivable experience; that is, more akin to metaphysics than to what is normally called "science" or "physics."

8. Concerning the question whether or not the model is compatible with special relativity (the, so called "relativistic invariance" question) it is clear—no long formulas being needed—that the above-described strange feature should make us more than skeptical towards a positive answer. For, in special relativity, when two events are spatially separated from one another, their temporal order is not intrinsically defined. It depends on the referential within which *we* decide to consider them. Then, if I am moving in such a way that for me the left-hand-side event takes place before the right-hand side one, with the same assumptions as above both measurement outcomes should, for me, be +. But if I am traveling the opposite way they should both be –. Now here we have to do with one and the same pair of measurement outcomes, registered by instruments that are both at rest with respect to the source. And remember that the model is aimed at describing reality-per-se—that is, the whole set of things and events—just as it *is*. Can a real, registered event be at the same time what it is and its opposite? The least that can be said is that such a view is hardly consonant with the idea that seemed at the start to be the most attractive feature of the model, the one of resorting only to familiar notions!

Admittedly, anomalies in models are as a rule liable to be corrected by appropriate changes made on them, so that what is shown here is that— as difficulty 4 already portended—making the model relativistic necessarily implies modifying it in some drastic way. However, what presently emerges from the research works carried out in this direction seems to be that, in striking contrast with the guiding principle of special relativity, it is, in the model, hardly possible to avoid postulating the existence of some privileged space-frame. In return, the model can then be generalized, along the above sketched lines, in such a way as to preserve all the *appearances* predicted by special relativity (the negative outcome of the Michelson-Morley experiment, etc.). But the physicists who crave for an ontologically interpretable theory are not inclined to rest content with just this.[4]

If now we come back to the imaginary person called Jack above, we observe that finally the reasons he had for adopting the Broglie-Bohm model—essentially those Sokal and Bricmont mentioned in their book—cannot be kept. For these reasons essentially rest on the fact that the said model is expressed by means of very simple concepts, whose success in ordinary life Jack took to be warrants of their validity. But difficulty number 4, corroborated by difficulty number 8, shows that this coming back to the exclusive use of simple, familiar concepts was, actually, illusory. This fact is significant concerning the structure of Jack's reasoning for it ruins the objection raised by him to Hume's skeptical standpoint towards realism. This objection, as we remember, was that since these simple concepts work very well in ordinary life not to trust them in other fields would merely reveal whimsical sophistication. Clearly this objection does not apply when, as here, the simple concepts finally turn out not to work in the "other fields" in question.

From the existence of these numerous difficulties most of us will presumably infer that it is hard to really believe in the validity of the model, in the sense of considering it to be endowed with a scientific asset comparable to that of commonly accepted scientific theories. On the other hand, however, it should be noted that the difficulties in question do not in any way imply that the predictions of the model are at variance with the data. They are not. In other words, at least within the nonrelativistic realm we should consider the model to be one possible, not convincing but quite consistent, way of accounting for the said data, including also the surprising ones. In practice this has two interesting consequences. The first one is that if, upon analyzing a problem that quantum mechanics raises, we encounter some apparently unsolvable conceptual difficulty and if we observe that, in the Broglie-Bohm model this difficulty does not arise, we are entitled to consider that it is not truly a basic one. In section 9-3-1 (point 3) we already saw an example of this and in section 10-3 we shall have an opportunity to make use of the remark. The other consequence is that it may be convenient to make use of the model as a touchstone relatively to such and such new concept. I mean, it may be useful to wonder what form the concept should take within the model and inquire whether or not, within this framework, it is consistent with what is other-

---

[4] These difficulties and others made David Bohm build up a novel theory, based no longer on the corpuscle notion but on that of an implicit order, quite distinct from the explicit one that shows up in phenomena. The two Bohmian approaches, though quite different, still have in common the fact of being oriented toward the edification of an ontologically interpretable theory, which bars radically opposing one to the other.

wise known. Let it moreover be stated that the somewhat metaphysical nature of the model would not justify bluntly discarding it. For, as we could already observe, as soon as we set radical idealism aside, present day physics forces us to take seriously conceptions lying so far apart from our usual experience—the scientific one included—that, to qualify them, the epithet "meta-physical" naturally comes to mind.

9-4 The So-Called "Modal" Interpretation

This is a comparatively recent interpretative scheme. It was put forward by the philosopher Bas van Fraassen (1972, 1974) who gave it this name in order to stress the fact that—like, to be sure, several others—it involves intrinsic probabilities, a notion belonging not to ordinary but to modal formal logic. Van Fraassen's idea was taken up by several groups of theoretical physicists, who developed it along different directions. At present there therefore exist not one but a whole family of so-called modal interpretations, which, moreover, are continuously evolving. Here we can only have a look on some essential features they have in common. One of them is that (although van Fraassen himself is, philosophically speaking, not a realist) they focus on some kind of realism. And another one is that they aim at dispensing with the notion of measurement collapse, normally associated, as we know, with any interpretation in realist terms of the wave function. For this purpose they dissociate the two roles that the elementary realist interpretation attributed to the wave function (or the symbols that generalize the latter), namely the role of describing reality and the one of specifying what outcomes a measurement of a physical quantity may have and their respective probabilities. They impart to it merely the second one, that is, they reject the view that the wave function of a physical system always constitutes the most accurate possible description of the reality of the system. Quite on the contrary they postulate the existence—in some well-specified circumstances—of finer states, in which precise values are attributed to quantities that, within the conventional quantum formalism, would not have any. In this, they resemble the Broglie-Bohm model and, just as the latter, they must be considered to be supplementary (or, in received terms, "hidden") variable theories. But, contrary to the said model, these theories are indeterminist.

When conceptually inessential technical details are disregarded the guiding idea, common to all these theories, may be schematized by means of the example described in section 4-2-2 and further in section 8-1-1. These theories assert that in such a process, after the electron-instrument interaction is over, there is a probability 1/2 that the considered system (in this case, the pointer) should *really* be in state A and a probability

½ that it should *really* be in state B. But they also assert that the quantum state (here, in other word, the wave function) is A+B.[5] Here, of course, the important word is "really." We do not have to do, as in conventional quantum mechanics, with purely mathematical pieces of information making it possible to predict what will be observed. We are here informed about properties of which it is claimed that, with the specified probabilities, the system actually *has* them.

But then—it will be asked—what about the "practically unfeasible" tests mentioned in section 4-2-2? In other words, what should be predicted concerning the outcomes of not easily feasible measurements such as those we examined in section 8-2? True, for practical reasons when macroscopic systems such as pointers are involved such measurements cannot actually be performed. But still, "theoretically" the physical quantities they would bear on are measurable. And, what is more, the here considered modal theories are supposed to apply, even in cases in which the involved systems are microscopic, that is, in contexts in which the corresponding measurements are quite conceivable, even in practice. Is it the case that, for such measurements, the modal theories predict results grossly at variance with those quantum mechanics predicts? This would amount to an immediate ruin of these theories for we know very well that the observable predictions of the latter are reliable.

Of course, the modal theories were devised with just the purpose of yielding a realist description avoiding this elementary pitfall, and the way in which they do so is easily grasped. Within orthodox—that is, hidden-variable-free—quantum mechanics, when we say that a system is in state A this is tantamount to saying that it is in quantum state A. It cannot mean anything else since, in this theory, there simply is no description of the reality of things finer than the quantum one (the one using wave functions or equivalent mathematical symbols). To claim that, in an ensemble, some systems are in state A and other ones in state B thus amounts to claim that the said ensemble is a proper mixture, in the sense defined in section 8-2. And, in the cases of interest, this description is at variance with some (verifiable and verified) predictions that we directly infer from the quantum formalism by performing straightaway calculations, without indulging into any such conjectural attribution of properties. But when we admit that there are variables that are even finer than the wave function the outlook radically changes. We may then consider a system to be in a state A—labeled by a given value of some supplementary variable, and called a *value state*—without implying anything concerning its quantum state. Now, according to the modal theories it is merely the latter—which is strictly identi-

---

[5] Let it be recalled that in chapter 4, for the sake of maximal simplicity, the electron was still assumed to be absorbed by the instrument.

fied with the quantum state of orthodox quantum mechanics (in other words, to the wave function or equivalent symbols)—that must be made use of for making observational predictions. The value state has no role in the latter Unsurprisingly, these theories therefore yield the same predictions as the calculations directly based on orthodox quantum mechanics, disregarding intermediate value attributions. The similarity between this "mechanism" and the one described in section 3-3-8 apropos of the Broglie-Bohm model and fringe production is worth noting.

The reason why a number of physicists consider these theories to be attractive and interesting is that they are ontologically interpretable Within their conceptual framework It may legitimately be considered that after any measurement process—be it even of the generalized type investigated here in sections 4-2-2 and 8-2—the pointer, before anybody observes it, really lies in one definite graduation interval (and a cat is either alive or dead). However, we may wonder (and the same holds true concerning the Broglie-Bohm model) to what extent this advantage is more than just a language convention. What legitimates this query is that, as we already noted, the strong objectivity and reality notions are rather tightly linked with those of prediction and counterfactuality. Our feeling that such and such a thing *really has* such and such a property appears to rest, to an appreciable extent, on the idea that, because of this property, this or that may happen to it, or that it would react in such and such a way to what events might possibly affect it. At least as long as merely inanimate objects are considered, it is difficult to set aside the view that ascribing to things properties that would entail no such implication whatsoever would be a kind of a gratuitous move and would be similar to a metaphysical assumption. Along these lines, we would expect that the fact, for the supplementary variables, to really have precise values should have specific consequences. As we saw, this does not seem to be the case.

Quite apart from this, the theories under review meet with a few technical difficulties that are still under investigation. Just as is the case with the Broglie-Bohm model, one of them bears on the problem of reconciling the theories in question with special relativity. Here, again the prospects are, at the time of writing, not really satisfactory. In the opinion of some of the main experts in the field, the nonlocal character of these theories is really at variance with the orthodox interpretation of relativity theory (Dickinson et al. 1998).

Another somewhat tricky point has to do with the problem of simultaneously ascribing definite properties to a physical system and its subsystems. When we think of a system and one of its subsystem (for example a car and one of its wheels), commonsense as well as conventional logic (and even also orthodox quantum mechanics) tell us that if at some time the subsystem has some definite property—specified by a definite value of some physical quantity—the overall system ipso facto also has it. Now, strangely enough, within the formalism of the modal theories it turns out that this is not always the case (Vermaas 1998). And the upholders of such theories therefore had to introduce quite a surprising logical distinction, that F. Arntzenius (1990) illustrated by means of a pleasant analogy. He suggested we should think of a table the left-hand side of which is painted green. This state of affairs may be described by stating either that the left-hand side of the table is green or that the table is green at its left-hand side. What Arntzenius pointed out is that, from a purely logical and formal point of view, these two propositions may be said to express two different properties, that of "being green" and that of "being green at the left-hand side," ascribed to two objects that themselves differ from one another, namely "the left-hand side of the table" and "the table as a whole," respectively. It seems unquestionable—some call it "the table paradox"—that to salvage the internal consistency of the modal theories this distinction has to be made, and that one of the two statements may be true without the other one being true as well. But obviously a realist may consider the subtlety of such a distinction to reach beyond what he feels able to tolerate.

Finally, these theories also meet with a few difficulties—to be merely mentioned here—concerning the analysis of the measurement problem. According to some authors there are cases in which the formalism in question predicts the indication yielded by the pointer will be at variance with the observed value. In view of these various difficulties and oddities the impartial analyst is bound to consider that such attributions of properties to systems are not extremely convincing. It is true however that, in some fields, the model may be considered inspiring (see section 10-3).

9-5 The Heisenberg Representation: It Does not, by Itself, Yield a Solution

Both the Broglie-Bohm model and—to a slightly smaller extent—the modal interpretation in many ways strongly differ from textbook quantum mechanics. We may wonder whether, for preserving realism, such momentous changes are actually necessary. Indeed, scientists who did not specifically investigate such problems might well, at first sight, seriously doubt they are and, in this and the following section, this opinion of theirs is to be examined.

In particular, some among them wonder whether the conceptual difficulties analyzed at length in chapters 4 and 8 might not be removed just simply by having recourse to the *Heisenberg representation*. The representation mode that has up to now been alluded to at several places in this book, and in which the wave function (alias the state vector) depends on time, is technically known as the *Schrödinger representation*. Within its realm the wave function has a twofold role. On the one hand it, indeed, evolves in time (according to the time-dependent Schrödinger equation) just as the physical system does, and for this reason, as already noted, it is often taken to be—between two measurements—a representative of the physical reality of the system. On the other hand it describes our *knowledge* of the latter, and it is in virtue of this role that it has to be suddenly "reduced" when, due to some measurement made on the system, our knowledge of it suddenly changes. At first sight it seems conceivable that the mentioned difficulties should have this duality of roles as their origin. On the contrary, in the Heisenberg representation, which is known to be operationally equivalent to the Schrödinger one, the two roles in question are quite clearly separated and ascribed to two different entities. Within both representations the so-called "dynamical" quantities (position, velocity, etc.) are set in correspondence with mathematical symbols called "self-adjoint operators." But in the Heisenberg representation these operators are time dependent (actually they obey the same equations as the corresponding classical dynamical quantities) and may therefore be considered to *be* the dynamical quantities. As for the wave function, it is time independent so long as no measurement is performed and suddenly changes when one is made, just as would befit a quantity whose role would merely be to label the state of our knowledge concerning the system. In view of the fact that the Heisenberg representation thus apparently distinguishes what takes place within the physical world from mere increases of knowledge, we might have the feeling that, in it (and within the realm of a realistic conception), the mentioned difficulties do vanish.

Unfortunately, upon further examination of the subject it turns out that interpreting the Heisenberg representation in such a way is not possible, as is presently to be shown. For this purpose, note first that, when a nondisturbing measurement of a quantity pertaining to some physical system is made, the number we read on the dial informs us of a value that, obviously, is contingent. It does not follow from the laws of physics alone but also depends on the initial conditions. Hence, for a measurement to be, not a *creation* but a mere *mental registration* of a value—as physical realism requests—it is necessary that, on the system, the said contingent value should somehow have preexisted to the measurement. And if the measurement is disturbing the same holds true, except that the preexisting contingent value need not be the same as the output one. Now, such a view is not compatible with the Heisenberg representation. True, within the latter a self-adjoint operator is associated to every dynamical quantity. But, contrary to what is the case concerning values of dynamical quantities within the classical physics approach, this operator is not, at any time, in a one-to-one correspondence with some definite *contingent* state of the system. It simultaneously and indiscriminately refers to *all* the possible values of the latter. And what, in the representation in question, most nearly corresponds to the contingent element specifying the state is the initial wave function. Consequently, the assignment of roles (between the self-adjoint operator and the wave function) is not, in this representation, the one that would make the above described realist interpretation possible. The role of the self-adjoint operator is too modest since this operator fails to label the contingent states. And correlatively the one of the wave function is too vast, since instead of being restricted to a mere description of our knowledge it also labels the initial state as it "really is." While, in classical physics, symbols such as E (electrical field), B (magnetic field) ρ (density), etc. represent, at one and the same time, physical quantities and the values these quantities have at a given time, in quantum mechanics nothing of the kind is possible, not even in the Heisenberg representation. And the foregoing analysis shows that this difference is the deep reason why the said representation fails to restore a realist outlook.[6]

9-6 Feynman's Reformulation and the Corresponding "Fabricated Ontology"

Concerning physical realism some high-energy physicists nowadays start asking themselves a few questions but, up to now, most of them have serenely kept, concerning such matters, to the views held by their "classical" predecessors. In other words, they consider physical realism to be self-evident and, as already noted, they do not believe for a moment that keeping to it should require turning to the Broglie-Bohm model or some equally revolutionary scheme. Their optimism—or conservatism—comes, I guess, from their intuitively adhering to the "fabricated ontology" the main features of which were sketched here in section 2-7. And, in turn, this adhering is naturally explained by the fact that the said ontology does indeed appear in most attractive colors. To clearly see why it is appropriate to enter into some details concerning it, and this is what will presently be done.

To begin with, let us note that it entirely proceeds from a reformulation of quantum physics that Richard Feynman discovered (1949). Within the high-energy field of research this reformulation is outstandingly efficient, which, of course goes in par with quite a crucial role played, in it, by

---

[6] A quantitative study of the way in which, in Heisenberg representation, a measurement process is to be described illustrates rather clearly the impossibility in question. It is described in section 10-8 of *Veiled Reality*.

some mathematical—or, better to say, computational—developments. To describe it here as if we were to actually make use of it would therefore drive us much too far. Fortunately, separating its conceptual aspects from those that are of an essentially computational nature turns out to be relatively easy. For grasping the nature of the fabricated ontology and check whether or not it is a *genuine* ontology, only the conceptual aspects are needed. Let us therefore try and find out what they consist of.

The first point to be noted is that the Feynman reformulation rests on three main ideas, or methods. They are (1) a mode of analyzing particle behavior focused on the use of probability amplitudes, (2) an entirely novel way of interpreting the antiparticle notion, and (3) a new concept, that of virtual particles viewed as carriers of interaction. All three ideas are—it must be granted—conceptually rather difficult ones, with which even the physicists who professionally deal with them find it somewhat hard to get accustomed. The essence of the one numbered 3 above was already described in section 2-7. Let us consider the two others.

1. To grasp the substance of the first one let us turn back to the so-called "Young slit experiment" already considered in chapter 4, in which a beam of photons or other particles impinges on a screen after passing through a diaphragm with two parallel slits. We saw that, then, fringes get formed on the screen, which leads one to suspect that the beam has wavelike features. The phenomenon is a fully general one. To interpret it most physicists (Feynman among them) set aside—without even stating they do!—the Broglie-like idea of particles driven by a wave. And, correspondingly, they point out that it would be quite meaningless to speak of "the probability that such and such incident particle passes through one given slit" (or, equivalently, of "the probability that the particle *is*, at some time, *within* one given slit"). For, if this way of speaking were appropriate this would imply that each individual beam particle actually passes through but one slit (even though nothing would predetermine through which one it does) and (as we saw in chapter 4) no fringes could then appear. In quantum mechanics such views are therefore set aside, as we know, and replaced by the wave function notion, the wave function being, in the one particle case, a function—of the three position coordinates and the time—whose modulus square yields a certain probability. Now, the probability of what, exactly? Not (of course, see above) of the particle to *be*, at the indicated time, at the place the space coordinates specify. But the probability that if a detecting device is—or were—at that place it will—or, respectively, would—yield the signal "particle detected." After it has passed the diaphragm, the overall wave function is the (algebraic) sum of the partial waves that passed through each one of the two slits and, as a well-known calculation shows, on the screen (then playing the role of the just mentioned detecting device) fringes appear.

As just recalled, the modulus squared of the wave function is a probability. Now, in various domains of classical physics (electrodynamics, etc.) such a relationship between two physical quantities is a familiar one, in that, in these domains, intensity shows up as being the modulus squared of some amplitude. For this reason, and since an obvious relationship holds between probability and intensity, the wave function is often designated by the name *"probability amplitude"* or, for brevity's sake, just *"amplitude."* In the Young slit experiment, if B is a point on the screen we then have to say, "there is a probability amplitude (just equal to the wave function) for the particle to be found at that place, the condition 'if it is searched for' being understood." But of course, since the screen *is* (taken to be) a detecting device its very presence entails that, at B, the latter condition is automatically fulfilled. In order to build up his remarkable reformulation Feynman, who was fully aware of the difficulties of interpreting quantum mechanics but absolutely reluctant to discuss them, made use of this "way of speaking" but not without strongly simplifying and generalizing it. Like a prestidigitator of superior rank, he chose to speak of "the amplitude for arrival" at some given point B,[7] and to do so without restricting (as we just did) the domain of definition of B to be the screen or any region where particle detecting devices are present. Of course, the wave function is mathematically well defined at any point C of space and, in particular, in small regions R around each one of the two slits, where, by assumption, no detecting device is present. Hence we may meaningfully speak of the value the probability amplitude (i.e., the wave function) has at any point C lying in one such region. But may we meaningfully speak of the probability amplitude *for arrival of the particle* at such a point C ?

From the point of view Feynman adopted—which, to repeat, centered on the quest for an efficient operational procedure—unquestionably the answer is "yes," and the fact that the Feynman method was such a tremendous success shows that the choice was a good one indeed. However, from a conceptual point of view the same answer may not be given, for, in fact, the verb "to arrive"—and the noun "arrival" as well—have just the same realist connotation as the verb "to be." It follows that, when we have to do with an intermediate point C lying in one of the regions R, we have a choice between two options. Either we consider that using the said verb (or noun) means that the particle really arrives there, that is, *is* actually there at some time, and then whether we talk of amplitudes or not, no

---

[7] In fact, but this technical point has no bearing on the question under study, rather than directly making use of the wave functions Feynman chose to use the corresponding Green's functions. Such a procedure makes it possible to express the formalism in a very neat and general way but, from a conceptual point of view, the two descriptions are equivalent.

fringe can appear on the screen. Or we consider that, in the context, use of the verb in question merely serves to formulate in a pictorial way what, in fact, is but an efficient recipe for calculating observational predictions. Since the fringes do actually appear on the screen, clearly the correct answer is the second one. We see therefore that Feynman's particle physics—just as quantum mechanics, the basic principles of which it keeps unaltered—is but of a predictive nature. It is not descriptive of anything. It is just predictive of observations. That it seems to be descriptive is merely due to a semantic shift concerning the word "arrival."

2. We still have to consider Feynman's very fruitful idea of considering antiparticles to be particles propagating toward the past, which will give us the—rare and philosophically most instructive—opportunity of observing *an idea being generated by a method*. Not actually by a method considered in the abstract. Rather, by the way in which a computing procedure may be applied to a datum. Here, the datum is in fact quite a specific, "technical" one. It consists in the fact that fermions (protons, electrons etc.) are described by an equation (Dirac's) that, along with positive energy solutions (which do describe the said fermions), also has "negative energy" ones that, apparently, correspond to nothing at all. Moreover, strangely enough, the quantum rules predict that if a "negative energy state" happens to be unoccupied (or "free"), a fermion has, in some cases, a nonvanishing probability of "falling" into it. Since this seems absurd, Dirac had, in the past, conjectured that, normally, all the negative energy states are occupied. As we know (section 2-6) this hypothesis (the "Dirac sea" one) was extremely fruitful since it opened the possibility of considering the particle-antiparticle pair creation phenomenon to consist of one particle popping out of this "sea"—and thus becoming "visible"—together with the corresponding "hole" in the "sea" (which is just the antiparticle) becoming visible as well. Historically, as we know, this hypothesis is the one that led physicists to the notion of particle creation.

The—just issued—statement that, according to the equations, the probability for an unoccupied negative energy state to become occupied is nonvanishing may of course be repeated concerning the corresponding probability amplitudes, since the former are the modulus squares of the latter. But the mathematical formalism the Feynman approach makes use of is flexible. In fact, it has the property that, within it, it is possible to remove the corresponding difficulty in a manner radically differing from Dirac's one. More precisely, in the equations of this formalism new terms may be plugged, which are void of any unwelcome effects and cancel the above mentioned meaningless probability amplitudes. Now, it so happens that the mathematical form of these new terms is remarkable. When compared with similar already interpreted mathematical forms it is such as to suggest imparting to the terms in question an interpretation (which, in ordinary language would, obviously, be meaningless) according to which they describe particles propagating backward in time. Within this new allegory, antiparticles are no longer "holes" in a "sea." They are ordinary particles traveling toward the past.

The considerable interest this idea has is best illustrated by considering the somewhat complex phenomenon of a particle-antiparticle pair being created at time t_1 and the antiparticle in the pair undergoing later, at time t_2, pair annihilation with some pre-existing third particle. Within the here described scheme this phenomenon may be very simply described. It suffices to consider that in fact only one particle, namely the just mentioned "third particle," is involved in the process, but that the space-time path of this particle suffers two bends. The "first" one occurs at time t_2 and results in the said particle starting to propagate backward in time. The "second" one takes place at time t_1 and makes the particle resume its normal behavior and travel toward the future. To convey the gist of this interpretation Feyman resorted to quite an evocative analogy, that of a plane flying low following some road. At some time the pilot sees two other roads appearing within his field of vision. And some time later he sees that one of them meets the one he has been following and that both of them simultaneously disappear. He then realizes that there is really but one road and that what he flew over was but a large zigzag in the latter.

Finally a point has to be noted that is both scientifically and philosophically quite important. As Feynman showed, with respect to observational predictions the Feynmanian reformulation reviewed here and the original quantum field theory are quite strictly equivalent.

REMARK

The Feynman diagrams approach is relative to particle theory, and hence to the "microscopic world." And, as we just noted, it is strictly in accordance with the basic quantum mechanical axioms. (In particular, in the nonrelativistic limit, contrary to other schemes to be considered farther in the book, it yields the Schrödinger equation without any small ad hoc terms.) Its inconsistency with objectivist realism therefore does not come as a surprise since the said realism rests on a tacit locality hypothesis that, as we saw, Bell's theorem refutes.

A similar observation may be made apropos of some theoretical developments regarding the relationships between logic and physics that appeared during the second half of the twentieth century. As is well known, for stating the quantum mathematical formalism in the most general manner, the notion of an infinite dimensional abstract space called the "Hilbert space" is required. With such a space a non-Boolean logic may quite naturally be associated. In the field of mathematical physics this fact gave rise to quite a large amount of remarkable works that, same as the Feynmanian approach though even more radically, were directed at reformulating basic quantum mechanics without altering its observational predictions. Now, like the Feynmanian approach, these new versions could at first sight seem to pave the way for descriptions bearing on the objects themselves rather than measurement outcomes. To put it otherwise, it could intuitively be thought that, thanks to them, objectivist realism would anew be made compatible with contemporary physics, as it was with classical physics. Needless to say, this hope did not materialize. Reformulating quantum mechanics in non-Boolean terms brings no significant change to the nonseparability or nonlocality problem, for the reason that nonlocality merely expresses an experimental fact (the violation of the Bell inequalities), which does not depend on any theory. Hence, while some of the formalisms grounded on quantum logic are mathematically extremely satisfactory it must be granted that they are not—not any more than the conventional quantum one—interpretable as describing microsystems per se.

Another approach is due to Griffiths, Gell-Mann and Hartle, and Omnès (with significant differences between these authors). It consists, as we know, in the use of the (closely related) notions of "partial logics" and "decohering histories." While differing very much from the above-mentioned one, this approach is also aimed at reformulating quantum mechanics without changing its observational predictions. And again, still on the same ground, it is not compatible with objectivist realism. Indeed, for reasons mentioned here in sections 3-3-5 and 4-2-1 and explained more in detail in *Veiled Reality* (chapters 11 and 12) the developments in question are also at variance with physical realism as defined in section 1-2. Their interest is great, but lies elsewhere.

9-7 A "Realism of Signification"

From the foregoing analyses it follows that to reconcile physical realism with the set of the experimental data is definitely quite a difficult enterprise. Nevertheless, realists are not strictly bound to admit defeat. As noted above (section 8-2-1) decoherence does account for the appearance[8] of classical things. They may wonder whether with the help of some philosophical thinking, they might take at least a small step further, in the sense of bringing such a notion of appearance somewhat closer, so to speak, to the one of reality.

---

[8] In the sense: "as opposed to 'reality' "; see note 20 in chapter 8.

To investigate this question let us first observe that, as already noted and contrary to other tentative interpretations, it is exclusively within the macroscopic realm that decoherence theory yields new pieces of information. Concerning the microscopic systems we may not expect to receive from it any indication supplementing the information yielded by standard (i.e. hidden-variable free) quantum mechanics, the core of which—what, in it, is unquestionable—is merely "weakly" objective, as we saw. This has the philosophically significant consequence that whoever aims at matching the latter theory with realism has to direct his thinking toward meaning rather than toward existence. For indeed, if we rely on decoherence and it alone for justifying objectivist realism what has just been noted implies that it is only within the macroscopic realm that we can hope to find the arguments for such a justification. And within a way of thinking focused on the notion of existence it would seem hardly conceivable that an object should be real without its constituents being real as well.

In view of this remark it admittedly seems reasonable not to expect from decoherence anything but the justification (within the macroscopic realm) of an objectivist *language*, of an "everything takes place as if." And, as we shall see in a moment, this will, when all is said and done, be our final standpoint. But we cannot firmly take it without having first studied another option and wondered about its consistency. For, a priori, a realist focusing his thinking on the meaning of *statements* still might set some hope on a definition of realism such as the one Dummett put forward (see section 6-6 above). He might expect to find a way of imparting to individual statements a meaning defined by "objective truth conditions," that is by conditions that "do or do not hold, quite independently of whether we are able to find out that they do or that they do not" (J. Bouveresse 1997).[9]

What is most significant in such an approach is that it focuses not on the notions of "substance" or "being" but just on the meaning of statements. So that, in it, there is no reason to claim that the question "realism versus antirealism" should get the same answer concerning propositions bearing on all types of systems. Adopting it should therefore enable an objectivist realist to push into the background the sort of paradox sketched above (strong objectivity regarding macroscopic systems, weak objectivity regarding microscopic ones). Concerning such a tentative realism—which may be called "of signification" or simply "macrorealism"—some sources of worry nevertheless remain. One of them has to do with the very notion of a macroscopic object, for this notion is ill defined. The most serious attempts at defining it make use of irreversibility. But the

---

[9] Here and in what follows we apply to decoherence observations that Jacques Bouveresse (1997) produced regarding Dummett's definitions.

latter is, itself, grasped only with the help of statistical mechanics considerations, resting on the fact that the observational abilities of human beings are limited. To a realist concerned about strictness and logical consistency this makes them quasi-unacceptable. Another, more general, problem hinges on the fact that a notion of truth totally disconnected from the one of possible truth recognition is quite difficult to accept. As Jacques Bouveresse noted (1997), even diehard realists do not go quite that far. Due to all this, objective conditions of truth are, within the approach under study, hard to define. How shall we, at the same time, avoid both the trap of considering them to be mere human-centered verifiability conditions (which would lead us to antirealism) and the pitfall of an explicit ontology (bringing us back to an objectivist realism whose quasi failure we noted)? The least unsatisfactory procedure presumably is to make use of the old trick of a Laplace demon. More precisely, it consists in using the, antirealism-inspired, notion of "ascertaining as being true" but in attributing the function of ascertaining to an imaginary being endowed with abilities considerably surpassing ours. To proceed along these lines we therefore must decide that the statements to be considered true or false are those that such a demon would so consider.

With these reservations and perspectives in mind, we may turn back to the question of finding out whether or not decoherence theory meets the expectations of physicists who believe in the universality of quantum mechanics and aim at grounding on it a consistent version of macrorealism. In view of our foregoing remarks it is clear that if such a macrorealism were understood in the trivial, existential sense the answer would be negative. To repeat, it is impossible to conceive of real beings the constituents of which would not be just as real! But it is also clear that if we decide to follow Dummett and center our macrorealism on the notion of meaning, such a "no" answer is not a priori logically compulsory. It is then by turning to physics and wondering if a positive answer is compatible with the objective conditions of truth inherent to such a type of realism that we may hope to settle the matter.

As just observed, within this approach the only way of getting a clear-cut answer seems to be to resort to some kind of a Laplace demon, that is, to the fiction of a being whose abilities at observing and measuring are limited by no considerations of a practical nature. Who, for example, would be able to simultaneously measure the positions and velocities of all the molecules in a classical gas; or, more generally, all the physical quantities that are operationally definable and mutually compatible in the quantum mechanical sense. We should imagine that the demon sets up, as we do, "measurement processes" in the sense defined in section 8-1-1 and performs on the ensembles of systems they involve all the measurements he is able to do. The answer to our question is then positive or negative according to whether, for him, the conceptual difficulty referred to in section 8-2 as the Schrödinger-cat problem is or is not removed, as we saw that, for us, it *is* removed due to our inability at making extremely complex measurements.

Unfortunately, the question whether or not this is the case is a delicate one. So delicate indeed that the theoretical physicists do not unanimously agree as to the answer. The main contentious point bears on the extension of the set of aptitudes the demon is supposed to be endowed with. Should we, or should we not, consider that these aptitudes are limited by purely factual and apparently contingent data? That, for example, the demon is unable to perform measurements necessitating instruments the complexity of which is so great that they must involve more nucleons than there are in the Universe? In fact, the physicists who most recently investigated quantitatively this problem took up an a priori standpoint that amounted to answering the question positively. And the (already hinted at) detailed calculations they made on this basis (Caves 1993; Omnès 1994b) then convinced them that the Schrödinger cat difficulty is removed "period"; that is, is removed also for the demon. For what they showed was, in fact, that even a Laplace demon, if subjected to the limitation in question, would be unable, for lack of sufficiently large instruments at his disposal, to observe the highly complex quantities that, in this context, are the worrying ones as we know. Hence it seems clear that, as soon as we take the said limitation to hold, according to the guiding idea stated above of the realism of signification we should consider these quantities to be just simply nonexistent, which, to repeat, removes the cat difficulty.[10]

The argument does carry weight. But, on the other hand, it may also be considered that when aiming at defining the very condition under which an object, or a quantity, may be called real, referring to contingent data (such as size) pertaining to some given real object (here, the Universe) is a vicious circle. Within such a standpoint (which will perhaps be taken to be the more reasonable one), the demon must be considered able to measure *all* the physical quantities that are operationally definable and mutually compatible,[11] irrespective of any "factual" contingencies. But within this framework it is impossible to demonstrate the non-measur-

---

[10] Note that in the limiting cases Hepp (1972) considered, such a "realism of significa-tion" is justified independently of the limitation considered here. Even within this realm, however, problems remain. They concern the order in which some data bearing on the sys-tems and some data bearing on the instruments are supposed to tend to infinity (Bell 1975).

[11] Of course, this twofold condition is in no way equivalent to the much stronger and in fact unacceptable hypothesis according to which all self-adjoint quantum mechanical operators would correspond to physical quantities.

ability of all the quantities referred to above as the "worrying" ones. Concerning some decoherence models it even could be proved (*Veiled Reality*, section 10-6) that some such quantities are definitely measurable (at least in principle, in other words "by the demon"), and hence real.

Finally, concerning locality one difficulty still remains. It has to do with the fact that the *mathematical* representation (by means of a partial trace) of an ensemble of systems *S* interacting with their environment (the "reduced state" we met with in section 8-2-1) is compatible with an infinite number of *physical* representations of the said ensemble. One of the latter, as we know (section 8-2-1), describes the ensemble as composed of apparently localized systems, as we intuitively expect should be the case. But the others do not (Joos 1987; d'Espagnat 1995, section 12-3). Why is the first mentioned one the "real" one? Or is it? Above, concerning this point, we introduced the idea that locality should be taken to be an a priori form of our sensibility and stressed that decoherence renders the quantum prediction *compatible* with this a priori condition. But it is clear that an upholder of realism —be it merely "of signification" and "macro"—can hardly rest content with considerations of this kind. He tends to demand that the statements we utter concerning the places where macroscopic objects lie should be true in virtue of state of affairs independent of *our* aptitudes as well as of the structure of *our* mind. It seems that decoherence does not, all by itself, fulfill such a request.

To sum up, it appears that, combined with the decoherence notion, the idea of a "realism of signification" does, in a way, open a vista toward a realist interpretation of physics. But it is clear that the said interpretation must be one in which such notions as truth recognition and a priori forms of sensibility (or at least of perception) still have quite a crucial role. On the whole, the idea that emerges from the content of this section is therefore that macroscopic physics describes an *empirical* reality, concerning which we now understand better than before why it displays such robustness but which still remains basically empirical, that is, relative to us.

9-8 Nonlinear Realist Quantum Theories

One attempt more at explicitly restoring realism in quantum physics is to be mentioned. It is grounded on the observation that linearity (in the mathematical sense of the word) is what essentially lies at the origin of the conceptual difficulties besetting quantum measurement theory. And, as a last resort so to speak, it consists in modifying in a more or less ad hoc way the quantum evolution law—that is (in a nonrelativistic framework), the Schrödinger equation—by plugging in it nonlinear terms. It can be shown that modifications of such a type may be made to it, that are weak enough not to appreciably alter the usual quantum observational predictions but still large enough to induce frequently occurring *spontaneous* reductions of the center of mass wave functions of macroscopic objects. Assuming such nonlinear terms to be present therefore makes it possible to attribute an ontological significance to the said wave function without being thereby committed to introducing the "ontological incongruity" of a mere measurement causing arbitrarily large physical effects at arbitrarily large distances. In effect, several theories based on somewhat different physical assumptions were constructed along these lines. In some of them, such as the Ghirardi, Rimini, and Weber one (1986) the nonlinear terms are purely ad hoc. Others, such as the Karolihasy one (1966), aim at relating them, in some way, to gravitation. Broadly speaking this is also the guiding line of a theoretical program that Penrose (1995) put forward and that may quite well be promising. These models are not to be described here (the first named one was analyzed in *Veiled Reality* with some detail).

Since, relative to microsystems, such theories yield, by construction, the same predictions as quantum mechanics they, of course, are nonlocal. In particular, they correctly predict the outcomes of the Aspect-type experiments. True, nonlocality does not look quite as conceptually troublesome in them as within orthodox quantum mechanics but still they too imply the existence of faster-than-light "influences" that do not permit signaling (Bell 1987a). Moreover, in both the technical and the conceptual realms these theories have to face other obstacles. Without going into details let it just be mentioned here that most of them only approximately satisfy the law of energy conservation and that they meet with quite serious difficulties in connection with special relativity theory. To the last objection their upholders retort that the same is true concerning orthodox quantum mechanics—quantum field theory included—when measurements are considered. However this answer is not unreservedly valid. More precisely it *is* valid in the eyes of whoever explicitly or implicitly adheres to the realist view according to which either the instrument pointer or, at least, the observer's neuron variables finally take up definite values, considered to be elements of reality-per-se. On the contrary, it is not valid in the eyes of those who—as has been done throughout this book—set the said view aside and consider the quantities in question to merely be features of empirical reality (see section 18-4-1 for details). This is one more illustration of a kind of general rule of which we already encountered various manifestations; the rule that it is the demand for realism that puts a check on making relativity compatible with quantum physics.

Finally it is worth noting that the theories in question are, in significant respects, formally comparable to decoherence theory. The observational predictions they specifically lead to exclusively concern macroscopic systems. So do those from decoherence theory. And moreover, in the purely operational realm there are, no—nor *can* there be any—significant qualitative differences between them. For here and there what is achieved is but a rational justification, grounded on a general theory, of the common notions of form, position, motion, etc. that lie at the basis of both our day-to-day behavior and classical physics. It therefore comes as no surprise that the mathematical formalism of those among the said theories that are elaborated enough to possess one is very similar to the one of decoherence. Essentially, the difference between the theories under study and decoherence is but of a conceptual nature. However, in this respect it is quite considerable. For indeed according to decoherence objects, etc., are but phenomena, that is, appearances common to everybody—and belong to the realm of empirical reality—while according to the theories in question they are objective in the strong sense and are parts of independent reality.

On the other hand it should also be noted that this comparison raises a philosophically significant question. For let us consider a situation in which, for some reason or other, a macroscopic system finds itself in a state in which its quantum features are manifest. Within the theories under study this corresponds to a case in which, after, e.g., a measurement-like interaction, the spontaneous reduction has not yet occurred. We remember (section 8-2-1) that then, due to decoherence, the system reverts in an extremely short time to a state in which the classical appearances are restored. And this evolution, as we know, takes place automatically, that is, in virtue of the "orthodox" quantum laws, without anything like nonlinear terms being involved. This being the case, it may be wondered what the hypothesis that such terms are present adds to the picture. Admittedly, one obvious answer may be given. It is that, due to them, "appearance" is changed to "reality." But—many philosophers will ask—if we dispose of a set of "appearances" that hold good for everybody, isn't that enough? Is it actually rational to strive to change, in an artificial manner, these appearances to ontological realities? Isn't this trying to build up some kind of a poor man's metaphysics, akin, some will claim, to pure verbalism?

In the second part of this book we shall have occasions to ponder over questions akin to this one.

9-9 Outlook

If, we discard the realist nonlinear theories on the basis of the foregoing objections—which, without being decisive, make their artificial flavor rather patent—and cast again a look at the expectations of the conventional realist, we have to admit that we are at a loss how to meet them. For even if we reconcile ourselves to unexpected and surprising variants of realism we finally have to grant that the conceptions people try to build up along such lines, either are quite unconvincing or just simply are, when all is said and done, merely *apparently* realist. And anyhow the claim—made by our imaginary Jack—that independent reality is, so to speak, plainly readable in the world outside, is at variance with the data.

This result should undoubtedly be considered a contribution from science, not only to thought but even to the methodology of thought. For it makes us clearly realize the depth of the gulf separating the a priori from the a posteriori, and what seems obviously true from what turns out, when all is said and done, to be correct. More precisely, it forces us to grant that the problem of realism is one of those that reason alone cannot solve. According to some pure idealists—and also according to Kant's followers—reason is able, all by itself, to prove that (conventional) Ontology is vain and empty. On the contrary according to others—remember Jack's reasoning here reported on in section 9-2—reason evidently shows that physical realism is *the* correct standpoint, all other views just being Byzantine plays of the mind. The philosophers strive to solve the riddle with just the help of good, hard thinking, and it should be granted that they do succeed in appreciably shedding light on it. Indeed, in the past they had, on such matters, daring ideas other people could not have dreamed of. And it now turns out that some of these ideas seem to be quite right, after all. But if, at present, they look right, this is not due to pure philosophy. It is due to physics. Admittedly physics does not discard all the possible versions of realism (as its name shows, the "veiled reality" conception is a variant of realism, and is not refuted) but it sets on them quite strict conditions. And it so happens that these conditions, as we saw in this chapter, rule out just the versions that people who consider themselves "serious," "clear headed," "pedestrian," "rationalist," "matter-of-fact," etc., call "reasonable." It should be granted that this piece of information is of considerable significance.[12]

On the other hand however, let us also observe that when the human mind is in quest of rational explanations, those of the "realist"—not to say (in a broad sense) "mechanistic"—type are the ones that it most naturally builds up and feels are the most understandable. Hence (as already

[12] As Hervé Zwirn (2000) noted: "It would seem that, for the first time in the history of philosophy, choosing to believe in the existence of a world both external to all observers and roughly consistent with what we observe has become impossible, short of taking up an irrational standpoint."

noted apropos of the Broglie-Bohm model) when a serious conceptual difficulty arises, if it turns out that an explanation of a realist type partly sheds light on it this should count as a positive element, even when, for general reasons, the said explanation is not actually convincing. It is easily realized that the paradox, here, is merely apparent.

CHAPTER 10

SCHRÖDINGER'S CAT, WIGNER'S FRIEND, AND VEILED REALITY

§⧉

10-1 Introduction

IN THE FOREGOING we could observe that, to account for the phenomena, present-day physics seems to favor some kind of a two-level explanatory scheme. The level that may be termed "the strictly scientific one" consists in explaining both the regularity of the observed phenomena and the observed intersubjective agreement by referring to laws. This, of course, is just the standpoint normally taken up in science, but with here the significant difference, motivating the notion of a possible "second level," that while in classical physics the laws were objectively interpretable—they supposedly described what exists—in quantum physics they are but observational predictive rules. Actually this somewhat unusual state of affairs induces two distinct questioning. One of them just concerns the fact that, at first sight, some of the predictions ensuing from the laws seem to be grossly at variance with most commonplace observations; as we know, such is the case concerning the Schrödinger cat paradox. The other one, which actually is the one motivating the view that a "second level" might well be needed, has to do with the fact that stating any *rule* whatsoever unavoidably calls for some justification. In the present case we would like to know "what makes" the quantum rules so much more successful than any other recipes we might a priori invent. In this chapter—which is meant as a transition from the, mainly informative, first part to the more speculative second part of the book—a clear dividing line is drawn between these two subjects of inquiry. Sections 10-2 and 10-3 deal with the first one and section 10-4 with the second one.

10-2 Of Pointers and Cats

In section 8-2 we observed that decoherence theory accounts rather well for the fact that the world appears classical. But in section 8-4 we noted that, strangely enough, it is only by considering statistical ensembles of systems that it succeeds in doing so, whereas our observations of course bear, as a rule, essentially on individual systems. And on that account we wondered whether, finally, the proposed explanation is not, in some way, incomplete or deficient. To be sure, the theory guarantees that if we were to experiment on a whole ensemble of cats, none of our observations would disprove the view that some of them are alive and the others dead. But, after all, the entities we normally deal with are not statistical ensembles of cats. They are individual cats. Similarly, we never have recourse to statistical ensembles of instruments. We use instruments one by one. And at this "level of the individual" it is difficult to dismiss the feeling that the "and-or" problem—the problem that "state" $aA+bB$ seemingly implies both aA and bB to be present, rather than *either aA or bB*—is not yet solved. So, finally, is it solvable?

The question may be dealt with in two ways that appreciably differ with respect to the ideas they refer to but both yield the answer "Yes." One of them—call it Approach A—consists in strictly confining oneself to the purely *predictive* conception of the nature of quantum mechanics that was taken up in chapter 4. This approach—we remember—stands quite at the opposite of objectivist, or even physical, realism. According to it wave functions (alias state vectors) describe no physical reality whatsoever. They are just tools, making it possible to apply to various problems the quantum mechanical predictive rules introduced here in section 4-3. Now, since the said rules serve to calculate *observational* probabilities, they of course involve, along with the time-dependent Schrödinger equation (describing the spontaneous evolution of the system), also the so called "generalized Born rule" that finally yields these very probabilities. And since the notion "probability" inherently involves the concept of *alternative* possibilities, otherwise said the idea the word "or" expresses, within Approach A no particular justification for the use of this word is, in fact, needed. Its presence is an inherent part of the rules that were adopted right at the start. At the stage at which we now are—that of explaining the *phenomena* just by means of the quantum rules—there would therefore be no point in striving to explain it.[1] Incidentally the view (somewhat underlying this approach) that anyhow no *scientific* explanation of the said rules can be obtained will become even clearer when (in chapter 15) we come to the conclusion that, at an appropriately deep level, the very notion of "explaining" turns out to be problematic.

[1] Elsewhere (d'Espagnat, 1997b) a very simple theory (called the "operational" theory) corresponding to this stage in our "explanation quest" has been set forth as a simplified version of the "partial logics" or "consistent histories" theory (Griffiths 1984; Omnès 1988; Gell-Mann and Hartle 1989). I think I substantiated there my claim that this simplified form becomes quite a natural one as soon as the (anyhow unreachable) goal of inserting the said theory within realism is given up.

The other way of dealing with the question in hand—call it Approach B—stems from the fact that, as pointed out in section 2-8, when viewed in the light of the strong completeness hypothesis standard quantum mechanics, albeit overtly operational, still, implicitly, has in it some remnant of ontology. It has, just due to the fact that positively *asserting* the nonexistence of hidden variables constitutes an existential statement, even though a negative one. Approach B cures this defect—to the extent that it is one—by adopting, not the strong but just the *weak* version of the completeness principle (see section 2-8). The hypothesis may then be made that supplementary—called "hidden"—variables should exist, specifying structures finer than the wave function, and the Broglie-Bohm model is no longer ruled out. Now, as already mentioned, this model does, in a sense, yield a self-consistent measurement theory. For indeed, let us consider once more an electron-instrument interaction such as the one investigated in chapters 4 and 8 (that is, a generalized measurement procedure in which the initial electron "state" is taken to be a quantum superposition, such as c, of those, a and b, the instrument was meant to discriminate between). Within the Broglie-Bohm model the pilot wave is just the wave function of the electron-instrument composite system in the configuration space corresponding to this system. And, in the case of a measurement procedure such as this one, at a time t subsequent to the electron-instrument interaction the region of the said space in which this wave function is appreciably nonzero splits in two neatly separated parts, R_1 and R_2 say, corresponding to aA and bB respectively. In other words, the pilot wave may be considered to be a superposition of two waves—call them O_1 and O_2—corresponding to aA and bB, the O_1 (O_2) intensity being appreciable only within R_1 (R_2). Since in the model, the modulus square of the pilot wave at a given place is, by assumption, the probability for the representative point to *be* at that place, the said probability, obviously, is appreciable only within these two regions.[2] It is therefore clear that, after the electron-instrument interaction has taken place the representative point lies *either* in Region R_1 or in Region R_2. And that therefore the instrument pointer may be considered to lie in either one or the other of the intervals on the scale corresponding to A and B.[3]

It should be realized that there is no incompatibility between the two ways, A and B, of answering the question in hand, and that therefore we need not, actually, choose between them. Approach A consists in asserting

---

[2] A reader interested in the detailed structure of the corresponding calculation may find it , for example, in chapter 14 of Bell (1987b).

[3] At least, this holds true to the extent that the material state of the pointer is considered adequately represented by just the representative point. In fact this is not strictly true. We should not forget that, in the model, the pilot wave is real and is also, so to speak, an inherent element of the reality of the objects. When predicting the future of the latter we

that we know nothing concerning "what is Real" including the, real or unreal, nature of the wave functions. That, concerning them, we merely know the role they, together with the generalized Born rule, have in yielding observational predictions. And that therefore their detailed mathematical structure can't raise any existential problem whatsoever. The structures of "the Real" are such that these predictions come out right. This is all that there is to know. Approach B may be viewed as illustrating this point. For it is just an example of how the said "Real" might conceivably be structured and it shows that, then, the "and-or" difficulty here under study does indeed vanish (which does not mean of course that all related difficulties also nicely fade out; we know that it is not the case!).

Hence it is clear that Approaches A and B are both adequate as long as the only involved systems are pointers, or, generally speaking, objects of which it is quite clear that nobody knows them "from the inside." But that much may hardly be claimed when the just stated condition is not fulfilled, or is not fulfilled in an obvious way. To understand why, let us temporarily set aside Approach B, the Broglie-Bohm model and such other would-be realistic attempts. And let us focus on the guiding line of our whole investigation, which consists in looking for a theory merely predicting observations while refraining from interpreting it as a descriptive theory. This, indeed is what enabled us above (when considering Approach A) to finally state that decoherence theory does solve the measurement problem. But observe that this result was obtained at the price of attributing all the predictions of observations to just a disembodied, collective We, vaguely conceived of as preparing systems and instruments and looking at pointers. If, instead of pointers—or along with them— we had to do with conscious beings, *they* also might conceivably make predictions. And this may well render the problem more complex. Such situations are therefore worth being considered for their own sake, and this is the subject of the following section.

10-3 Wigner's Friend

Of course, if states of consciousness are but states of matter, there is no point in distinguishing between animated and inanimate objects. Since in scientific circles and within the enlightened general public this way of looking at things is almost a "received view" (a critical examination of it will be taken up in chapter 18), it may be felt that the foregoing analyses generally apply, whether or not the involved systems are animated. However, a

---

may not totally disregard it. It is assumed however that our senses only inform us about the representative point.

moment reflection shows such a reasoning to be hasty. This is because the analyses in question were carried out within the realm of a conception in which the basic notion—the one on which the other ones rest—is that, not of "matter" but of "being aware of" or "registering sense data."

True, it might be pointed out that the need of ultimately referring all what we assert to the evidence of the senses was long since recognized, and that, still, this did not prevent most biologists and neurologists from implicitly reducing consciousness to matter. But this was because, up to quite recent times, the great principle "everything takes place as if . . . ," held dear by Berkeley as well as by Kant and the neo-Kantians, was believed to be true. Everything takes place—it was thought—as if the classical axioms held true at any scale. And since the latter are ontologically interpretable everybody thought it legitimate to argue *as if* we had an access to ontology. As if, therefore, the ultimate reference to conscious and perceiving beings could just simply be left unspoken.

Today, the "great principle" in question was proved false, as we saw. The legitimacy of such a "leaving unspoken" procedure therefore appears questionable. And indeed, if we look back on the content of the foregoing section we realize that (as noted in its last paragraph) when animated beings are involved in the process the reasoning carried out in it raises a feeling of uneasiness liable to develop into a full-fledged objection. For, in fact, we can hardly avoid believing the cat has its own views about whether it is alive or not. Moreover, if, following Descartes, we are bent on considering the said cat to be a mere mechanism, nothing forbids us to replace it, in thought, by some human being—a friend, Wigner suggested (Wigner 1961)—who gets such and such impressions under the conditions in which Schrödinger's cat dies or survives. The question then is: within this framework, what about our two approaches A and B of section 10-2? Are they still valid? More precisely, is it consistent with quantum mechanics to attribute at any time definite states of consciousness both to the external experimentalist and to the "friend"?

To study this problem we must first clearly grasp the nature and the origin of the feeling of uneasiness noted above. To this end, there is no other way than to imagine the experimentalist disposes of a million copies of some simple "experimental setting" and sets in, with each one of them, a measurement process bearing on an electron lying in "state" $c = a+b$ (in the above notations). Moreover, in an extremely idealized way let us fancy these "experimental settings" to be both animated and inordinately simple (neither pointers nor "friends" but microbacteria, say). We may then consider that, when this whole process has come to an end, on the thus constituted ensemble of electron-bacteria systems the experimentalist is able to perform extremely fine measurements of various kinds. And, in particular we may imagine she makes ones by means of which she checks that the ensemble is not a proper mixture of systems some of which lie in "state" aA and the other ones in "state" bB. Our uneasiness then proceeds from the fact that the idea the "friends" (here the bacteria) have "states of consciousness" (are aware that they are in state A or in state B) just seems to suggest that the said ensemble *is* such a mixture (even before the experimentalist performs her checking).

For investigating this question the best is to first analyze it within the framework of Approach B, for, as we noted, this approach provides us with some sort of a concrete picture of what actually takes place. In it, the validity of the Broglie-Bohm model may be assumed. Now, in this model the representative point of the whole composite system lies at time t within either one of the two configuration-space regions R_1 and R_2, the notations being those of the foregoing section. Hence it is indeed true that the "friend" (the bacterium) is (and feels to be) either in one of its possible states or in the other one (either in state A or in state B). Within the strictly "orthodox" quantum theory, where strong completeness is assumed (no hidden variables), this fact would unavoidably entail the inconsistency trouble the specter of which worries us. An ensemble made up of such elements would be, perforce, not a pure case but a mixture. But within the framework of the Broglie-Bohm model the situation is altogether different. This is because, in the model, the fact that the representative point lies within some well-defined region—R_1, say—has no influence on the wave function. The latter remains, at all times, the total wave function that just results—without any "reduction" taking place—from applying to the initial wave function the Schrödinger time-evolution law (generated by the Schrödinger equation). The later behavior of the considered ensemble—hence also the outcomes of measurements possibly carried out on it—are therefore consistent with the predictions yielded by the pure case. No inconsistency pops in.[4]

Let us parenthetically take notice here of a point that will later prove useful. It is that, due to decoherence, if the "friends" are macroscopic it is quite possible in practice to *do as if* the ensemble were indeed just a mixture. Here "in practice" means "with respect to all the measurements that human beings would be able to perform on the elements of the ensemble (every such elements being here composed of an electron, an instrument and a "friend"). Under these conditions we may indeed treat the ensemble as if it were a mixture some elements of which are describable by aA and may therefore be associated to the reduced wave function O_1 while the other ones are describable by bB and may therefore be associated to O_2. (Of course the relative proportions of these elements is then

---

[4] The similarity between this and the content of section 3-3-8 is worth noting.

yielded by the generalized Born rule.) Obviously, such associations result in imparting to the state of consciousness of each individual "friend" a, practical, predictive power that it completely lacks when the friend is not macroscopic (the microbacteria case).

Readers with a taste for analogies may note the similarity between the situation here-analyzed and the Broglie-Bohm-like interpretation of the Young-slit thought-experiment with gas. If we fancy the particles involved there to be endowed with consciousness, associations of the above-described type are equivalent to substituting—for each individual particle—a fictitious wave function having passed through the slit the particle passed to the actual wave function. If the gas is dense (which corresponds to the case when the friends are macroscopic) our thinking particles run, due to such a substitution, only a negligible risk of making erroneous predictions. For indeed the actual distribution on the screen of their points of impact then practically coincides with the one the said substitution leads to predict.

We still have to consider the problem within Approach A. The foregoing observations show that, in its realm, a problem arises. For what characterizes this approach is, as we noted, that it is purely predictive and free from any descriptive or "ontological" elements. Hence the knowledge the experimentalist has of, in particular, the system wave function is meaningful only with respect to the prediction possibilities it provides him/her with. Now, this knowledge is a state of consciousness and, on philosophical grounds, it may seem tempting to "extrapolate" this remark and consider that states of consciousness are meaningful only in virtue of the predictions they make possible. The problem then is that, within Approach B—which, to repeat, is but an illustration of Approach A—at least if the "friend" is not macroscopic his or her states of consciousness are deprived, as we saw, of any predictive function. It may therefore be feared that, in Approach A, the very notion of a state of consciousness of the "friend" should be of questionable validity. The specter of solipsism becomes menacing.

The notions of consciousness, state of consciousness, becoming aware and so on, are to be investigated with some details in chapter 18. But let me note already here that, in my opinion, the reason why the *scientific* meaning of the state of consciousness notion seems to be limited to the predictive realm is just that science itself is intrinsically so limited. And that this does not prevent the states of consciousness—of the existence of which we have an immediate knowledge!—to exist independently of whether or not they are predictive [in this connection see also d'Espagnat (2005)]. If this is granted it becomes clear that attributing to the Schrödinger cat or to the microbacterium the certainty of being alive—even when it is an element of a pure case ensemble of the $aA+bB$ type—implies no consequence of a scientific nature. It therefore entails no inconsistency with regard to the measurements the experimentalist makes or will make. And of course the same holds true concerning the friends' consciousness of having observed this or that.

It is worth noting that such a standpoint shows some outward similarity with the so-called "modal interpretation" discussed in section 9-4. In it the—essentially predictive—wave function corresponds to what, in the said interpretation, is called the "quantum state," while the impression the "friend" has corresponds to what is there called the "value state."[5] In fact the said similarity appears even more clearly when the details of the modal interpretation are considered. For indeed, the symbol $aA+bB$ that in the above scheme, represents the cognitive state of the experimentalist has the distinctive feature of containing no cross-terms such as aB or bA (of course this is why, from what he/she observes, the experimentalist is able to infer something about the electron). Now, within the modal interpretation it is precisely when cross terms are nonexistent that "value states" exist. Nay, it might even be claimed that within the framework of an extension of the modal interpretation to the Wigner's friend problem, the paradoxical nature of what we called "the table paradox" gets weakened. For we may decide that what corresponds to the statement "the left-hand side of the table is green" is the claim that the friend is in one, precisely defined, state of consciousness, this claim then having, for the friend, a well-defined truth value (it is true or false). And we may decide to set statements such as "the table is (is not) green at its left-hand side" in correspondence with claims that, in fact, do not concern one individual system but just statistically predict the impressions an outside experimentalist taking a look at the preparation of an ensemble of systems would get. Admittedly, such a parallelism is a rather "far-fetched" one. But the impression of oddness that, in the modal theory, we derived from the distinction about the table gets somewhat alleviated nevertheless. For, within the conception considered here it is quite clear that the two considered types of statements bear on quite different "objects," namely, the impressions of the friend and those of the experimentalist.

10-3-1 Talking with the "Friend"

If, when the measurement procedure is over, the experimentalist turns to his "friend" and asks her what she saw, the latter's answer will obviously

---

[5] This similarity was stressed by Michel Bitbol (2000). Note that, though he is not a realist, Bas van Fraassen still demands that some kind of a world view presented in a realist language should be possible. I suspect, therefore, that we are here departing to some extent from van Fraassen's views.

be that she observed this or that, i.e., had some well-defined impression. Suppose now that (in order to remove the wiping-out action of the environment) we, again, put some microbacterium (assumed endowed with consciousness, nay, even, with language!) in place of the "friend" and her instrument. Here again, we may assume that the experimentalist makes millions replicas of exactly that experiment and asks the question to all the bacteria involved. He will get, now the answer corresponding to one of the two possible outcomes, now the answer corresponding to the other one. In other words, he will get the (by the way, entirely correct) impression of having to do with a proper mixture. But of course there is no incompatibility between this and the fact that, by assumption, the ensemble was initially a pure case (the pure case $aA+bB$) since, for the experimentalist, the very fact of asking a question to a bacterium amounts to making on it a measurement by means of some instrument, whose interaction with the bacterium may well disturb the state of the latter. Admittedly, to account for the fact that the friends claim they had definite impressions already before the question was asked them (though after they looked at the dial, of course) is not so easy. But in fact this problem is, in substance, identical to the one investigated above about Wigner's friend (or the cat), being, already just after the "measurement procedure" has taken place, in a definite state of consciousness. And our proposal for solving it must therefore be just the one explained above. Consider first Approach B. Observe that, within it, the difficulty vanishes since, just after the procedure, the representative point already lies in one definite region. Attach to the said point a state of consciousness of the "friend." Note that, theoretically, it is not predictive. And, finally, transfer this notion of "nonpredictive states of consciousness" to Approach A.

10-3-2 The "Several Friends" Case

A generalization of Approaches A and B to the "several friends" case is worth having a look at. Specifically, we may consider situations in which several people (several "friends") look together at the same instrument pointer. In this case we of course expect that, subsequently to her observation, each one of these "friends" finds herself in a well defined state of consciousness (even when the initial conditions did not predetermine the latter: this point was dealt with above). But we also expect all these states of consciousness to coincide (all these people will see the pointer at the same place on the dial). And just the same holds true within somewhat more complex situations, such as those prevailing in Aspect-type experiments (section 3-1). There again, when the two "friends" situated at a great distance from one another have oriented their polarisers in one and the same direction, their observations set them in the same, well-defined, even though nonpredetermined, state of consciousness.

Within the, purely predictive, framework of Approach A, such inter-subjective agreements are implied, as explained in chapter 5, by the generalized Born rule for, as noted in section 5-2-5, this rule is impersonal. It deals with the probability, not that such and such a person should have a given impression but that *we,* collectively, should have it. For example, in the "Aspect-type" experiments it is supposed to apply to measurements made by individuals standing very far from one another and, to repeat, it and it alone is what predicts the correlation effects that turn out to be experimentally verified. Now, we may well wonder what justifies it. For indeed it seems it postulates some sort of a puzzling, immaterial link between persons who are arbitrarily far from one another.

Such being the case, it may seem tempting, at first sight, to take up the model sketchily described in section 5-2-5 that is, to substitute for the said nonlocal, abstract probability law an individual one, to be independently made use of by each involved person. In this approach, the Born rule is supposed to hold true for each person individually, so that any one of them may in this respect consider himself or herself to be the only mind in existence. As we know, in order to predict (in probability) both what she herself will observe and what other people will tell her they observed, this person, in principle, just needs apply the Schrödinger equation to all the existing physical systems, be they conscious or unconscious, without performing any "reduction" whatsoever. She then just needs use the Born rule for calculating *her own* observational probabilities (including those relative to the, printed or verbal, messages she may receive), without being in the least concerned by the senders' states of consciousness (if any!). What makes this approach worth considering is of course the fact that all of us are equally qualified for carrying out this sequence of operations. And that, the quantum rules being what they are, no public disagreement can ever arise between us concerning observational data, even though our internal states of consciousness may well differ.

These ideas may be applied to the problem here in hand. In other words, while we may go on considering that the Born rule yields probabilities that hold true for a wide, impersonal We, alternatively we may as well take the said rule to be valid for each observer separately. Within this option the states of consciousness of the various Wigner's friends are not any more necessarily correlated. Each friend then lives under the delusion of an intersubjective agreement that has no existence "in reality." As we see, this "several friends" problem thus sets us on the horns of a rather strange philosophical dilemma that discursive arguments may well be unfit to remove. Concerning it, it is however appropriate to observe that the second alternative might well lead to representations of quite an inordinate strangeness, since each involved person would tend to attach to her own world-picture a specific, macroscopic time evolution differing from the ones the other persons would predict. Hence, at the present stage of our knowledge on such matters I, for my part, still tend to hold fast to the conservative view that the states of consciousness of the various observers are truly correlated, as the "impersonal" Born rule implies they are (see section 5-2-2 above). However, in the present state of the investigations in this field I do not rule out the possibility that the alternative idea according to which the private states of consciousness of the various observers actually are not correlated but are deprived of (complete) predictive power should be correct.

10-3-3 States of Consciousness and Predictability

We noted above that, theoretically, the friend's state of consciousness is not predictive. That within a "generalized measurement" process such as the one we considered all along, it is only the quantum state—the overall wave function—that may serve at making observational predictions. However, we also noted that, in virtue of decoherence, the "friend," provided she is macroscopic, may, in practice, make use, for this purpose, of her own state of consciousness without any significant risk of error. Does this possibility extend to the observational predictions that the various Wigner's friends might want to bring to one another's notice? In other words, can it constitute the basis of an intersubjective agreement of some kind, within such a "community of friends"?

To find the answer to this question remember first that within the framework of a measurement of the $aA+bB$ type we could show (section 5-2-2)—using the conventional Born rule—that two friends looking at time t at the same pointer necessarily get the same "impression." Remember also that, as we saw in section 10-2, in the electron-instrument configuration space the overall electron-instrument wave function at time t is a superposition of two quasi-non-overlapping waves O_1 and O_2. For the purpose of making observational predictions, each one of the two friends will of course build up a wave function (the wave function called D in chapter 4) which, under these conditions, should at least approximately coincide either with O_1 (if she had impression A) or with O_2 (if she had impression B). Now, both having had the same impression the wave function they will thus build up will of course be the same for both.[6] And since

---

[6] Here we purposely avoided all technical complications. Actually the electron-instrument, which is a highly complex system, is normally to be described by means of quantum mechanical symbols more complex than just wave functions. Correspondingly, the quantum

it constitutes their predictive tool it follows that the probabilities of future events that the two friends thus calculate will be the same, which means that the answer to the question is positive. In virtue of the quasi-nonoverlapping of O_1 and O_2 and decoherence, they run no risk—as we saw in the case of the friends considered individually—of making any error they could detect. And this, in spite of the fact that, from the point of view of the experimentalist who planned the whole experiment and keeps an eye on its evolution, the wave function in question is not the "good" one.

This of course may be generalized to an arbitrary large number of "friends." And the notion of such a "Wigner's friends community" may perhaps turn out to be interesting from the point of view of speculative philosophy. For it is indeed conceivable that our whole human community—or, say, the set of all living and sentient beings—should be of this type. In chapter 18 we shall take a brief further look at this idea.

10-4 The Veiled Reality Hypothesis

We are here at a turning point. In the foregoing chapters we noted that the observational predictive rules of quantum mechanics work beautifully in a great number of domains and were never found at variance with experimental outcomes. Moreover we could establish—which actually was far from being self-evident—that the assumption according to which they are universal is in no way refuted by a few commonsense notions that, at first sight, seemed both to be at variance with them and to obviously reflect plain facts. We now would like to ponder on the question why these rules are there and where they come from. This problem will interest us all along the second part of the book. It is a delicate one on several grounds, and in particular because it does not lend itself to a neat and univocal formulation. Some—the "diehard instrumentalists"—go as far as considering it is meaningless. In such a general domain, they claim, the very idea of "explaining" is inappropriate. We dispose of predictive rules that work well, and there is nothing more to ask for. Others, following a somewhat Kantian line of thought, grant that some explanation of the said rules is to be looked for but expect to find it within the human realm. Finally, still others do not rest content with either of these two approaches. Along the lines of *Open Realism* (section 1-2) they are on the look for a mode of understanding grounded on the notion of "what

---

mechanical symbols that the friends build up are also more complex than just wave functions. And they are likely to differ from one another in many aspects. It may however be shown that concerning the problem at hand these aspects are irrelevant, so that the wave function description, while oversimplified on the technical side, still yields a correct account of what takes place in the situation under study.

truly exists"—alias, "the Real"—while being fully aware of the insuperable obstacles contemporary physics sets in the way of all too naive such ideas. The latter approach is the one that I took and strove to justify in several works. In the second part of this book we shall compare it with the two other ones and investigate to what extent it does or does not fit with various philosophical standpoints. But before going into this it seems necessary that some broad survey of what it actually consists of should be given. It is appropriate that, while so doing, we should also sketch the main lines of the reasoning that leads to it. But this, of course, will be done without entering at this stage in the elaborate philosophical debates to which it may be, and will be, submitted in further chapters.

10-4-1 Open Realism and Extended Causality

By way of introduction let us first note that to properly grasp the magnitude of the present-day problems concerning knowledge we should keep in mind the width of the gulf that divides, in this respect, the bulk of the philosophers from a vast majority of scientists. The said gulf consists, to repeat, in that most of the formers lean toward modes of thought more or less connected to Kantianism or radical idealism, a standpoint which seems to be rare among the latter. The formers stress the, quite obviously decisive, role that our sensorial and intellectual abilities do play in the building up of our representation of the World. From it they infer that there are no "objects-per-se," or that if there are ones, they are utterly unknowable. This table, this book, this computer, all these elements of our daily life are but "objects-for-us," and the laws that account for their behavior merely describe our human ways of systematizing our own collective experience. The latter—I mean, the scientists—reply to them by appealing to obviousness and commonsense. They point to the bones of a dinosaur and ask their opponents whether they actually believe that this 65-million-year-old animal existed just simply "for us." As for the philosophers, the only criticism most of them aim at Kant consists in claiming that he was not daring enough. He should have just simply rejected the notion of a reality-per-se, instead of keeping it while labeling it unknowable. On the contrary, most scientists stress that—in both magnitude and nature—our whole basic knowledge immensely changed since Kant's time. And they call obsolete the reasons Kant had of denying that human knowledge constitutes a reliable account of what truly *is*.

For my part, I endorse this judgement of theirs, but merely on the condition of not interpreting it to mean more that what it literally asserts. Unquestionable is the obsoleteness of most of the reasons Kant put forward though, by the way, he himself was fully justified in considering them adequate since he knew neither of evolution (of life and the Universe) nor of our mathematical physics. But, precisely, these are pieces of knowledge that we now possess. And, for reasons to be developed in chapter 13, I do consider that indeed they render obsolete the arguments of Kant and most of his successors. However (and this is, in my view, the important point) I also consider that arguments quite different from Kant's ones, grounded on the very knowledge that we now have (extended to the most recent data), lead to views that, roughly speaking, coincide with the hard core of Kant's ones. More precisely, I think this is true if the hard core in question is taken to be Kant's claim that the purpose of science is not a knowledge of "the Real" but just that of phenomena. For it seems to me that the content of the foregoing chapters does prove that much.

Let it be quite clear however that to thus go beyond traditional realism does not amount to rejecting any notion of a human-independent reality. Neither Plato (for whom Ideas are Beings) nor Kant (who clang to the thing-in-itself notion) adhered to the said rejection, though neither of them was an objectivist realist. And for reasons the reader partly knows (they are made more precise in what follows) I came to judge that, in this respect, they were not wrong. Hence I found it appropriate to take up, as a starting point in my reasoning, a minimal realist postulate, the one of "Open Realism" (section 1-2). I then convinced myself—through arguments that were discussed at length above—first, that realism of the accidents is to be barred out and second that the attempts at overstepping it while remaining within the, more general, physical realism framework are most unsatisfactory on various grounds.

It follows from this that indeed physics cannot be interpreted to be a faithful description of "the Real," and this reduces it to merely be concerned with the phenomena, in the philosophical sense of the word. In other words, it is, at best, a description of what has been called "empirical reality" in section 4-2-3. Like a large number of physicists I however considered it as significant that some major mathematical laws of physics, such as the Maxwell equations, remained remarkably stable and relevant in spite of the numerous changes that, in the course of time, their interpretations underwent. In other words, the, wary, conjecture that these laws indirectly furnish some not altogether misleading glimpses on the general structure of "the Real"[7] not only is not at variance with the data but seems not to be altogether unwarranted. It may also be expressed by stating that, beyond Kantian-like causality—which takes place between the phenom-

---

[7] Clearly this assumption is much weaker than the *(strong) structural realism* one, according to which the great physical laws do describe the said general structures. The fact that, in such a conceptual context, one of these laws, the one that no influence is propagated with superluminal velocity, appears as being violated (because of nonseparability) speaks *against* such a structural realism.

ena—there should exist an "extended causality" consisting of influences that are exerted by "the Real" on phenomena (and can't be approached in any quantitative way). This idea I may, admittedly, put forward only as a supposition, which means that the importance I impart to it must necessarily be considerably smaller than the one I attribute to the above-mentioned proof that objectivist, and even physical, realism(s) are erroneous. Nevertheless, it seems to me that whoever is not a "radical idealist" in the sense explained in section 1-2 may well consider that some physical data render the idea in question plausible.

I expressed the said idea in a metaphorical way, by asserting that "the Real" instead of being a radically unreachable "pure x" is merely veiled. Now, as a rule, metaphors are ambiguous, and this one is no exception. Some quick readers thought it meant that particles are localized but merely approximately! Fortunately, most of the criticisms addressed to it were distinctly more to the point. The main ones are here analyzed in chapter 17 but let me already stress that, in my mind, the metaphor was in no way aimed at suggesting some vague similarity between "the Real" and our everyday experience.[8] In particular I cannot ignore the fact that—in view of Bell's theorem and Aspect's and others' experiments—the conjecture that "the Real" lies in space-time implies nonlocality. This basic fact incites me to consider the said conjecture to be greatly lacking plausibility. I am therefore inclined to think that "the Real"—alias human-independent reality—is not embedded in space-time. And, indeed, I go as far as speculating that, quite on the contrary, the nature of space-time is (as were, in Kant's view, the ones of both space and time) not "noumenal" but "phenomenal," that space-time is a "reality-for-us."[9]

As can be seen, there is a rather high degree of convergence between the views I arrived at and those entertained by a large number of philosophers, even though, to repeat, I got to mines through channels different from theirs. These views, common to them and me, are quite at variance with a "received view" (physical realism) that is still widely accepted, and

[8] Nor is it meant to suggest that the symbols our theories call into play are rough approximations of what the Real "is made up of." In my opinion, wave functions (for example) are not—not even "very rough"!—approximations of "constituents of the Real."

[9] For completeness' sake let me however mention here a stimulating suggestion from U. Mohrhoff (2000, section IV; 2001). This author aimed at preserving the notion that the physical space is strongly objective while ascribing to it some at least of the features of phenomenal space. In view of experiments of the Young-slit type he granted that all of the latter cannot be kept. And the one he chose to give up was the idea that the various regions of space should exist by themselves independently of the objects that may or may not be in them. In other words, it was the idea that physical space is intrinsically partitioned into infinitesimal regions (and can therefore be represented by a transfinite set of real number triplets). According to him the place where an object is is comparable to the color of the object, in that the latter cannot conceivably be separated from the very existence of the said object. As an alternative to my views Mohrhoff's suggestion is certainly worth studying.

I think they are essential for attaining at a well-balanced conception of the relationship between human beings and the World. Compared to the weight of this agreement between us, the differences between my standpoint and the one of the said philosophers are relatively secondary. Apart from a somewhat subtle one (veiled reality rather than "pure x") to be scrutinized later, they mostly concern my "open realism" postulate, which separates me from the philosophers (neo-Kantians, etc.) that I, perhaps somewhat approximately, grouped under the label "radical idealists." Now, concerning the said postulate I couldn't think, of course of proving it (this is why I called it a postulate!). But still I believe I can show that— contrary to an opinion that, in philosophers' circles, is widely held—it is by no means arbitrary. That serious arguments (some of which we already saw) speak in its favor. That these arguments successfully hold out against criticism. And that therefore the postulate is, when all is said and done, hard to discard.

What are then, in detail, these arguments of mine? In the course of time, and following various exchanges of views, they somewhat evolved, but, on the whole, not considerably. In the second part of the book critical analyses of them are to be made, but for clearness sake it is necessary that they should first be concisely stated. Here they are, without comments.

1. The notion "existence" is prior to that of "knowledge." To be sure, the phrase "if thought did not exist we would have no knowledge" is a truism. But it is one that is worth being stated for it reminds us of the fact—lost sight of by so many thinkers!—that, logically, existence comes first. With all due respect to Descartes what, in my view, the *Cogito* proves is neither the existence of a personal *I* nor the primacy of thought on matter. And not even that the former might be independent of the latter. It is the "experimental" validity of the notion of existence (the existence of at least one entity, namely thought) and its primacy with respect to those of experience and knowledge, which make sense only in virtue of it. With the corollary that, while it is quite proper to make the idea that a particular thing exists depend upon the notion of possible knowledge of the said thing, it is inconsistent to refer the very notion of existence (in itself) to the one of possible knowledge. Neither Plato nor Kant indulged in such inconsequence, and quite obviously they were right.

2. In physics it often happens that fully consistent and beautifully rational theories are falsified by experiment. They have, in too many domains, verifiable consequences that turn out to be at variance with what is actually observed. Something says "no." This something cannot just be "us." There must therefore be something else than just "us."

3. As we saw (chapter 5), contrary to our intuition neither the "no-miracle" argument nor the one based on "intersubjective agreement" jus-

tifies—or even backs up—any view resembling realism of the accidents. But (as explained in sections 5-1-3, 5-2-3, and 5-2-4) these two arguments nevertheless strongly support the view that the physical laws do not depend exclusively on us, which of course calls for the notion of something else that is not "us."

4. Science normally makes use of descriptive concepts. In Kant's time the ones on which it rested—Euclidean space, universal time, determinism, and so on—were so close to (schematized) familiar notions that this philosopher could consider them to be a priori elements of human knowledge. And this idea, in turn, enabled him to reject the notion that science has to do with some (conjectural) reality-per-se viewed as the source of such concepts. Today however, science has recourse to new notions—space-time, curved space, indeterminism, etc.—that are anything but familiar. Whoever, therefore, proposes to follow Kant with respect to both the use of descriptive concepts and the just mentioned rejection has to face awkward problems. Is it meaningful to speak of sets of a priori notions that, in virtue of increase in knowledge, would undergo a time evolution? Should we have recourse to the notion of a "reserve" of a priori notions, on which each age would draw the ones that best suit it? Can a concept that, at a given time, is an a priori of knowledge not be one any more a century or two later? For my part I find it difficult to answer these questions otherwise than negatively. In my mind it therefore looks as though we are forced to either come back to physical realism or rest content with just predicting what we shall observe. Now, as we saw, present-day physics bars out the first one of these two conceivable ways out. We must therefore take up the second one. But then Argument 3 above shows these predictive laws signify that the "independent reality" notion can hardly be dispensed with.

Admittedly, in the philosophical literature objections to these arguments are easily found. In the second part of the book they will be made explicit and we shall analyze and discuss them. Those of them that were specifically directed to my views will of course be studied with a special care. Their critical analysis is the subject of chapters 17 and 19.

10-4-2 Two Misunderstandings

The Veiled Reality conception is—I granted—open to discussion, and the content of the following chapters will pave the way to the latter. Unfortunately the said conception also gave rise to pure and simple misunderstandings. Here two of them are examined (and a remark of a more general nature will then be made).

MISUNDERSTANDING 1

It is to be found within a generally favorable, well-disposed analysis of my standpoint put forward by Roland Omnès in one of his books (1994b, chapter 12). This author had the excellent idea of paralleling my views with those of Nicholas of Cusa (*De Docta Ignorantia*, first edition 1440). As far as the end outcome of my quest is concerned this clearly is a good remark, and I willingly grant that the title of this book expresses the said outcome concisely and beautifully. But on the other hand I am afraid I cannot agree with Roland Omnès at the place where he stated that the quoted author's arguments and my own are "remarkably similar." For indeed, first, my admiration for the title of the book does not prevent me from considering—after having read its content—that Nicholas of Cusa's deduction procedures are, to say the least, well . . . disconcerting. And second, the said author lived in the fifteenth century so that drawing a parallel between his reasoning and mine implies ignoring the whole set of my "negative arguments," which consist in proving that a nonveiled—objectivist or Einsteinian—form of realism is incompatible with *today* physics. I, on the contrary, consider these arguments to have great weight. For indeed I think it is clear that whoever would choose to remain within the realm of pre-quantum physics would naturally be led to develop a conception of Reality modeled on those of d'Alembert and Laplace, in which the very notion of "learned ignorance" would be totally out of place.

However, still more disconcerting in a sense is, for me, the account that, in the said analysis, Omnès gave of my definition of physical realism. Especially puzzling is his statement that I discriminate between "a 'strong realism' when Reality is known in itself and a weak realism, when this knowledge is only partial."[10] If I ever had recourse to such expressions, this must have been very seldom (as a rule, I make use of those of strong and weak *objectivity* and, concerning realism, I use other qualifiers). But never mind the choice of epithets! The main thing is that I feel rather sure not to have anywhere evoked the idea of an "only partial" knowledge of reality; and that indeed my veiled reality conception is altogether different from what Omnès believed it to be. This is quite clear from the example he made use of. He considered the energy levels of the matter composing the object "the Eiffel tower" and pointed out (and proved) that we are utterly unable to know whether or not the said matter is, at a given time, in such and such quantum state, corresponding to a well-specified energy level. Such, then, is his notion of a necessarily incomplete knowledge. And, according to him, what the words "veiled reality" ex-

---

[10] *Loc. cit.* p. 513. It is alas somewhat "in the nature of things" that the author of some work or other should consider accounts made of it by others to be unsatisfactory. But, on the other hand, occasions of rectifying misunderstandings of this type are so rare that when an opportunity presents itself of doing so an author should not let it pass.

press is just the impossibility of supplementing this knowledge and making it complete. It is important to realize that this is actually a misunderstanding. The foregoing chapters and sections—as well as the content of my former books—clearly show the veiled reality conception is not that one. It is not the view that, within Reality-per-se, there are a number of "things" and "facts" (laws, structures, the fact that the Eiffel tower is in Paris) that we truly know as they really are "by themselves," along with other ones that we are unable to really know. It is the—radically different!—conjecture that "the Real"—alias Reality-per-se—has some structure, that our great physical laws are (presumably hardly representative) emanations of the latter and, last but not least, that objects are in no way "things-in-themselves."

The source of the misunderstanding might well lie, in part, in the fact that most scientists do not worry much about the distinction Kant introduced between transcendental and empirical realism. Intuitively, just like Mr. Average Man, they have the first one in mind. They believe in plain physical realism. But when faced with some conceptual, or even merely technical, difficulty they often fall back, in their explicit analyses, on the second one, empirical realism, without realizing the magnitude of the conceptual jump they thereby implicitly make. Since, in my eyes, this magnitude is quite considerable, I, for one, directed my efforts toward showing that physics itself, philosophy apart, now demands that the jump in question be made. As for Roland Omnès, he, doubtless, is aware of the existence of the jump but, clearly, he is not haunted as much as I am by its magnitude. The expressions "independent reality" and "empirical reality" are not parts of his vocabulary. Maybe, therefore, he did not fully grasp the nature of the set of problems I have in mind. Having noticed, however, that, in my writings, a distinction is made between two forms of realism, he may have coined this "weak realism" notion (that is, the idea we merely have partial knowledge of the details) in order to account for it. But in fact I think I made it clear that it is the distinction between independent and empirical reality that lies at the core of my veiled reality conception. So that the latter notion is no more the knowable fraction of the former than, with Kant, phenomena are knowable bits of noumena. All this, even though, to repeat, I do not bar out the idea (to be discussed later) that our basic mathematically expressed physical laws do vaguely reflect some great structures of "the Real."

MISUNDERSTANDING 2

Since the expressions I used, "veiled reality" or "uncertain reality," are purposely suggestive of "something that escapes us," some may feel tempted to link up to them *any* somewhat vague idea. And since, in this realm, the possibilities are infinite it comes as no surprise that some upholders of esoteric visions should have purposed to relate their views to mines. It goes without saying that I endorse none of these speculations before having examined them and that such an examining, when I have a possibility of performing it (when I know about the attempt), most often leads me to a negative conclusion.

That much being said, it remains true that exceptions may occur. At any rate, there are some "intermediate cases," that is, approaches in which correct and interesting views are intermingled with speculations that I consider risky. The latter often take the form of lofty—but shaky—considerations, quite attractive in view of the evocative words they make use of but the strictness of which seems to me to be questionable. To me, such intermediate cases pose, as is easily guessed, a delicate problem. For example, how could I not feel some affinity for authors such as Thierry Magnin (1998) who imparted a decisive role to the notion of levels of reality (in further chapters I shall explain with some details why I agree that this notion is important). But on the other hand, how could I approve of such passages of his book as "unpredictability and chaos restore the place of time and open the way for its constructive role in building up an 'uncertain reality' (Bernard d'Espagnat)"? As the content of all the foregoing chapters makes it clear, it is not (and it never was) on the "constructive role" of time—and even less on chaos—that I ground my view that independent reality is "uncertain" (veiled). Similarly I find it hard to accept this author's statement that "to dare speak of weak objectivity, as, for example, Bernard d'Espagnat does, is also to dare redefine the concept of the scientist's neutrality." A redefining the author links with the idea that the choice of what hypotheses are basic are "often guided . . . by the social context of the time when they are put forward." The question whether or not choices guided this way do or do not have an effect on the objectivity of science is considered in the forthcoming chapter. But let me here already forcefully stress that—as, again, the whole of what we have already seen clearly shows—my notion of but a "weak" objectivity has strictly nothing to do with the notion of a social guidance of any sort.

A General Remark

In a totally different domain, it occurred to me that this introduction to the veiled reality conception should be supplemented by a warning. The point is as follows. I was informed that the conception in question gave rise, in the mind of readers of my foregoing books, to a conjecture that, I think, has no real substance. This is the view that the said conception yields the proper framework for an explanation—and thereby for a wider recognition—of, true or conjectured, *anomalous phenomena*. To be sure, all of us have heard of phenomena seemingly defying the laws of science. Since, in the mind of most people, science identifies with a mechanistic conception of nature, it is not surprising that when an author—I or somebody else—shows mechanism to be outdated, people who believe in such phenomena should look for a confirmation of their views in the theses this author puts forth. It should be noted however that, as far as I am concerned, such an association of ideas is erroneous. Concerning, for example, nonlocality, the short discussion put forward in section 3-2-4 clearly shows, I believe, that the hypothesis according to which it explains strange (magic?) influences-at-a-distance is totally artificial, and hence not credible. More generally, far from doubting of the robustness of the physical laws, I consider it most likely that all the predictions they lead to are correct and universal. And, even more generally, I consider that to situate the interest of the veiled reality conception in a conjectural explanation of hypothetical anomalous phenomena—"phenomenon" meaning "what is observable"—is to go completely astray. The interest of the conception in question is, in my view, exclusively of the nature of understanding and, more precisely, *philosophical* understanding.

PART 2

A PHILOSOPHICAL ANALYSIS

CHAPTER 11

SCIENCE AND PHILOSOPHY

11-1 The Impossible Split

A DOCTOR OF SCIENCE is a Ph.D. This old acronym reminds us that until the seventeenth or eighteenth century science and philosophy were hardly separate, that people such as Descartes, Pascal, and Leibniz were brilliant in both, in short that most prominent thinkers did not shun mixing up scientific and metaphysical research. But we also know that in the course of the eighteenth century, due to the growing specialization of the sciences and the critical thinking of philosophers such as Hume, such a mode of research came to an end. The fields of science and philosophy progressively became very clearly distinct. And in, say, the early nineteenth century the unspoken rule governing this splitting could indeed have been simply stated. It consisted in entrusting philosophers with problems bearing on the nature of things, and scientists with those concerning their behavior. Of this, Fourier's theory of heat propagation remains a vivid example. In Fourier's time several theories were in competition concerning the nature of heat: the theory of caloric, the one—already!—of molecular restlessness, and several others as well. Strictly in line with the rule just stated, Fourier abstained from taking sides in the controversy. Rather than speculate on the *nature* of heat, he chose to investigate, with strict quantitative methods, its *behavior*. This way, he could write down the heat propagation equation. And, in fact, this equation will forever remain valid, precisely because it is independent of any speculation concerning the basic nature of the "object" under study and merely deals with its evolution in time.

Such a distribution of roles was intellectually satisfactory and in quite a number of fields it still remains appropriate. It must be acknowledged nevertheless that, to day, it is far from suiting all research fields. For we must realize that, in the foregoing example, the focalization of the scientist on just the *behavior* of the entity under study was possible only due to the fact that everybody knew what "hotter" and "colder" meant in practice, so that the theory could spare the trouble of defining these words. But when what is at stake is a quantum field, or the metrics of curved spaces, or similarly unfamiliar notions, this is no more possible since, obviously, pretending to deal with the behavior of such entities without having first defined them would be absurd. With respect to them the physicist has therefore to choose between two standpoints. Either he gives up the rule of keeping to just a study of behaviors, that is, he agrees to take into consideration definite views concerning existential realities. Or he greatly waters down the substance of the said rule by taking up the view that, when all is said and done, the word "behavior" merely refers to a sequence of observation results registered in our memories. However, this dilemma means that, finally, the physicist is forced to take a glance at the—so delicate!—domain of the philosopher: for in the first case he touches upon ontology and in the second one he comes close to operationalism.

Hence, just as the philosopher who takes interest in the problem of reality may hardly ignore what the physicist has to say, similarly the physicist aiming at being more than a technician in physics nowadays can hardly escape having to cope with philosophical questions. Consequently, in this second part of the book priorities will be reversed. It is in the philosophical questions considered for their own sake that we shall take an interest, and we shall have recourse to what we know in physics in order to try and throw light on them.

11-2 Epistemology in the Late Twentieth Century

It is appropriate that such a study should start by a quick survey of philosophy of knowledge with special interest for that of scientific knowledge. Let us therefore have a look at epistemology.

One point that, in this field, easily attracts attention is the fact that the galaxy of contemporary epistemologists is composed of authors of rather diverse opinions and trends. Formerly this was not the case. Some forty years ago the logical positivists, as we all know, almost filled up the epistemological scene, equipped with a theory of knowledge that may be questioned but whose elaborated, unified and consistent structure is undeniable. Since then, sharp criticisms of varied sorts—some justified, others less so—made them fall, temporarily perhaps, into discredit.[1] And their successors, the epistemologists who, during the last decades of the twentieth century, were most in fashion, grounded their works on philosophical standpoints that hardly have these characters. If we were bent neverthe-

---

[1] Let the main reason for the said discredit be briefly noted. It consists in the failure of two attempts, that of justifying the verifiability principle and that of building up an inductive logic; as well as in the "discovery" that an isolated statement cannot be tested (Duhem-Quine thesis).

less on finding features common to many among them I guess we could find two. The first and most conspicuous one is an attitude of extreme severity towards science, combined with a radical skepticism ("everything goes") with respect to knowledge in general. The other one is the fact that their analyses impart an essential role to notions such as the one of "paradigm" that, without being explicitly bound to a realist approach, still are closely akin to those (strict conceptualization, in particular) that lie at the basis of objectivist realism.

I already pointed out (section 7-2-1) the reasons that I consider the second one of these two features to be one of the main causes of the first one. It is a fact that, in physics, realist interpretations of experimental data underwent quite momentous changes in the course of history, and clearly this does not prompt us to rely on those that, to day, are grounded on somewhat similar notions. But, to repeat, we can but be surprised by the stubborn silence of the epistemologists concerning two points that, in the eyes of the physicist, are both obvious and basic. One of them is the fact that, notwithstanding the changes that their interpretations underwent, the fundamental equations of mathematical physics have remained endowed with a considerable both heuristic and predictive value (think, for example, of Maxwell's equations). And the other one consists in that the power of physical science at predicting phenomena steadily increases. Admittedly, it is possible to discuss these two points at length, develop learned commentaries concerning them, brand truisms and so forth. But the facts cannot be denied. Irrespective of all sorts of conceptual changes, once established the equations go on yielding correct (and, quite often, extremely useful) predictions. And the domain in which we are able to make reliable predictions is an ever-increasing one. These are two facts that, personally, I take to be quite remarkable and that, it seems to me, most contemporary epistemologists did not take into due consideration.[2]

This being said, there is no question of denying the existence, in the works of the latter, of highly valuable contributions. The most illuminating one probably was the observation they made of the phenomenon Thomas Kuhn called "paradigm change." Formerly, thinkers thought scientific knowledge advanced in a continuous manner. Roughly speaking, it was believed that sciences rested on theories whose generality steadily increased, and which, as time went on, explained more and more facts, just because their laws and concepts grew ever more powerful. The episte-

---

[2] It goes without saying that this observation is not in the least at variance with the fact (point 2 of section 10-4-1) that experiments often refute theories. The theories that experiments disprove are (sometimes) descriptive, realist ones (such as the geocentric theory) or (more often) clever ideas aimed at further developing, or replacing, existing theories. However, what is at stake is then but the very short term. In such cases the experimental disproof is an inherent element of the normal progressive building up of science.

mologists of the last half-century showed that this conception was very much idealized. Basically they were historians of ideas. They gathered quite an extensive documentation and, on its basis, could rather convincingly show that, as a rule, things did not take place this way. That major advances resulted, in fact, from former theories being "defeated" by new ones, based on altogether different viewpoints. Gaston Bachelard (1949) was one of the first philosophers who brought such phenomena into light. He gave them the name "epistemological ruptures." But it was the works of thinkers such as Thomas Kuhn (1962) that made this conception quite generally known. Among the philosophers of science community its interest was then immediately acknowledged. But it is on the main "opinion makers" that it had the most important impact. In this group indeed it created a considerable stir, the consequences of which are still with us.

That it had such an important effect was partly due, of course, to its intrinsic significance. In particular, Kuhn pertinently stressed that, in periods in which new data show the "received" theory not to be really adequate, in general the said new data are yet insufficient for establishing the validity of an alternative conception. So that, the new theory must originate, at least in part, from other, less objective sources and, in debates, its upholders have therefore to rely very much on just persuasion. But the large audience such views received also came from the circumstance that Kuhn did not hesitate to stress and tacitly extrapolate the sociological nature of the just mentioned process as well as the fact that it implies some arbitrariness. Unfortunately, when taken at face value these two observations finally are—I claim—misleading ones

In fact, what, prima facie, makes the first one look so far-reaching (and thereby seems to undermine the reliability of science) is something akin to a shift in meaning. It is of course true that some sociological conditions are more favorable to scientific research than others. It is also obviously true that fashion—the fact that at a given place and time such and such ideas are in the air—may orient research along specific directions and divert it from other ones. But from this it does not in the least follow that the *outcomes* of well-defined scientific inquiries are themselves dependent on such factors. That, when a specific inquiry has been started and carried out with scientific methods, the end-result should depend on sociological or cultural data. Are such outcomes under such dependence or are they not? Answer is to be found by looking at facts. Unfortunately, while prominent epistemologists put forward numerous and convincing cases of sociocultural circumstances having had a bearing on the *choice* of a research theme, it is not an easy matter to find, in their writings, examples bearing, in a specific way, on the *outcomes* of given, well-defined investigations.

As for the concomitant assertion that the outcomes in question are to some extent arbitrary, the reason why some epistemologists apparently believed it to be true is, I think, to be found in the fact they were historians. For it incited them to take interest in what took place on such and such occasions, otherwise said on short-term events. Now it is quite true that, considered under this light, sciences, mainly in their most productive periods, undergo somewhat chaotic evolutions. Conflicts are bitter, prejudices are persistent, new experimental findings are not, as yet, either corroborated or disproved, new ideas seem to contradict one another, which later will often be found to express the same truth through different means, and so on. All this is quite unquestionable but it is just as undeniable that gradually everything becomes clearer. At this stage the question whether or not, in the long term, objective criteria make up the decision must therefore still be considered an open one.

A negative answer to it was, not bluntly stated but, somehow, suggested by Kuhn and was later explicitly given by philosophers on his side. It rested essentially on two remarks. One of them basically was the observation that no winner group would ever admit that its victory is no stride forward. A remark meant presumably to imply that scientific revolutions, which are always claimed to be advances, might sometimes not be genuine ones. But is this not a trifle sophistic? For the real point is this: is it true or false that the "winner" theory is the one that accounts for both the facts the former theory already accounted for *and* some additional ones the former theory failed to explain? Now, concerning this question, remarks of a purely psychological nature such as the just mentioned one are obviously irrelevant.

The other remark on which the said epistemologists' denial hung looks somewhat more formidable. It just consisted in answering the foregoing question negatively. In other words, it was the claim that when a theory gets replaced by some other, different, one, grounded on some new set of concepts, as a rule some valid elements of explanatory power thereby become ignored, hence as good as lost. If this were actually the case it would be very worrying indeed. It would confirm the validity of a thesis that the works of this whole school of thought has the effect of spreading, namely the view that there are advanced pieces of scientific knowledge, as valuable as the ones we have or perhaps even more valuable, that were irretrievably lost. But, fortunately, it is easy to see that the claim in question is erroneous. It suffices to analyze the examples that were put forward in its support. Consider for instance the one that Larry Laudan (1977) drew from seventeenth century physics. It bears on birefringence. It consists in noting that this phenomenon was well explained by Huygens' wave theory and that this theory got replaced by the corpuscular one of

Newton, which did not yield the explanation in question. Here is a clear-cut case—we might think—of loss of explanatory power.

True, with respect to short-term events Laudan is right. But this is not what is at stake. What really matters is what results on the long term. The sole significant question is whether *today* our theory of light still fails at accounting for birefringence, or yields of it an explanation poorer than the one Huygens gave. Now, to this question the answer, clearly, is "no." For indeed, the whole informative, explanatory and valid content of Huygens' theory has now been incorporated within a much wider and subtler one—quantum electrodynamics—that takes into account both the corpuscular and the wavelike features of light and, in particular, accounts for birefringence. Whoever would claim Huygens theory is "lost (or ignored) knowledge" would just simply be talking nonsense.

Admittedly, since most epistemologists reject universality they still enjoy the possibility of retreating on safer grounds. They may claim that the foregoing refutation of their views merely concerns the realm of physics, and that the said views remain cogent in other sciences. Even though, for reasons explained in chapter 6, I myself believe that science is, in a sense, universal, it is difficult for me to harry them there. For (as shown at the end of section 5-4) the type of universality I think holds good still leaves place for some quite appreciable specificity of the various sciences. I cannot therefore completely discard the idea that in some of the latter an everlasting loss of predictive power occurred. Conceivably this may have taken place within sciences that are on the borderline of technology, for the idea that some (mainly craftlike) procedures got irretrievably loss is a plausible conjecture. But the question remains open, for the examples Laudan put forward in support of his thesis are, for me, unconvincing in two respects. On the one hand, same as the above one, they concern but temporary states of affairs, and on the other hand their author tends, I think, to mix up predictive power loss and temporary disinterest for such or such types of problems.

Finally, let the weakest point in the views of followers of Kuhn and Feyerabend be recalled. Because they adhere, some of them to objectivist realism, others to constructivist views built on concepts akin to those of objectivist realism, whenever concepts get replaced this amounts, in their eyes, to a dramatic crumbling down of the theory that made a manifest use of the said concepts. The well-known claim of some of them that physics is noncumulative seems to be based on this idea. Now, it must be forcefully stressed that, at least concerning the epistemology of physics, this is a radically erroneous view. For, to repeat, we must keep in mind the crucial role equations have in the very formulation of physical theories and their remarkable stability upon changes of wordings and outward interpretations. And we must also remember the steady predictive power increase that we owe to such constructions. We shall then recognize that, in theoretical physics, going over from a set of concepts to another one often does not at all amount to destroying the theory in hand and replacing it by another one. It merely amounts to generalizing it and simultaneously changing the allegorical picture by means of which we express it in words. We shall then realize that physics *is* cumulative.

To sum up, what I personally derived from taking cognizance of these epistemologists' works was a mixture of contentment about details and skepticism—not to say more—concerning the results they get. It seems to me that they combined most pertinent and interesting pieces of information relative to methodologies, research traditions and so on, with views concerning the nature of knowledge that are simplistic, arbitrary, and denote serious misreading of the advanced parts of the sciences.[3] But anyhow, many authors have published a number of quite elaborate critical studies of the Kuhn, Feyerabend, etc., theories. Presumably, all the arguments, both pro and contra, have been developed and it goes without saying that the foregoing rapid survey is in no way aimed at taking the place of these pertinent—and long—analyses. By putting it forward I just meant to explain the reasons why I, personally, am more than reticent with regard to these views, and why, therefore, I take interest in other, totally different, approaches.

11-3 A Critical Glance at Some Claims

Although, during past decades, the Kuhnian conception raised considerable interest it did not monopolize attention. In fact, during the last half-century a variety of viewpoints appeared. There can be no question of giving here a systematic review of them all. In this section and the following one let me just, less ambitiously, put forward some critical remarks of a general character.

1. I shall, I hope, be forgiven if I stress the necessity we are in, when we read texts from philosophers (epistemologists included), of keeping our critical mind very much awake. That, in philosophy just as in any other field of study, a professional language should be used, is, of course, entirely normal. But within the said other fields what makes such professional languages useful is mainly the fact that they prevent shifts in mean-

---

[3] In particular it seems that these thinkers failed to recognize the force of the bind created within each discipline by the great laws constituting it. For example, after having read books dealing with scientific subjects philosophy students will naively ask, "who is the author of such and such formula?," without in any way realizing that the said formula is but an elementary consequence of equations expressing quite general laws.

ing from occurring. It would be excellent if the same were true in philosophy. Unfortunately we can but observe that in a number of philosophical writings, professional language, far from preventing such shifts in meaning from taking place, quite on the contrary generates them. In some cases the shift is even quite intentional. Its author apparently expects that, as in poetry, it should somehow express the ineffable. But most often it is not introduced purposely. It just arises from the writer indulging too freely in the use of ambiguous terms or having too strong a taste for mannered style, wrongly identified with keenness in analyzing. Anyhow, when reading philosophical works it always is most important to be very much on one's guard, to avoid being lured by words, and to systematically judge by oneself whether or not the text is at all meaningful. This is a counterpart to the irreplaceable discovery—which we owe to prominent philosophers—of how broad the spectrum of possible modes of apprehension of the world actually is.

2. Some epistemologists deal with the various elements of scientific knowledge in a very roundabout way, enter into great details concerning conflicts and their psychological and sociological consequences. And we finally realize that they kept quasi silent relatively to the one basic problem, the question of determining what science really deals with and what information it truly yields concerning it.

3. Jean-Jacques Rousseau's adversaries often criticized him for having arbitrarily raised to the level of a basic truth an opinion of his—the view that human beings are good by nature—and having then correctly drawn from it, especially in history (we now would say prehistory) most questionable conclusions. Unfortunately, philosophers of science sometimes behave in a Rousseau-like way: choosing an idea, making it a basic truth, and finally inferring many things from it, but forgetting in the meantime that it is only from an *idea they had* that these brilliant things get deduced. There are many examples of such a behavior. Empiricists formed the—a priori quite reasonable—hypothesis that our whole knowledge comes from our senses. Then they reasoned hard on this basis. And indeed they so widely did so that in their minds the—quite interesting—starting hypothesis progressively became unquestionable truth. Same with the positivists and their verification principle. They decided to put it forward, they explored at great length its consequences, and they came to raise it to the level of a dogma. So much so indeed that the physicists in their ranks have come to simply abhor, in a purely a priori way, the hidden variable notion. We can easily see that they take it to be irrational, whereas it is just at variance with an axiom stated by fiat. Same also with the realists who, on the contrary, make it a dogma that external reality is knowable by us, at least theoretically, as it really is, its contingent features included. And

who made so much use of this idea that they cannot even imagine it could just be an assumption.

4. By now, epistemologists have realized that relativity theory must be quite seriously taken into account. But they are very far from being appropriately on their guards against the devastating consequences quantum mechanics may have concerning the validity of some of their views. For indeed quantum theory bluntly falsifies some of the great principles a few of them choose to take as the starting points of their quest. Of course, the realist epistemologists are those who are most in danger. As we saw, some of them believe they can safely assert that any entity we can manipulate and make use of is an object endowed, in any circumstance, with individual existence. Others—or the same—state that, by definition, particles possess continuous trajectories, and so on. We may wonder whether a non-negligible part of contemporary epistemological literature is not to be set aside, just simply on the ground that the ideas it puts forward are incompatible with present-day knowledge.[4]

5. Precisely, concerning quantum mechanics we saw in chapter 4 that its objectivity is of the type we called "weak." And a sentence from Bohr quoted here in section 4-1-4 shows that this is the kind of objectivity that the founders of quantum mechanics actually had in mind. The supporters of objectivist realism are aware of the fact but they often claim that the difficulty—if it is one—is but of a historical nature. To remove it, it suffices, according to them, to observe that in the 1920s, at the time when quantum mechanics was discovered, positivism (Machian or logical) was at the top and the idea therefore prevailed that only what human beings can verify has a meaning. In their eyes, the quantum mechanics founding fathers simply went with the stream. I think it should be stressed that, grounded as it is in collective psychology, this explanation explains in fact nothing at all, precisely because it entirely rests upon social psychology. Today the "fashion of the time" has considerably changed. In the eyes of many scientists, discoveries such as the one of the genetic code, those relative to neurosciences and those made in astrophysics unquestionably restored—to some extent at least—the legitimacy of the intuitive trust

---

[4] A close relationship is to be noted between such gaps in specialized knowledge and the fact that most present-day epistemologists bluntly reject the views of their positivist predecessors. Generally speaking, the normal way in which pure philosophy develops is that new philosophical "schools" emerge that consider the philosophical reasoning of the then prevailing philosophical circles to have unsatisfactory aspects and reject the latter's standpoints just for that reason. Some epistemologists (not all!) therefore naturally tend to believe that the same may be done within epistemology. That, for example, we may switch from positivism to realism on purely philosophical grounds. But this is not so! What science proper has to say is quite crucial in such matters. And it is a fact that, by raising strong objections to realism, it indirectly supports views showing admixtures of positivism.

they, like everybody else, a priori put in the realist approach. If the psychological explanation examined here were the correct one, now that psychology has changed the physicists should have no difficulty at all at solving the quantum interpretation problem. They would simply take up the, now prevalent, realist approach and smoothly produce a new interpretation of quantum mechanics along its lines. Unfortunately, as we saw, this does not work. True, models were constructed reproducing the quantum predictions and aiming at both scientific credibility and ontological interpretability. However, as we could observe, those that are endowed with scientific credibility are not ontologically interpretable and those that are ontologically interpretable lack scientific credibility. The above-considered "historical" thesis is therefore valueless.

11-4 Physics and Linguistics

Föllesdal's definition identifying meaning with the joint product of all the evidence available to people who communicate is well known. At first sight it does not look so surprising. But upon reflection it is startling for it suggests that anything that does not hinge on communication is meaningless. That, therefore, language is prior to both our logic and our apprehension of things in general (scientific knowledge included).

Such an ascription of a central role to language explains why, throughout the twentieth century, very many philosophers, and a few physicists as well, got deeply interested in it. According, for example, to Cassirer (1923-29), language is indeed one of the main modes of imparting a meaning, for it quite decisively contributes to the production of the *symbolic forms* human beings elaborate in order—as P. Uzan (1998) put it—to have access to significance. The stress laid on such a notion as the one of "access to significance" is clearly related to the idealist thesis that, within any domain whatsoever—physics included—meaning is *the* central notion, to which all others are subordinate. And emphasis of the role language has in the process distinguishes the present approach from the one of Kant in that meaning is here taken to proceed not from innate—hence frozen— forms but from a genuine elaboration process.

This is no proper place for entering into a general discussion of these ideas. Concerning the relationships between logic and language let us simply note the pertinence of Jean Largeault's (1980) twin remarks that "thought is not a directly observable public phenomenon," and that "the logic we know in the West relates to thought merely through the medium of language." This, he explains, is why many linguists "gave up the idea that languages are just clothes that would superpose to pure thought as a form to a content," and replaced it by the view that language is what molds the very core of our thinking, that is, logic. The idea is a perplexing one however, for languages do not all, by far, have the same structure. If it were true that our logic is linked to the one of the Indo-European language (and that all other cultures superficially borrowed it) we would therefore have to do with an extreme form of relativism: the necessity of giving up the very idea of the unity or reason.

In this case, as in the one of the Kuhnian criticism of science, it presumably is, to an appreciable extent, the revolutionary aspect these views have with respect to the received ones that made them successful within the circle of educated, cultured people. On the other hand, let it be noted that, here also, the examples the linguists put forward in favor of their thesis lay themselves open to criticism. When, for instance, the latter point to the advances Aristotle's introduction of the word *potentia* made possible (e.g., by allowing for a distinction to be made between potentiality and actuality) we, at first, are impressed. But then, on thinking the matter over we come to wonder whether the cause-effect relationship is really the one from language to thought. In science, at least, it seems that it goes the other way round. Advances in research quite often make it clear that a given word is not accurate enough. That, in fact, it happens to designate two notions between which, before some data were known, it was not possible, or convenient, to distinguish. In such cases a new term is, of course, invented and the ambiguity is thereby removed. Why shouldn't Aristotle's mind have worked that way? Conceivably, he first realized the pertinence of a new notion, differing from that of actuality, and then created a new word in order to designate it. Seen under this light the invention of a new word appears to be secondary with respect to that of a new notion rather than the other way round. That a suitable language makes thinking much easier is, of course indisputable. But the phrase: "language creates thought" might well, after all, be but an "axiom à la Rousseau."

Hence we are hardly surprised when we find out—for example on reading Largeault (*loc. cit.*)—that this belief in dependence of thought on language is shared neither by all linguists nor by all philosophers. Some of the first named do discard the view that we spontaneously think by means of language. They conceive of language according to the above-described scheme, that is, as being a means of communication enabling the speaker to code a nonverbal thought and the listener to decode the message. Some of the second put forward arguments leading them to the (intuitively plausible!) conclusion that a set of formal rules corresponding to a language radically differing from ours (one, for instance, in which the sentences could no be decomposed into subjects and predicates) could not be called a logic.

Although the question is complex and somewhat abstruse, still physics may be of some help in shedding light on one of its aspects. For in it—as we saw in chapter 1—it proves necessary, not only to make use of quite unfamiliar notions but even to part with the view that the two sets of concepts corresponding respectively to objects and to predicates are totally distinct and never mix. In line with the "primacy of language" idea it might have been expected that the necessity of watering down this seemingly basic linguistic distinction should have led, if not to the failure of logic, at least to the need for a completely new one. But, in fact, it turned out that this consequence could be avoided. True, ways of formulating quantum mechanics were put forward that do involve formal systems parting with classical logic. But making use of them is in no way necessary, as is clear from the fact that most textbooks on the subject do not utilize them at all.[5] So that, indeed, some of the main promoters of these systems, such as J. M. Jauch (1973), wisely chose not to impart to them the great name "logic," and to use the less prestigious expression: "propositional calculus." This choice they motivated by stressing that even a physicist who systematically makes use of such a calculus in fact has to utilize ordinary logic when he *talks about* the latter. In other words, they pointed out that the meta-language has anyhow to obey the rules of universal logic.

Conceivably, this might constitute the embryo of an answer to the more general question raised by the linguists and philosophers whose views we are considering. The idea would be to distinguish—within what we call "logic" or "logics"—on the one hand what has to do with the basic structures of thought and on the other hand the combinatory thinking rules that concern specific physical or mental process. We might then grant that these rules largely emanate from human beings and their—evolving—aptitudes at conceiving and describing things and effects. But we would maintain that a basic logic—that of metalanguage—exists, which *is* universal. And to our opponents we would simply point out that they themselves cannot avoid using the latter logic whenever they strive to formulate their own thesis. In this spirit we might even go as far as to grant to B. L. Whorf (1956) that we carve nature along the lines our mother tongue laid out, but we would then make it clear that what thus depends on language is merely the set of the above-mentioned combinatory rules. And that therefore, between our customary ways of inferring and those some Amerindian tribes use, or quantum propositional calculus implies, the differences merely concern the rules in question. That, as far as metalanguage is concerned, but *one* logic makes it possible to use it.

Indirectly we would then come close to Bohr and his "basic truth" (for indeed, in his opinion it was one!) that everyday language (suitably refined

---

[5] We already noted that, contrary to what some expected, these systems did not make it possible to build up a realist theory obeying local causality and compatible with the observed facts.

by admixing technical words) finally is the only nonambiguous means of communication that we have at our disposal. For indeed, if everyday language really is the ultimate, unavoidable, hence universal reference, this obviously implies that the ordinary logic on which it is construed also is endowed with the character of being *the* ultimate resort, hence of being universal. To this argument it might conceivably be objected that everyday language merely applies to the macroscopic objects (and more generally to classical physics), in other words, to a description of what we called "empirical reality," and that this *limits* the domain of validity of basic logic. But the answer to this is simple. If basic logic so nicely applies to the description of such objects, it might well be because, for the purpose of organizing their experience, our remote ancestors progressively constructed the concepts of space, locality, objects and so on—in short all those that compose our (human-centered) empirical reality—according to the very rules of this logic. Within such a conception, in the triplet "logic, language, empirical reality" (basic) logic would have the fundamental role. Contrary to the claims of many philosophers of language it would be prior to language and contrary to the views of objectivist realists it would be prior, not, of course, to mind-independent reality but to empirical reality, that is, to what *they* call reality. This would account for the fact that for studying phenomena not belonging to the macroscopic realm the most appropriate method consists, not in creating new logics (quantum or others) but in making use of predictive rules formulated within the framework of conventional, universal logic.

11-5 Sociologism

The picture of science Kuhn's followers made popular generated a conception that may be called sociologism. What I mean by this is a tendency to consider that the various pieces of scientific knowledge we dispose of have no intrinsic significance. That (as has been claimed) all of them must be thought of as being imbedded in anthropological situations that are constituent elements of them. This veritable sociological *creed*—that, in this or other forms, appears in the writings of many "post-something" thinkers—is, in my opinion, a subtle blending of verity and error and, on the whole, a caricature of truth. This hybrid nature it has gives it its persuasive force and thereby makes it redoubtable.

   The (small) part of truth it contains consists, as we already know, in the fact that nowadays not only philosophical reasons but also arguments most firmly grounded in physics have rendered objectivist realism practically indefensible. True, available data do not forbid the physicists who take realism to be a genuine demand of reason to try and justify their position by producing an ontologically interpretable theory endowed with some degree of scientific cogency. But such endeavors are, as we found out, unconvincing and anyhow no contemporary "realist" physicist can derive from his scientific knowledge positive arguments enabling him to really convince others that realism is the correct philosophical approach. It is on *this* point, if only they put it forward, that the anthropocentric reference of the adepts of sociologism would be pertinent. But what they claim is altogether different.

Let us reconsider their views. True, some of their statements are so general that they might conceivably be interpreted along the just sketched lines. But others are precise. This is, for instance, the case of the above mentioned claim that all scientific assessments must be thought of as imbedded in anthropological situations that are constituent elements of them. The words "anthropological situations" (note the plural) cannot be understood otherwise than as referring to *contingent* situations. To circumstances that hold true at some places and times and not at others. Which, as soon as we leave the realm of broad, general views, leads to disconcerting questions. Should such assertions be understood to mean that scientific statements of any sort, including those that predict what is to be observed under specified conditions, may well be true at some places and false elsewhere? That the impossibility of building perpetual motion machines or the one of sending orders at superluminal velocity are sociocultural facts, akin in nature to fashions? That, had Lorentz, Poincaré, and Einstein been differently brought up, we might be able to send at distance (to space probes, for example) immediately effective orders? That if, during the twenties, the German and Danish societies had been culturally different from what they actually were we could violate Heisenberg's indeterminacy relationships? Since these are sheer absurdities, if the assertions in question do have such meanings their pertinence is, to say the least, questionable! Unfortunately, in view of their wording it is quite difficult, nay, it seems almost impossible to attribute to them any more reasonable sense that would clear them from such implications.

Luckily, sociologism practically did not affect the scientific circles themselves.[6] But it worked havoc in the humanities world and, through this channel, its thesis became broadly known and quite popular. In circum-

---

[6] True, a cursory reading of Heisenberg's *Ordnung der Wirklichkeit* (Heisenberg 1989) might give the opposite feeling. For indeed its author went so far as to claim that, at one time, the Olympian gods *truly* ruled the Greek world. It should, however, be observed that in the quoted book a sociological interpretation of science is nowhere proposed. Admittedly Heisenberg refused to split our diverse "patterns of reality" into subjective and objective ones; into "truths" and "appearances." But still according to him the succession in time of those of the said patterns that were linked with science was due to causes that were purely internal to knowledge.

stances such as this one, hoping to bring public opinion back to sounder views by mere reasoning is quite hopeless. Other channels are to be used. Fortunately, the physicist Alan Sokal realized this. The well-known trap (Sokal 1996) he set for the sociologists was in no way an unfair one and, had they been serious, they automatically would have escaped it. That they did not hence casts crude light on the ludicrous character of some of the assertions to which sociologism may lead and, more generally, on the lack of critical mind it generates. Since then sociologism is somewhat less the fashion. Unfortunately Sokal himself drifted to the other extreme, the one that consists in bluntly rejecting any view parting from physical realism. We already know the considerable difficulties this approach leads to. They were extensively analyzed in chapter 9.

11-6 The End of Certainties?

Nowadays, in papers and books, the phrase "the end of certainties" often appears. Does it adequately characterize the present-day state of affairs? I do not think so. For what do we actually mean by the word "certainty"? What pieces of knowledge do we consider should be sure ones? In Leibniz' and even Kant's times determinism was taken to be one of the basic "possibility conditions" of science. And indeed some people still think this is the case. According to them, outside the deterministic framework no scientific certainty is at all conceivable, so that, in their eyes, quantum indeterminism just simply rings the knell of certainty. However, the fact that the golden age of physics coincided with the advent of quantum mechanics clearly shows that determinism is not a necessary condition for the appearance of new pieces of knowledge to which the epithet "certain" unquestionably fits. And a similar remark obviously applies to the views of people who ground the "end of certainties" on the discovery of chaos. For others, to whom indeterminism is not a problem, no certainty is conceivable outside the framework of objectivist or at least physical realism. Also they, once they have taken cognizance of the foregoing chapters, will have to consider that certainty is lost, since physical realism is now greatly in trouble. In my opinion, however, this is unjustified pessimism for, as we repeatedly noted, once they have been duly established within some domain, our observational predictive laws never are refuted on any significant point within this domain. They work all right, irrespective of the changes their interpretations may undergo. We truly know that under such and such conditions we shall observe this or that. Or we know what the probability is that we shall observe this or that, and then we also know that, by repeating many times the operation, this probability will be experimentally confirmed. Isn't this "certain" knowledge? For my part I answer it is. I do not therefore join the choir of thinkers who announce, at present, the end of certainties.

On the other hand, it goes without saying that I agree with those who are aware that familiar concepts are deceptive and stress, for this reason, the vanishing of *illusively simple* certainties.

CHAPTER 12

MATERIALISMS

§

12-1 Introduction

A S ALREADY NOTED, drawing a precise dividing line between the philosopher's and the physicist's respective domains has by now become awkward. This, we saw, is due to the fact that the old rule: "to the philosopher the study of the nature of things, to the physicist that of their behavior" progressively became inapplicable. The questions concerning the behavior of light, which in the late nineteenth century became crucial ones for physics, vividly illustrate this evolution, for it seemed practically impossible to keep them totally immune from any "ontological" consideration. Admittedly it might be maintained that, in a sense, Poincaré did so, just by asserting that the question whether ether exists or not is merely one of convention (Poincaré 1902). But note that this very statement of his amounted to taking up a standpoint that partook very much of philosophy. And, of course, another clear-cut example of the evolution we are here considering is the fact that, during the same period, also the investigations concerning the behavior of matter quite naturally led the physicists to ask themselves questions about the nature of the latter.

This state of affairs most presumably contributed to a renewal of the view (which had been the prevailing one during the previous centuries) that, after all, science might possibly convey some amount of information of an ontological nature. It is therefore likely that it played its part in the growth of materialisms. The word "materialism" is here written in the plural for this philosophy took and still takes, as we shall see, different forms. Below, three of them will be considered: dialectical materialism, scientific materialism and a third one, for which the name "neomaterialism" seems adequate.

12-2 Dialectical Materialism

At this place it would of course be unthinkable not to mention dialectical materialism. But I hope I shall be forgiven if I do not produce anything resembling a critical analysis of its content. For various (and not all equally pertinent) reasons this doctrine is no more considered, by now, as a very inspiring one. However, the true reason why we shall not expatiate on it here is not its present discredit but just the fact that, for about a century, it was made the subject of so many books that it would just be impossible to deal with it in a few pages. Let it merely be recalled here that while, as already noted, some relationship may well be detected between Bohr's thought and dialectics in general, it would be quite illusory (although some tried!) to strive to link the said thought to dialectical *materialism*. The reason simply is that the Bohrian approach is crucially human centered, so that calling it materialism would imply dramatically—and arbitrarily—changing the meaning of the latter word.

12-3 The So-Called "Scientific" Materialism

The version of materialism that is here labeled "scientific" (just in order to distinguish it from other ones) is a venerable doctrine, much older than science itself since, in its "atomic" version, it goes back at least to Leucippus and Democritus, that is, to the sixth and fifth centuries B.C. Great minds defended it. One of its distinctive features is that it is a genuine philosophy, that is, a world-view that lays claim to universality. This is why, for instance, Cartesian mechanism is no materialism: in Descartes' view it bore exclusively on matter, not on thought. The same remark approximately applies to Galileo's and Newton's views and, more generally, to those of most seventeenth-century scientists, even though they were, on this point, less explicit than Descartes was. Apparently, it was not before the eighteenth century that the (presently widespread) idea began to dawn according to which thought might conceivably be included within a materialist approach.

"Materialism"—or "mechanism"—does not necessarily means "atomism." As we know, Descartes himself did not believe in the atomic theory. And in fact ancient atomic theory, the one of Democritus and Epicurus, underwent an eclipse of a surprisingly long duration—almost two thousand years!—before it reappeared, first episodically with Gassendi in the seventeenth century and then in an objectively grounded way at the beginning of the nineteenth century with Dalton's works. Nevertheless, science never took atomism to be of the nature of a universal principle. Even at the time of its apex, about at the end of the same century, scientists, materialists and nonmaterialists alike, set the concepts of ether and fields outside its realm.

This being so, it is all the same true that nowadays, in most people's mind, the word "materialism" calls forth a predominantly atomic view of nature. More precisely, it evokes a version of *objectivist*, nay even of *near* realism, whose central notion is derived—through abstraction and simplification—from our experience of solids. The philosophical atom—in other words, the particle—is, in it, conceived of by means of a thought extrapolation process. Implicitly we think of a sequence of smaller and smaller sand grains or dust speck, and imagine the limiting case of an infinitely small, but still finitely massive, such grain or speck. Such corpuscles are thought to interact through forces and, in view of the success of the Newtonian theory, it became intuitive to think that not all of them are contact forces. Nowadays it is commonly granted that some do act at a distance, although with an intensity that decreases when distance increases. And it seems that, in the general public they are most commonly conceived of as being emanations from the particles. As a rule, the whole world—our brains included—is, in the said public, taken to be composed of such "atoms," and the phenomena taking place in it—including thought—are considered to merely follow from the existence of these atoms and the forces that connect them.

Admittedly, such a world-view is grossly overschematic. Calling it "scientific" would be entirely inadequate, unless it were specified that this epithet refers, not at all to what a genuine scientist would say but merely to the ideas that, due to the oversimplifications they read or hear, people in the street[1] commonly form concerning science. This being said, it must be granted that, schematic as it is, in some domains such a scientific materialism did provide scientists with a conceptual framework in which highly complex and most fruitful disciplines could grow. This for example, is the case concerning molecular biology, which essentially developed on that basis (indeed it could, for a simple reason sketched here in section 1-1). So that we may well understand (without endorsing it of course) the attitude of the still numerous scientists who bona fide consider materialism, taken along the above sketched lines (with the refinements the complexity of nature makes necessary), to be the necessary framework of scientific research in any domain.

A fortiori, we may easily grasp that the successes of such a materialist-atomistic world-view entailed its general spreading. To the extent indeed that within well-developed societies it has now become the instinctive, received "ontology." I mean that now it practically constitutes our almost

---

[1] But, unfortunately, not they alone. In fact, it seems that among contemporary thinkers—and those, in particular, who call themselves "physicalists"— many would endorse J. R. Searles' naive claim (1995) (reported by M. Bitbol [2000]) that we live in a world entirely composed of physical particles embedded in force fields. In view of the information reported on in the first part of this book, such a conception is obviously obsolete and most seriously misleading, and the fact that prominent philosophers still argue on its basis may indeed be considered worrying. (To some extent they may, however, be excused for doing so in view of the fabricated ontology that some physicists believe in, as we saw, and disseminate.)

automatic way of thinking of anything whatsoever, be it a natural object, a machine, a living being, its brain, and hence, transitively, the thoughts that go with the latter and that we take to be produced by it. But, however psychologically justified this process may have been, its end result finally is erroneous. We already noted the scientific deficiency of philosophical atomism (see section 3-3-7). We know that an atomistic materialism reducing the whole world to a set of atoms, particles and so on, interacting through distance-decreasing forces, is an experimentally disproved conception. However great the attraction is that it may exert on some minds, such materialism is just simply false.

Admittedly, one might hope to salvage it by somewhat modifying it, and some authors strive to do so. Roger Penrose, for example (who describes himself more as a physicalist than a materialist) keeps at a distance both from atomism and from standard quantum theory. As we already know, he (Penrose 1995) chose to trust a conception grounded on wave functions and, within the latter, a theory that slightly alters the Schrödinger equation (by taking conjectured gravitation effects into account). As for Bricmont (1994) and Sokal (Sokal and Bricmont 1997), they made use of the Louis de Broglie (1928) and David Bohm (1952) model we already know of, which, to be sure, outwardly has a strong similarity with atomism. However, as we also know, the similarity in question is, to a great extent, misleading since, along with corpuscles, the model involves either a nonlocal entity or distance-nondecreasing forces or both. In fact, such conceptions are indeed possibilities but marginal ones. They do preserve strong objectivity but the first one is presently considered to be a research program more than a full-fledged theory and the second one has to face the above-mentioned difficulties regarding its matching with relativity. Making use of either one for the purpose of rescuing materialism amounts to granting that the said materialism cannot any more be presented as it once was, that is, as the most reasonable world-view brought about by objective knowledge and thus somehow forcing itself upon any enlightened mind. And of course, needless to recall here that if a materialist scientist chose the alternative way of trying to find support for his views in the leading ideas of *standard* quantum mechanics, without adjunction or modification, he would immediately be blocked. For, as we saw, within this framework the notions that make up the conceptual basis of materialism, those of atoms, particles etc. can be but elements of a description of *empirical* or *epistemological* reality, that is, elements of a "carving up" of reality that *we* operate, according to *our* mind structure. But then the thesis that mind is a mere "epiphenomenon"—that thought emanates from brains that are themselves composed of atoms—obviously becomes inconsistent. Within it, the objects that are supposed to explain thought turn out to only exist relatively to thought!

Anxious not to condemn materialism unduly expeditiously, a number of scientists set themselves on the looking for arguments susceptible of salvaging it. They explored several ones that might conceivably be hoped to work. One is, of course, the general reasoning we already commented on above, grounded on the fact that quantum mechanics took shape at a time when the theories of knowledge most in favor were of a positivist type, stressing the basic role of perception and the meaninglessness of such "metaphysical" views as transcendental realism. As we remember, the argument in question consisted in claiming that this concomitance was the reason why, from the beginning, quantum mechanics was stated in terms of measurement outcomes rather than properties of mind-independent reality. Admittedly this reasoning is based on a true fact, but we noted it does not go through. We observed that, in order to get an adequate view of the question in hand it is, in fact, necessary to go into the analyses dealt with within the first part of the present book. That, in other terms, one has to investigate whether or not, among the ontologically interpretable theories reproducing the quantum mechanical predictions there are any that are really satisfactory. And that the answer is known. Ontologically interpretable theories do exist (and indeed their mere existence even makes it easier to analyze some conceptual problems having to do with measurement). But, besides drastically parting with atomism, such theories are not endowed with a real convincing power since they all meet with quite serious difficulties.

A different and somewhat more significant promaterialism argument (Bitbol 1998) rests on the notion of *value*, itself closely linked to the one of *traditions of research* that Larry Laudan (1977) put forward. Both are rooted in the simple truth that research presupposes a desire to know. As a rule, the latter has its source in some primeval intuitions that, for our mind, count as values. Values that it aims at clarifying, that, by the same token, it strives to justify, and that, finally, it sometimes finds itself forced, to its regret, to modify. What follows from this is a decisive role played, in research, by the said traditions, concerning which it must be observed that, admittedly, they evolve, but slowly enough for enabling beginners in science to find in them the conceptual supports they need. In Laudan's eyes, such traditions are neither research programs nor definite theories but rather tissues of theories and commitments of various kinds, both methodological *and ontological*. Many have a long history behind them. Laudan produced several examples, most of them taken from the human sciences (Marxism and capitalism in economics, behaviorism and Freudism in psychology etc.). But he also mentioned Darwinism, mechanism, and so on. He stressed—in his capacity as a historian—that every evolving tradition of research includes a whole set of, sometime competing and mutually inconsistent, theories. And that, contrary to the theories they involve, these traditions neither explain nor predict anything, and are not directly testable. Rather, he claimed, they are sets of quite general assumptions bearing on the entities and processes of the considered domain as well as on the methods best suited to the analysis of the corresponding problems and the elaboration of theories concerning them.

It is true that in chapter 11, on the issue whether or not significant elements of scientific knowledge sometimes get lost, we found Laudan's approach not to be fully convincing. But we have here to do with quite another set of questions, of an essentially historical and psychological nature, and I must say that, in this field, Laudan's "traditions of research" notion seems to me to be a most significant one indeed. Following Michel Bitbol we may well call these traditions "values" for, more than calculations of any kind—and more, even, than any "disembodied" rationality—, they are what, in the eyes of the scientist, imparts a meaning, and a "backbone" so to speak, to his quest.

Now, among them we find materialism. That materialism is indeed one is clear, for it is quite easy to discover in it all the characteristics that, following Laudan, we found traditions of research have; including even the one of incorporating theories differing very much from one another. For what features do Democritus' atomism, dialectical materialism and the "neomaterialism" we shall meet with in a moment have in common? None, apparently, except just a state of mind, made up of ontological realism, a robust belief that *what is* is intelligible, and an innermost rejection of everything that their adepts brand as "spiritual dreaming." Such convictions create links of a considerable psychological force, which, nevertheless, should not prevent us from asking what their objective validity is as regards present knowledge. With respect to scientific materialism as described at the beginning of this section, Michel Bitbol (1998) put, in this connection, an interesting argument forward. He pointed out that, however significant the criticism summarized above of this conception (or similar ones) may be, still it rests of arguments that—following (Laudan 1996)—he called "ampliative." What this means is that the arguments in question enlarge the corpus of the motivations for making a choice and extend it beyond the borders of strict empiricism. Above, for example, our setting aside the Broglie-Bohm model was grounded on the conceptual difficulties this model meets with concerning interaction at a distance, particularly in connection with relativity theory. It might as well have been based on the fact that it is "uselessly complicated," in the sense that it introduces basically unobservable physical quantities, taking account of which makes calculations unduly laborious. Arguments of such a nature always played a considerable role in physics. In fact, it would be impossible to do completely without them. That the one centered on complication is merely ampliative is of course clear, but this is also true regarding the one that refers to the "relativistic troubles." For, after all, nothing compels us to leave the (Galilean or Einsteinian) relativity principle untouched if we dispose of a theory such that experimentally *everything takes place as if* this principle were satisfied (or if we have some substantial hope of finding such a theory). Bitbol pointed out that, as a consequence, some researchers may choose to perpetuate, in spite of everything, an ontologically interpretable model such as the Broglie-Bohm one, in the name of meta-scientific values. He is right except that, even though traditions of research (and the values they carry) are more stable than the descriptive theories that, within them, succeed to one another, still there is no guarantee that they are to last for eternity.

To this it should be added that salvaging objectivist realism is not the same thing as salvaging *atomistic* realism. When Bitbol pointed out that the Broglie-Bohm model can be rescued (if "the price for that is paid") he more or less seemed to imply that the atomic theory of matter would thereby be saved. Now, these are, in fact, two different questions. As we noted, the said model resorts to the notion of myriad localized corpuscles. But it also involves the one of a wave function (or, with Bohm and Hiley [1993] of a "quantum potential") that is just as real as the latter and is not at all localized. Moreover, the need of taking high-energy phenomena such as (at least apparent) particle creation into account seems to imply, as we saw, that the idea the bosons are genuine particles should be given up and suggests that the same might be true concerning fermions as well. Finally, let it be noted that the physicists who most thoroughly investigated the model—Bohm himself foremost—discarded the idea of identifying it with any version of materialism. True, these physicists stressed that the model makes it possible to understand physics without referring—even implicitly—to consciousness, and that consequently, as far as physics is concerned, it is appropriate not to introduce the latter into the picture. But they considered that the changes quantum physics introduced into our knowledge of matter force us to give up all the arguments that, in classical physics, naturally led to discarding any attribution of a "mental pole" to what the said physics described. On the contrary, they claimed that these changes speak for the view that such a pole exists (Bohm and Hiley 1993). All this shows that the sophistication we may feel tempted to introduce within atomic materialism to make it consistent with contemporary physics, not only makes its "atomism" vanish but may even change it into a conception not having much in common with materialism any more.

Were a conclusion needed, it might be that, when Descartes put forward his two leading ideas, a clear splitting between matter and mind and mechanism concerning matter, he could not imagine the more-than-two-centuries-long error to which he thereby gave the "starting flick." During this time his followers respectfully—much too respectfully!—kept to the second idea—and progressively came to reject the first one. The end result was universal multitudinism and reduction of all being to machines or—at best!—computers. Today, we know that with the idea of the unity of Substance, Spinoza, on the whole, was better inspired.

12-4 "Neomaterialism" and Physics

Dialectical materialism and scientific materialism are not the only doctrines to which the label "materialist" may be attached. Another one exists, especially investigated by the philosopher André Comte-Sponville and that, for future reference, is to be called here *neomaterialism*. Its foremost feature is that it gives up most of the guiding ideas and concepts that constitute the main supports of scientific materialism, including several of those that—like many others—I, for one, took to be essential elements of any type of materialism whatsoever. For example, in some articles and books I pointed out that, today, the word "matter" usually evokes such properties as permanence, solidity, impenetrability and so on, and that it readily serves as a kind of intuitive explanation of them. And I stressed that, in view of contemporary physics, such an "explanation" is misleading since, even though the said properties are indeed quantitatively and quite satisfactorily explained by physics, it is not at all by referring to the idea of matter. It is by making use of formulas and mathematical notions such as antisymmetry viewed as primeval; and, correspondingly, *not* grounded on some more basic notion from which they would proceed. To this, Comte-Sponville answered in a most interesting footnote. "Contrary, he wrote, to what Bernard d'Espagnat keeps on claiming, in order to be material (in the philosophical sense of the word) matter needs not be 'what is conserved, what is permanent,' nor be 'what I can touch.' The notion of a non-permanent, intangible matter is by no means contradictory" (Comte-Sponville and Ferry 1998). And he noted moreover: "Bernard d'Espagnat is of course right in observing that the 'thing-ism' that could be found in the ancient atomic doctrine no more corresponds to the actual state of physics. But what does it basically change?" He then explained that the real question is not to know what the inward structure of matter is, "for example"—he wrote—"whether it is substance-like or energy-like, separable or non-separable. . . ." It is to know "whether it is of a spiritual or ideal nature . . . or of a physical nature (comparable, though of course not identical, to the experience we have, at the macroscopic level, of the bodies and forces that we call material)." And he then added that, according to him, this question *couldn't be answered by physics*. Indeed he went so far as to write: "Physics cannot even tell us whether the World exists or not. How could it tell us whether it is wholly material or not?" Similarly, somewhere else in that same book he stated, still concerning physics: "It deals, not with Being but with experience." And elsewhere: "A science never is either materialist nor idealist. But why should this fact prevent us from drawing from it a support for our opinion when we opt for one of these two standpoints?"

We shall come back to these important claims. But let us first fully realize how much the materialism Comte-Sponville conceives of differs, not only from the one of La Mettrie, Diderot, and d'Holbach but also from the above-mentioned "scientific" materialism (which still is implicitly at the root of the thinking of many researchers in hard and soft sciences). In particular, the fact that it incorporates nonseparability radically sets it apart from these traditional versions of materialism. For, as the content of the first part of this book and of the foregoing pages makes it clear, this fact renders relative all the explanations and deductions to be found at the core of arguments elaborated within the said materialist versions.

At places, for defining his conception and perhaps also for showing it is a genuine form of materialism, Comte-Sponville made use of somewhat surprising, all-of-a-piece formulas, such as: "It is the doctrine that claims there are but material beings." Let it be granted, however, that we should not overhastily criticize the apparently circular (or at least imprecise) character (a material being is what?) of such statements, which, after all are but shorthand expressions. We had better look at the passages in which Comte-Sponville gave a detailed description of his thesis. And, in particular, at the ample footnote the main lines of which have just been reported.

To study its content we have to separately analyze its various items and examine how they match one another. The order in which we do so is not what counts most. Let us start by a somewhat "formal" remark bearing on Comte-Sponville's assertion that the real question is to know whether matter is of a spiritual (or ideal) nature or of a physical nature (comparable to our experience of the macroscopic domain). What, there, makes me uncomfortable is unimportant by itself but constitutes in my view a sort of premonitory sign that shifts in meaning might well take place in what follows. It is the fact that an entity is evoked to which, before we know anything concerning its nature (for this is the very subject of the quest in hand) the author chose to give the suggestive name "matter." As for me, when, as Comte-Sponville here, I had to introduce, first in my thinking and then in my books, a yet-to-be-characterized entity, I chose to make use of no image-carrying words such as "matter," "God," or "spirit." I used the purposely neutral, nonsuggestive expression "mind-independent reality." Even then, some people (most of them idealists or related to idealism), in the eyes of whom reference is prior to existence, criticized me for having thus imparted consistence to an idea that is "primeval" to such a degree that it cannot be given a definition. I explained myself on this point (see the end of section 6-6) and it is clear that this justification also applies to Comte-Sponville's thought process. But, with respect to terminology, this does not prevent me from considering that making use, right at the start, of the word "matter" somewhat partakes of prejudgment.

Next, let us turn to the Comte-Sponville sentences reported above just after that one. And, first, to the one where he explained that the question in hand couln't be answered by physics since a science is neither material-ist nor idealist. And in which he added: "But why should this fact prevent us from drawing from it [a science] a support for our opinion when we *opt* [emphasis mine] for one of these two standpoints?"

This sentence is important in two respects. First, it involves the verb "to opt," which Comte-Sponville supplemented there with the likable phrase "to philosophize is to think further than one knows." Most materi-alists naively believe that they and they alone are the holders of rationality and hence of truth. They generally consider all other conceptions to be either chimera or intellectual aberrations. Through the quoted sentence Comte-Sponville overtly acknowledged that, quite on the contrary, mate-rialism is an *option*. This point should obviously be kept in mind. Comte-Sponville rightly rejected both scientism "which mistakes sciences for a philosophy" and positivism "which takes sciences to be sufficient." Mate-rialism, he stressed, exists only under the condition that it should avoid falling into these two pitfalls.

The second reason that makes the sentence in question significant comes from the phrase "drawing from it ['it' being 'a science'] a support" that it contains. Through this phrase Comte-Sponville made it clear that his opinion is, nevertheless, not—or at least not fully—arbitrary. And a question then comes to mind, namely, from what pieces of scientific infor-mation does Comte-Sponville actually draw a support for his opinion? In the footnote in question, as in the rest of the book, Comte-Sponville did not directly answer this question (which, of course, he knew is tricky). However, his implicit answer was clear and unambiguous. For remember that he gave himself—and gave us—a choice between two assumptions only. One of them is that matter is of a spiritual or ideal nature. And the other one that it is of a physical nature that is (according to the above quoted sentence), "comparable, though of course not identical, to the experience we have, at the macroscopic level, of the bodies and forces that we call material." Knowing that he is an adept neither of spiritualism nor of idealism it is clear that the assumption in favor of which he opted— like all other materialists—is the second one, the one that the inner nature of matter is comparable to our experience of the macroscopic domain.

True, on the part of philosophers of yore this second assumption had some degree of reasonableness. Besides, it is understandable that, even now, it should attract those among thinkers who cling most strongly to the age-old axiom that a genuinely creative philosopher should extract his "system" exclusively from his own mind, without letting himself being disturbed in any way by the whirl of discoveries. For, as we already noted, a priori it indeed looks plausible and natural that the inner structures of "matter"—in the: "ultimate reality" sense—should be qualitatively comparable to those of our experience of the macroscopic world. Let us therefore refrain from too harshly criticizing the philosophers who go on thinking along such lines, and especially those who, like Comte-Sponville, clearly make quite meritorious efforts aimed at keeping these "lines" acceptable. Nonetheless, it is true that a posteriori—however surprising it may appear—the lines in question prove to be a downright error. And this, it must be stressed, not only in view of one, given, generally accepted theory (of course I have quantum mechanics in mind) but also in view of all the alternative theories (the Broglie-Bohm one and others) that do not conflict with observed facts. Including even yet-to-be-discovered ones, for let us remember that nonseparability is theory independent and entails that ultimate reality, whatever it is, anyhow has no similarity with our experience of the macroworld. In fact, the "lines" in question are just simply at variance with the data.

In reality, as we see, the whole difficulty stems from the fact that Comte-Sponville gave himself—and gave us—a choice that is restricted to spiritualism or radical idealism on the one hand and a brand of realism inspired from our experience of the macroworld on the other hand. I think I showed that there is a third possibility, that of an "open realism" in which the notion of an ultimate reality either fully unknowable or, at best, "veiled" (and, in a sense, prior to the matter-mind splitting) is considered meaningful. And I claim that if the idea we both, Comte-Sponville and I, believe in, namely, that existence is conceptually prior to knowledge, is to be kept, open realism represents in fact, because of nonseparability, the only admissible conjecture. Besides, concerning the latter (nonseparability), in a more recent book (Comte-Sponville 1999) Comte-Sponville granted that it does seem mysterious. But, for the purpose of salvaging materialism he pointed out at that place that a mystery "cannot refute anything and, first of all, not a system of thoughts that raises its own finiteness to the level of a principle." "If"—he added—"matter is mindless, why should our mind necessarily be able to understand the whole of it?"

The least than can be said is that here we are far from the claim previously made that matter is comparable with the experience we have of the macroscopic world. Consequently, we may have the feeling that the choice is between two possibilities. Either Comte-Sponville's thinking is not fully self-consistent or (and this hypothesis is by far the most plausible one) it has recently undergone quite a significant evolution.[2]

However, let us refrain from issuing too rash judgements for the question is ticklish. To appreciate it correctly it is appropriate that we should analyze the way in which Comte-Sponville describes materialism as a philosophical position, that is, independently of any specifically scientific considerations. This is what is to be done now.

12-5 The Purely Philosophical Aspects of Neomaterialism

To fully understand what Comte-Sponville means by the word "materialism" the best is to turn to the chapter "What is Materialism?" of his book *A Philosophical Education* (*Une éducation philosophique*) (Comte-Sponville 1998). We all know that any sufficiently elaborate thought system must normally cope with a few difficulties, conceptual or otherwise, and in the said chapter Comte-Sponville gave a fair account of those that beset his own standpoint. He started by critically analyzing the way of defining materialism that consists in stating it is the thesis that "everything, including thought, is material." Comte-Sponville appropriately stressed that such a definition leaves undefined the very notion of matter (and, correspondingly, the word "material") so that, to impart a meaning to it, it is necessary to supplement it, either positively, by stating what the nature of matter is, or negatively, by setting matter over against thought. At this stage he expressed a doubt concerning the pertinence of the first of these two procedures. (It would imply, he noted, detailed elements of knowledge that materialism could dispense with for a long time and apparently still dispenses with.) He therefore chose the second one but, concerning it, he immediately noted that we cannot simultaneously, on the one hand consider that, with the sole exception of the vacuum, everything is matter including thought and, on the other hand define matter negatively, as being everything that is *not* thought.[3] This observation led him to grant that the Marxist thesis of the "primacy of matter" (otherwise said, the claim that mind is a *product* of matter) is, within the monist framework in which any materialism is, by definition, situated, logically inconsistent.

---

[2] Just like—it must be said—that of most of us. Besides, it seems clear that, at the present stage of evolution of ideas, the appearance of genuinely novel data forces all our contemporaries to evolve in this respect.

[3] For then thought would be and would not be matter. The no-contradiction principle would be violated.

At this stage Comte-Sponville did not deny that he is facing an aporia. In order to be able to go on with his quest he was therefore led to take up a rather subtle standpoint, of which some indeed would say that it verges on equilibrism. On the one hand he granted that, after all, we can hardly avoid resorting to some *positive* definition of matter, implying what he called a "gnoseologic realism." Matter, he wrote, is neither unknowable (for if it were we could not assert that it is not of a spiritual nature) nor reducible to the knowledge we have of it (for then it would be mind itself). Hence he acknowledged that materialism "does involve a theory of matter" and remains therefore "dependent . . . on the development of the natural sciences." But on the other hand he claimed that this is by no means essential, that materialism is, first of all, a theory of refusal and of fight, and that, by virtue of this, it is first of all a "theory of the mind." According to him its purpose primarily is to "explain mind by something else than mind itself, and, particularly, to account for such and such mental, cultural or psychic phenomena by means of material processes"; in order, he added, to "defeat religion, superstition and illusion."

These last lines are surprising. Not that Comte-Sponville should be criticized for having, at the start, something in mind. Who doesn't? We all know that aiming at reinforcing a thesis we believe is true, trying to defend it with the help of all the arguments we take to be correct, is one of the most powerful incentive for research and inquiry in general. But if the thesis is (for example) the illusory nature of religion, may we, for proving it, build up a theory—materialism—the validity of which we demonstrate on the basis of its usefulness in the fight against religion? Isn't this just plain circular reasoning? Isn't it tantamount to reversing the logical order, which would be first to show the thesis to be true, irrespective of its possible uses, and then possibly make use of it for a fight against some error? Were we here spiteful and ill intentioned we might go so far as to say that the argument in hand partakes of rhetoric more than research. In it we recognize Comte-Sponville's view that materialism is basically an option, or, as might also be said, a value. And indeed it comes within the scope of the rhetor's art to defend a value. What counts, for him, is the value itself, and making the latter striking is much more important than looking for objective reasons to accept it, acceptance being anyhow granted, and beyond question.

Such might well be, when all is said, Comte-Sponville's real position. Once and for all, he chose materialism. Still, it remains true that we cannot opt in favor of some standpoint before this standpoint has been defined. And if Comte-Sponville attaches importance to the *existence* of a theory of matter it is—as he himself wrote—because the latter makes it possible to define matter and thus avoid the aporia described above. But finally it doesn't much matter what the theory actually is, provided only there is one. Its existence allows for a nonempty definition of materialism and this is sufficient since Comte-Sponville has decided that, once defined, materialism will be true.

Now, what, after all, should we think of this whole thought process? There are two elements of an answer that readily come to the mind of a scientist when he considers this question.

The first one is to take notice of the unpretentiousness of the purpose actually aimed at. To enunciate a statement is always easier than to prove it, and even to make it plausible. A priori, in the text here under study the impartial reader could expect he would find arguments in favor of materialism, showing it is better than its rivals, spiritualism and idealism. And in fact, the clarity of the style, along with the intrinsic interest of the tackled subjects, entails that somebody who would rapidly peruse the document might have the impression of actually finding such arguments in it. (This is linked to the fact that when an author intelligently explains what an idea consists of, he thereby automatically makes it attractive.) But we just found out that, in fact, no such arguments are given. And moreover, we must do justice to Comte-Sponville by granting that he no-where explicitly claimed he would produce some. In other words, the content of the said chapter is literally faithful to its title, "What is Materialism?" Its author defined—as best he could—materialism; he did so in such a way that this materialism stands indeed in opposition to both spiritualism and idealism. But he nowhere produced reasons that would show the former to be more plausible than the two others. Obviously therefore the content in question cannot be rationally opposed to spiritualists or idealists, who just choose different options. Admittedly it is stated in it that the two doctrines in question are illusions to be fought against. But a statement is not an argument.

The second reflection that comes to mind is that even the unpretentious aim of just *defining* materialism is, after all not entirely reached. This is because, for the above-mentioned aporia to be really removed it does not suffice that we should dispose of any physical theory whatsoever, to which the name "theory of matter" would then be imparted. For example, a purely operationalist theory would obviously be unfit to such a purpose. As Comte-Sponville himself pointed out, what is needed is a theory in which matter is assumed described as it really is; in which, in other words, reality is taken to be knowable. For, if it is unknowable, asserting that it is matter rather than mind would obviously be meaningless. It is true that if thought is imbedded in ultimate reality—in the heart and substance of things, as Comte-Sponville puts it—there is no reason that it should be able to entirely grasp this ultimate reality. And apropos of nonseparability we saw that even the materialist Comte-Sponville considers it to be, in some respects, mysterious. It is therefore appropriate that, concerning the subject in hand, the existence of some mystery should be acknowledged and Comte-Sponville is right in doing so. But then, what is his justification for imparting to this mysterious "heart and substance of things" the name "matter"? True, ascribing names to things and concepts is arbitrary. We may therefore just *decide* that, from now on, the word "matter" is to be made use of for designating the said heart and substance of things, which largely escapes us. But then, to repeat, what sense does it have to assert that the heart and substance of things is matter and merely matter? Is not such a statement indisputably a tautology? Similarly, the statement "mind emanates from matter" is then to be translated as "mind emanates from the heart and substance of things." But if the heart and substance of things is unknown such a statement just means that the human mind has an origin, and such a statement is vague enough to be compatible with the most extreme variety of philosophical and religious standpoints. Indeed, as already pointed out, it seems that the only way Comte-Sponville could possibly have removed the difficulties he had to cope with would have been for him to take up my *veiled reality* conception. That, in other words, it would have been to grant that the heart and substance of things he referred to could be labeled neither "material" nor "spiritual" for it is conceptually prior to the splitting between matter and mind.

As a conclusion, let me note that Comte-Sponville and I stand in opposition to one another in some respects and agree in others. We stand in opposition, first of all in that I take some of his firm beliefs to merely be preconceptions. Not that I favor an "imperialism of reason" that would crush impressions, feeling, and ethics. But still I consider that, particularly when the question is whether some conceptions are to be rejected, inner conviction is not enough. Reason has something to say. Hence, unlike Comte-Sponville I certainly would not state that spiritualism and idealism are illusions that must be fought against without having, at least, produced some arguments in favor of my assertion. Correlatively, to give me a liking for materialism, producing a faultless definition of it would not be sufficient. It would be necessary that I should moreover be shown— by means of up-to-date information—that it rationally overrides spiritualism, idealism and particularly the veiled reality conception. Up to now indications along these lines are missing.

Another point concerning which Comte-Sponville and I have conflicting views has to do with the fact that, for the reasons we saw, I do not consider his essentially philosophical definition of materialism to be self-consistent. I would more easily grasp the content of a definition that would explicitly be grounded on an ontologically interpretable physical theory, provided of course that the latter be compatible with present-day knowledge. This last condition obviously implies that, to repeat, Comte-

Sponville's referring to our experience of the macroworld has to be rejected. In fact, what is needed here is a theory that, like the Broglie-Bohm model—would aim at a genuine unveiling of Being, while remaining consistent with observed facts. In the first part of this book we discovered the strange features all such theories necessarily have, and we took note of the reservation they therefore arouse in the mind of a great many other physicists. If, nevertheless, I were to become a mechanist (an unlikely hypothesis!) it could only be under the effect of such a theory, only freed from the deficiencies all the existing ones actually show. Finally, a third difference is that Comte-Sponville is extremely keen on making mind secondary to matter as he conceives it. I, on the contrary, consider that, once (by making use of ampliative arguments[4]) the ontologically interpretable theories have been discarded, the data described in the first part make such a thesis inconsistent, for, to repeat, they show the microobjects—particles, etc.—have existence and well-defined attributes only relative to knowledge, and hence to the mind. Indeed they show that such entities are but elements of an *empirical* reality.[5]

On the other hand, Comte-Sponville's and my own standpoints are rather close in that we both take the notion "existence" to be prior to any others, and in particular to the notion "knowledge." In other words we both consider that to be knowable by human beings is not a necessary condition for existence. This is indeed a basic point. In one of his books the philosopher Ferdinand Alquié (1950) used the expression "nostalgia of Being" for referring to the disfavor the notion of Being progressively felt in within the philosopher's community and the indefinable impression such an evolution may well induce in our mind. But this word "nostalgia" evokes ideas of loss and regret. Neither Comte-Sponville nor I acknowledge a loss, nor therefore have any regret. For this reason I would feel somewhat close to Comte-Sponville's neomaterialism if only it were improved in three respects. First a distinction should clearly be made between the two notions, "independent reality" and "empirical reality," and the same word "matter" should not be used for referring to either one of them. Second, mind should explicitly be considered to proceed from the first rather than from the second one. And third, for designating the thus defined conception the word "materialism" should, correspondingly, be given up.

---

[4] Ampliative but quite strong nevertheless; see chapter 9.

[5] Hence the same holds true concerning their compounds—that is to say, macroscopic objects—unless we accept to consider that the number of degrees of freedom of the latter is infinite and rest content with just the "realism of signification" introduced in section 9-7.

12-6 Materialism and Wisdom

Of the three just suggested "improvements" the one that the materialists will find the least acceptable is presumably the last one. From a purely logical point of view this may look paradoxical since it is the only one that is purely verbal and merely a matter of convention. But this appearance of a paradox vanishes when we remember that materialism is a Laudan-like "tradition of research" and that, in this author's eyes, such traditions are neither explanatory nor predictive, nor directly testable. That they may even include mutually contradictory theories. In a sense, they are flags, and nobody willingly gives up his flag.

On the other hand, they are far from being exclusively that. In a recent book Michel Bitbol (1998) developed wise views that do shed light on the present state of affairs. It is a fact, he noted, that notwithstanding our knowledge of quantum mechanics we, physicists, cannot refrain from making use of the formal notions of particles and their properties, even though it is "at the price of adulterating them far more than what would have been considered acceptable in the classical physics era." In fact we absolutely must keep these notions because we are quite unable to act, experiment, imagine new theoretical links and so on, within a world of pure equations merely referring to our observational predictions. We ascribe to such words as "particles," etc., novel meanings because we are aware of a *historical* continuity between classical and contemporary physics and because it is only by means of such an extension that we can, without logical inconsistency, consider the latter to be a continuation of the former. In the so-called "particle theory" domain the appearance, mentioned here in chapter 2 and analyzed in chapter 9, of a *fabricated ontology* is a vivid illustration of this psychological process. Now, broadly speaking such observations, which hold concerning atomic theory, are also valid concerning physical realism. We may therefore carry over to materialism the set of conclusions put forward by Bitbol relatively to the atomic world view. This means that, after all, the natural tendency of some scientists to set themselves, in thought, within a materialist perspective may be considered acceptable, nay even fruitful. And those among them who feel unable to conceive that rationality can be exerted independently of a materialist world-view should be encouraged to keep to the latter. The only request that should be addressed to them is that they should refrain from converting their choice into an illegitimate doctrinal creed.

CHAPTER 13

SUGGESTIONS FROM KANTISM

§§

13-1 Introduction

WE SAW THAT, on the whole, present-day physicists are doubtful concerning the way their science is to be interpreted. They wonder what it is, after all, "trying to tell us." Under such circumstances they would, clearly, make a mistake if they bluntly brushed aside the idea of having a look at philosophy. As Whitehead once noted "Each . . . science, in tracing its ideas backward to their basic notions stops at a half-way house. It finds a resting place amid notions which for its immediate purposes and for its immediate methods it need not analyze any further" (Whitehead 1933). However, in the course of years purposes and methods evolve and new, unfamiliar notions sometimes become necessary. Philosophy may then help.

Moreover, the physicists in question would also be wrong if they ignored the fact that philosophy has a history behind it. Science also has one, of course. But, much more than science, philosophy is, so to speak, co-extensive to its own history. In philosophy, to try making present-day ideas known without in any way referring to past ones would be very much counterproductive. And considering the whole information we reviewed it is clear that the Kantian standpoint, ancient as it is, still is likely to shed light on the problems we met, since it disconnected science from reality-per-se—which we acknowledged is hardly reachable—and assigned to it the study of phenomena. From comparing our factual conclusions with Kantism we may therefore hope to derive some new insights. But for this move to be effective we should first form a well-balanced evaluation, compatible with present-day data, of Kant's philosophical views concerning our relationship with the world. Such is the purpose of the next section.

13-2 A Look at Kantism

As shown by the content of the chapter of the *Critique of Pure Reason* devoted to the antinomies of pure reason, the objective Kant had foremost in mind was in no way to rule out the reality-per-se notion (that is, "independent reality," alias "the Real" in our language). If such had been his purpose, he would not have kept—nay even justified—the "thing-in-itself" concept. His intention—which we should unreservedly approve of—essentially was to define the domain in which the concepts of understanding may legitimately be made use of for building up genuine knowledge. This is why he repeatedly (and critically) mentioned the "presumption" of pure reason, which, he wrote, ignores its true objective and prides itself on its own penetration and knowledge at places where there is neither penetration nor knowledge any more. And correlatively this is also why he approved of the empiricists inasmuch, he wrote, as the latter's guideline is just a maxim recommending moderation in our claims and caution in our assertions, and instructing us to widen our understanding with the help of our only reliable master, experience. And (a point the reader is begged to remember) he added that when taking up this standpoint we should not be forbidden to indulge, for our practical benefit, into some speculations and accept some beliefs. But, he commented, with the proviso that the latter should not be presented under the pompous label of scientific or rational views (*Critique of Pure Reason*, II, 2, III). Below we shall meet again with this idea, to which my own views somewhat resemble.

That much being said, the starting point of Kant's and of all idealists' critical analyses is known. It is often described as follows (Luc Ferry 1987). When we look at any object, say a table, we are aware that we have, within us, a representation of the table. But we never quite stop at that. Beyond this representation we think of the table itself, that is, we think of it as an object lying outside us. Along with—and after—a number of thinkers (Descartes in particular) Kant wondered about the relationship between our representation and the object itself. Is our representation adequate to the object? And he stressed the obvious: to this question it is impossible to give a purely factual answer, exclusively based on observation. To be able to compare the object-in-itself with its representation it would be necessary to separately and directly dispose of both, whereas we only have access to the second.

Presented this way the argument is, as we all know, irrefutable. There exists no way of making certain that the said adequacy holds. However, taken in isolation the argument does not *demonstrate* that we have no reason whatsoever of conjecturing our representation to somehow be an approximate description of the object. A fortiori it does not prove that the elements that serve to compose our representation have *nothing* in common with the thing-in-itself. That, in Kant's wording, they are totally a priori.

As we see, the question of the a priori arises here. It is an important theme concerning which, in fact, it seems today quite difficult to entirely agree with Kant. Kant indeed claimed he could demonstrate—by means of a purely philosophical reasoning—that, for example, the concept of space is a priori. That it is in no way an empirical one derived from our experience of some "external space" existing quite independently of us. And, as we shall see in a moment, this claim played an extremely important role in his reasoning. His argument went as follows. For me to be able—he pointed out—to refer my sensations to something external to me (that is, something situated at some place other than where I am); and, correlatively, for me to be able to picture to myself the things as being "outside there," and lying next to one another, and hence as being, not only different but also situated at different places, it is necessary that I should already have some idea of space. Hence this idea cannot be experimentally derived from observed relationships between external phenomena.

Admittedly the logic of the argument is impressive and looks convincing. But is it really? When reading the foregoing argument it is impossible to completely brush aside the trivial remark that a similarly constructed one would prove it is impossible to learn how to swim, or ride on a bicycle. That only the people, if any, who happen to have an inborn, a priori, aptitude at riding on a bicycle will ever be able to go cycling. For if somebody does not have it how could such external objects as the bicycle, the road or even other people's advices ever impart it to him? In the example the loophole in the argument (and loophole there is, for experience shows it *is* possible to learn how to ride a bicycle) obviously consists in its ignoring the power we have of learning by apprenticeship. Apparently neural systems are able to "take notice" of which associations of small gestures were useful and which ones were prejudicial, and progressively favor the first ones. We, who have good reasons (which, admittedly, Kant had not!) for believing in the evolution of species, may we logically discard the assumption that a "real space" always existed, and that, within living beings, a similar process of trial and error progressively gave rise to the—useful—*notion* of space? As for me, I think we cannot.

To the foregoing it might conceivably be retorted that at least we have the innate power of apprenticeship, so that, if not space itself, still the power of apprenticeship concerning space is genuinely a priori. However, this is not what Kant wrote. And moreover, under the (presently objectionable but quite compatible with eighteenth-century physics) assumption that both space and time exist "in themselves" one grasps neither the necessity nor even just the rationale of such a referring to an *a priori* power of apprenticeship. For why should not life (conceived of, correlatively to this assumption, as having gradually developed within "real" space) have progressively gained—through a Darwinian process and along with others abilities—the ability at learning? Still within the realm of this assumption it seems clear that, in proportion as consciousness developed within the more advanced species, it augmented its abilities at apprenticeship and thereby gained a clearer notion of spatiality.

Besides, when we try to discriminate between what Kant did prove and what he did not we must remember that this philosopher explicitly aimed at *certainty*, and that this made him reject any hypothesis that could not be changed into sure knowledge, irrespective of its high or low plausibility. Several passages of *Transcendental Aesthetics* make this clear. In particular the one where Kant stressed that since the specific conditions of sensitivity are conditions not of the possibility of things but only of their manifestation, we may well claim space contains all the things susceptible of externally appearing to us, but not all the things-in-themselves, be they perceived or not and independently of who perceives them. If "to claim" is taken in the sense "to affirm" this observation is quite right. In its light we better realize that quite strictly speaking a pupil of, say, Laplace could not, on the basis of what he knew, claim it is certain that all that exists is imbedded in space. But it does not show that, in Laplace's time, such an idea was unwarranted. On the contrary, we may well consider that such a simple assumption, which was then at variance with no known fact, was indeed extremely plausible because of its very simplicity, and that therefore no serious objection could be made to the people who put it forward. As a consequence of all this it would be a serious mistake to consider that *Transcendental Aesthetics* yields a *demonstration* of the transcendental ideality of space.[1]

Of course, and as the reader has undoubtedly realized, the foregoing is neither a proof nor even an indication in favor of the so-called "realist" view that space and time both exist per se. Hence care should be taken not to infer from the above that Kant was *wrong* in what he stated concerning space, and that the latter does have some objective existence independent of our own. This would be an entirely unjustified inference and indeed we saw that recent arguments drawn from contemporary physics tell against such a view. What follows from the foregoing analysis is just that as long as we restrict our investigation to the purely philosophical realm we cannot consider the thesis that space is a priori to be demonstrated. In other words, in the present problem it is important to clearly apprehend the different facets of the question. We must disagree with the

---

[1] It is true that in favor of this view Kant put forward, in other works, a different argument. We refer to the one, commented on by Jules Vuillemin (1955), that is grounded on Galilean relativity. Is *that* argument a proof as we mean it here? There are reasons for answering negatively. See section 2-3 again.

realists who consider the absoluteness of space and time (or of space-time) to be "proved by reason and experience" and ground on this view their rejection of all forms of idealism. But we must also disagree with the adepts of critical idealism who claim that Kant fixed once and for all the conceptual framework within which any serious thinking has to keep. When separated from those possibly proceeding from other sources (quantum physics in particular) his arguments are definitely insufficient for settling some questions, the one of the nature of space in particular. In prominent philosophical circles the opinion that spatiality is but a framing of phenomena by the mind is so widespread that, apparently, everyone there believes it is a demonstrated truth—of which only "people not in the know" are ignorant. But on the basis of the above it appears to be but a "received view."

Such a skepticism (which, incidentally, outside the philosophical circles, seems to be shared by many) concerning the a priori character of concepts lying at the roots of knowledge unavoidably leads to questioning one of the most important aspects of Kantism, namely the alleged necessity of what Kant called his "Copernican revolution." The arguments on which he grounded the said necessity are well known. As most enlightened people of his time, he thought human beings had reached genuine certainty in at least two domains, namely geometry and Newtonian physics. Now, these sciences were founded on some great constitutive concepts such as (Euclidean) space, (universal) time and causality (in the sense of determinism) that (as we just saw concerning the former) Kant took to be a priori or, in other words, innate. If these sciences described things-in-themselves (as their conjectured certainty suggested and as, before Kant, most people thought) such innateness would have meant that innate constitutive concepts apply to things-in-themselves, which would have raised the above-noted enigma. How could it be possible to understand such an exact fitting to things-in-themselves of concepts that are innate (i.e., *not derived* from any experience of these things)? Apart from involving, à la Descartes, the divine providence notion, such a preestablished harmony looks miraculous and incomprehensible and is a scandal for reason. The difficulty was inextricable and Kant judged that the only way out of it was to give up one of the premises the conjunction of which created the trouble. But then, since he considered the a priori character of the constitutive concepts to be beyond doubt, he had no other choice than to assume that sciences do not describe things-in-themselves. He therefore took up the view that they only describe human representations, in other words, phenomena. To this change of outlook he gave the name "Copernican revolution" in analogy with Copernicus' discovery. Same as, according to the latter, the Earth revolves around the Sun rather than the Sun around the

Earth, phenomena, according to Kant, get regulated on human knowing possibilities rather than the reverse, as before thought.

So, as we see, the a priori character of the fundamental concepts lying at the root of human knowledge plays a basic role in Kant's demonstration that his "Copernican revolution" is necessary. Since, for the reason stated above, I, for one, do not consider Kant's proof of the said a priori character to be binding, I do not feel bound to believe right away in the necessity of the revolution in question and hence to accept it without further ado. If the conceptual foundations of physics—its "constitutive concepts" in particular—had remained the same from its beginning up to now, I would therefore feel inclined to set the Kantian view aside and take realism of the accident (or more generally physical realism) to be very likely true. As a matter of fact, however, the situation is much more complex. During the two centuries that elapsed since Kant's time new pieces of knowledge took shape that are significant concerning the question in hand. One of them is, as we just saw, the discovery of evolution (of species and the Universe). But another and most important one is the upheaval that took place within the conceptual basis of physics and thereby of science in general. Actually, physics itself proves that many of the basic concepts on which the former physical science rested—Euclidean space, universal time, precise localization at any time of the center of mass of several objects relative to one another, etc.—though remaining necessary for human action at the human scale, may no more be taken (as a realist would have it) to correspond to basic elements of Reality. In other words, even if, in view of the foregoing considerations, we decided to disbelieve Kant and claimed that the alleged "first concepts of understanding" proceed in fact (through apprenticeship) from our ancestral experience, we would not be entitled to believe that they yield access to "the Real."

REMARK

Concerning causality and its restriction to phenomena, Kant's assertions are not quite so clear-cut as they are sometimes said to be. True, concerning (discursive) knowledge he asserted that causality only takes place between phenomena, that is, merely between representations. But in view of what follows it is worthwhile to note that, at places, he referred to the *cause* of the said representations. Admittedly he did so just in order to stress that this "cause" remains unknown to us and that we cannot perceive it as an object (since, he pointed out, such an object could be represented neither in space nor in time). But still, he did. In the *Critique of Pure Reason* he even went so far as to specify that we may call 'transcendental object' the "purely intelligible" cause of phenomena in general.

13-2-1 From Kant's Views to the "Operationalist" Theory

We noted that a longing for certainty seems to have played a considerable role in the genesis of Kantism. Apparently, Kant thought he had gained the said certainty by means of his Copernican revolution and as a compensation, so to speak—as Abner Shimony (1993, I) suggested—for having renounced any hope of a scientific description of the thing-in-itself. Today, the advances in both mathematics and physics that took place during the nineteenth and twentieth centuries (think, for example, of Gödel's theorem) tend to make us skeptical in this respect. But does this means we should give up the Kantian ideal of a science that would be certain? The considerations developed in the first part of this book suggest that this is not really the case and that, in fact, we do dispose of a science that is both certain and universal. But, contrary to the one Kant had in mind, it does not—or, at least, not basically—bear on "objects for us" susceptible of being described by means of the *objectivist language*. This "science" is the "operationalist" theory here described in chapters 4, 8, and 10, that is, quantum mechanics restricted to its "hard core," in other words to its observational predictive rules. Consequently, what it essentially deals with—what it actually does describe—is just the set of the said observational predictive rules. This science is universal since theoretically it applies to any phenomenon, whatever its scale. It is certain in that, on the basis of sufficiently well specified initial data concerning the preparation of the system in hand, it unambiguously yields the probabilities that, upon measurement of such and such physical quantity on this system, the outcome of the measurement be such and such. (The "law of great numbers" then makes it possible to experimentally verify these predictions with an arbitrarily great precision.[2]) Such a certainty is equivalent to the one the phenomena were supposed to have within the physics Kant had in mind. In both cases it follows from the fact that what is dealt with is not an element of some external reality but just the apprehension of some sense data. Indeed, this feature is even more clearly apparent here than in Kant's conception since, contrary to the latter, such an "operational physics" does not, in principle, take the roundabout way of using the intermediate notion of "objects for us."

As it stands, however, the latter statement is slightly too simple and should be somewhat watered down. In practice, we still have some need for the objectivist language, that is, for the language of strong objectivity (take good note: just for the *language*, not for the *notion of objects-*

---

[2] Concerning a few highly theoretical restrictions that are to be brought to these considerations, see section 14-6.

per-se to which this language usually refers). We need it in order to, on the one hand, describe our whole equipment (in a broad sense) and, on the other hand, synthesize in just a few short phrases an information that could be expressed in the observational predictions language only by means of prohibitively long, complicated sentences. But this does not imply any basic duality within the very principles of knowledge since it is now proven (remember, e.g., the section 8-2-2 content) that the observational predictive rules of everyday life and of classical science follow from the general quantum ones. Incidentally, note that, besides the usual notions, the language thus introduced also involves other ones (virtual particles, curved spaces, and so on) that are not a priori notions of human sensibility (the main ones are derived, as we know, from mathematics).

What, in all this, is especially worth noting is that, through a thought process differing very much from Kant's one and grounded on information this philosopher did not dispose of, we finally arrived at a position that, basically, is not very different from Kant's one. Due to its bearing but on phenomena, science as Kant conceived it may be called weakly objective. Due to its bearing exclusively on observational predictions, the same holds true concerning our quantum centered physics. Moreover, in both cases it is just this weak objectivity that makes our knowledge a sure one (and, in particular saves it from the paralogisms Kant called "antinomies of dialectical reasoning"). Compared to the magnitude of those that separate Kantism from objectivist realism, the remaining differences are not considerable. The main (already noted) one is that the empirical reality notion is not the same in the two approaches. In Kantism it involves all of the "objects for us" and more generally all phenomena at any scale of magnitude. For, in it, our a priori concepts universally apply within the whole field of scientific knowledge and so, therefore, does objectivist language (even though care must be taken not to interpret the corresponding statements as bearing on "the Real"). This is why Kant frequently made use of just the word "reality" for referring to empirical reality. Indeed in his eyes, except for the unknowable "thing in itself" there can be no other reality. In our present-day atomic and subatomic physics the situation is quite different since, as just recalled, many of its statements—basic ones included—have the form of universal, observational predictive rules inexpressible in the objectivist language. In virtue of this, the empirical reality notion only applies in domains that are defined—though even then, only vaguely—only with reference to human dimensions and human actions.[3]

---

[3] As several authors and especially Cassirer (1936) pointed out, the difference between what Kant called "empirical reality" (or just "reality") and what this expression may nowadays be taken to mean becomes especially clear when we consider what Kant called the

13-2-2 On Kant's "Refutation of Metaphysics"

It still remains to be seen whether or not within an explicitly purely philo-sophical approach it is possible to do more than just describe phenomena. Whether, for example, the conjecture that mind-independent reality ("the Real") might, after all, be merely veiled rather that totally hidden may be considered "conceivably acceptable." As we may remember, the conjec-ture in question essentially is that our great physical laws might, after all, be not-altogether-deceitful traces of "the Real" (while, on the contrary, the contingent properties, alias the "accidents," that we observe within empirical reality are mere Kant-like phenomena). Such a conception typi-cally falls into the category of those the possibility of which Kant, arguing on the basis of the science of his period, thought he could disprove. What we must therefore presently examine is whether or not his arguments to this effect can be transposed within the framework of contemporary scien-tific knowledge as described in the first part of the present book. Here therefore we have to do with a question concerning, not just plausibility arguments but impossibility proofs.

Now, what are these alleged proofs? Above, we saw an example of them, namely, the Kantian proof that the notion of space is a priori. We might think of transposing this proof to contemporary physics and applying it to the great mathematical laws. Following Kant, we would then stress that, for synthesizing the whole set of the phenomena we found no other way than to make use of mathematical structures, and that the latter are therefore a mould imposed by us, as a pure product of the mind, upon the multifarious sense data. Which amounts to say that these struc-tures had to be laid down right at the start and that they therefore cannot have been derived from our experience of relationships between alleged "external" phenomena. There is thus no reason that they should corre-spond to any independent external reality.

"law of complete determination." Any object, Kant stated (in the *Critique of Pure Reason*) is subject to this law, according to which: "If all possible predicates are taken together with their contradictory opposites then one of each pair of contradictory opposites must belong to it (the object)." It is clear from his writings that Kant tightly linked this law (which of course, within his conception of the nature of predicates, is necessarily related to counterfac-tuality) with his notion of reality. And from everything that we saw it follows that the law in question is quite strictly inconsistent with quantum physics. It is interesting to note that ii is also inconsistent with some of the so-called "ontologically interpretable" models. For example it is inconsistent with the Broglie-Bohm one, at least as long as, in Kant's spirit, we identify "predicate" with "what should, in principle, be measurable." For, the momentum p of a particle is a predicate in that sense and still, in the model, it is in general not true, concerning some still unobserved particle, that its momentum p either has value a or has not value a. Indeed, as we saw in section 7-2-2, a statement of this type is correct only concerning the momentum π. But π is not observable.

Here again, the logic of the reasoning is impressive. But, concerning Kant's reasoning with respect to space, we could observe that in somewhat complex matters simple logical constructions only referring to very general ideas are liable not to encompass all the data. More precisely, relatively to the said reasoning we noted that it ignores the possibility that living beings should have progressively—through trial and errors that is through their experience of the external world—mastered the concept in question A somewhat similar remark seems here to be in order. First, not all mathematicians—by far!—consider mathematics to just be constructs of the mind (some believe they constitute an *external* reality). But, for the sake of the discussion, let us assume they are. Even then, it is a fact that they swarm with concepts and theorems, only a small fraction of which is made use of in mathematical physics. From the point of view of a pure mathematician the latter have no special features. If the structuring of our sense data were but a process internal to our mind there would therefore be no reason that just the concepts composing this tiny fraction should have been chosen. Hence it cannot be claimed that the precise structure of our mathematical description of the great physical laws is a priori entirely given to us by the very structure of our own mind. This selection process, at least, is not. Consequently, not only it is possible but it even looks reasonable to assume (as we instinctively do as scientists!) that the choice of the mathematical formulas by which we describe observational predictions is, partly at least, governed by some Reality external to ourselves. I think this remains true even in view of the fact that, as Einstein himself stressed, the relationship between this Reality and mathematics remains mysterious (particularly when time is taken not to be an external arena since, then, the explanation by trial and error becomes questionable).

To sum up, what has been shown in this section is that, account being taken of the content of present-day knowledge, there are, in Kant's reasoning, no still-valid elements making it possible to disprove—or even just make incredible—the open realism postulate and the veiled reality conception.

13-3 Facing the Refusal of the Independent Reality Notion

Kant, to repeat, accepted the "thing-in-itself" (alias "reality per se," alias "the Real") notion.[4] Presumably he would not have objected to my open realism postulate being taken as a starting point, since this postulate as-

---

[4] This whole section is an analysis of the approach of one particular author (Cassirer). On first reading it may be passed for its content does not have much bearing on what comes after.

sumes nothing with respect to the thing-in-itself accessibility. The same is not true concerning most of Kant's followers. Neo-Kantians did not any more accept the said notion, except just as a "limiting concept." They refused to take it to correspond to some existing, though unknowable, entity. In his, above quoted, preface to a reprinting of the *Critique of Pure Reason* Luc Ferry (1987) took up a similar standpoint. He wrote "The thing-in-itself shall be understood, not any more as a reality external to the representation but as the very fact of the representation (as the fact that representations are given to us; that we do not produce them)."

Given by what or by whom? Neo-Kantians and their adepts of course must answer: "by nothing and by nobody" for, to accept the existence of a "giver" would be tantamount to granting that there is some reality prior to the phenomena, which is precisely what they deny. Unfortunately in everyday language the very notion of a gift presupposes that of a giver. In other words, we are instructed here to take the expression "given to us" in an altogether new sense, which we are not informed of. . . . We thus immediately realize that the discussion will not be extremely simple.

13-3-1 Cassirer and the Theory of Concept

To try and understand, let us turn to the philosopher who may well be the clearest of all neo-Kantians, namely, Ernst Cassirer. Cassirer's reasoning, remarkably described in his book *Substance and Function* (Cassirer 1910), began by contrasting two possible procedures for defining what is meant by the word "concept." To define a given concept, the one, say, of "tree," the first method, which goes back to Aristotle, is to start from the notion of things—individual objects—taken to be primeval. It consists in considering a collection of objects presenting both similarities and differences, in keeping the former and in ignoring the latter. The method is aimed at being efficient (the higher-level concept is expected to increase the intelligibility of the lower level one) but Cassirer pointed out that the procedure offers no guarantee that it should be. He noted for instance that if systematically making use of it induced us to group together the concepts "cherries" and "flesh" because such objects have in common the properties of being edible and red, we would thereby construct, not a valid "higher-level" concept but just a meaningless verbal combination. And he stressed that what, in fact prevents Aristotle's followers to fall into such erratic practices is the implicit—but metaphysical!—idea that, far from being a mere summing up of features shared by the objects composing some set or other, a valid concept expresses a facet of their inner reality.

Cassirer's criticism of this method bore first on its efficiency. He observed that the more the generality of the concept increases—from the oak to the tree, then to the plant, to the living being, and so on—the more its informative content decreases. In the limit we would dispose of a concept exhibiting a complete generality but that would not carry any information whatsoever. Now, he claimed, what we are expecting from a scientific concept is just the opposite. From it we expect that to the initial indetermination and the plural nature of the representative content it should substitute a strict, univocal determination. To the first method, implicitly grounded on the metaphysical idea of substance as we just saw, he thus greatly preferred the one that is made use of in mathematics and that refers to the notion of "function of a variable." A mathematical function represents—he pointed out—a universal law that, in virtue of the successive values the variable takes, underlies all the special cases it deals with, which entails that here the generality (the one of the law) in no way wears away the individual particularities. For example, while the concept of second-degree curves is more general that those of circle or ellipse, still, when we go over from the concept "ellipse" to that of "second-degree curve" we lose no information since we can get back to the ellipse just by specifying the range of values of some parameters. Not only—he claimed—do such concepts not wear away the special features of the specific cases to which they apply, they even express them in full. In this approach the higher-level (i.e., the more general) concept also is the one with the richer content. Our understanding—Cassirer noted—of our three-dimensional Euclidean space, far from being weakened gets more acute and precise when, thanks to contemporary geometry, we come to apprehend higher-level spatial forms. For it is only then that the axiomatic underlying the structure of our own space comes into full light.

13-3-2 Empirical Reality According to Cassirer

According to Cassirer the advent of this new conceptualizing method is an important step and should even lead to a renewal of our logic, which, he claimed, is still too much grounded on the "ontological" method. And he pointed out that, in fact, we do already make use of it in everyday life, without even noticing it. He noted that when, for instance, we bring together, under the concept "metal," the concepts "iron," "copper," "gold," etc., we do not set aside as much as we think the features that distinguish these various metals from one another. In fact, we keep them but as "variables" in the mathematical sense of the word. That is we think of them as constituting, in concert, a kind of variable x enabling us to think of a whole sequence of possible specifications. Viewed under this light the concept "metal" refers, according to him, less to the notion of substance than to the notions of "sequence" or "function."

All this applies of course also to physics. The physicist does not rest content with just noting properties, and when he/she generalizes it is, of course, not by wearing away such and such feature of the phenomena he/she deals with. It is by discovering a general law that brings together without obliterating them a whole sequence of less general specifications. Now, the physicist does not just read such a law within the observational of experimental results. True, for validating the law in question experimentation is made use of. But, Cassirer stressed, for the validation to be at all possible it is necessary that the very principle of the sequences, the points of view according to which the various elements are to be gathered and compared, should be specified and motivated by the theory. The chaos of the sense data is thus translated into a system of numbers and these numbers, in turn, get their interpretation, hence their specific signification, merely from the content of the concepts elaborated by the theory. In this, he claimed, lies the logical concatenation without which we could not impart an objective value to the operation that changes the impression of the senses into a mathematical symbol. To confirm its own validity the symbol cannot just rely on such and such element of perception. It can do so only by referring to the law that defines the concatenation of its various constitutive parts. And the latter is what will reveal itself with an ever-increasing clarity as being the very carrier of the "empirical reality" idea.

We here see appearing one of Cassirer's basic notions, the one, precisely, of an "empirical reality" that he conceived of as being essentially grounded on the concept construction described above and that constituted in his opinion the only valid acceptation of the word "reality." Somewhere else in his book he stated this even more clearly. The practical necessity, he claimed, in virtue of which we wrench ourselves free from the realm of separate, unlinked impressions and work our way to the idea of continuous objects linked together by strict causal laws is, after all, a logical necessity. Reason and the universal validity of its rules hence constitute the framework within which we sketch the very concept of Being.

Such views and connected ones also expressed in the book show that Cassirer disagreed with each one of two mutually opposite ideas. On the one hand he discarded, and even radically refuted by means of his theory of concepts, a phenomenalism that grounds science exclusively on sense data, considers they only may be called "real" and thus comes near Berkeleyan idealism. And on the other hand, he relied on the same theory of concepts for also discarding the views of conventional realism, in other words for stressing (in his own words) the limits of any sort of "transcendence" whatsoever. Hence, of course, as already noted, he rejected the thing-in-itself notion. But the great difference between his ways of dealing with these two approaches is worth noting. In fact, it is on the first one that he centered almost the totality of his critical reasoning. His objections on this point being hardly relevant to our subject, let us merely note here that they focused on the idea that direct perception yields but scattered fragments. That the "seen and heard" criterion only provides us with inconsistent and temporally unorganized perception groups while, on the contrary the concept "object" needs a complete filling up of the temporal sequence. And that he forcefully stressed that it is on the *logical* demand of a "connection of the given" that science bases its definition of nature and of the natural object. Radical therefore was his criticism of traditional empiricism and its claim that nothing is initially given to us except the self and its impressions, and that empirical reality is entirely constructed starting from this. Indeed the "logical necessity" notion was so central for him that it played in his book the role of a primary datum, conceptually prior to any possible derivation, nay even definition. This clearly appears in, for example, his description of what he took to be an elucidation of the "mystery of induction." In his opinion, considering a number of identical cases just leads nowhere. It is already at the stage of just one case that, be it only implicitly, a potential has to be assumed that leads beyond its limited and fragmentary aspect. In other words the logical function in virtue of which we are able to continue the analysis of an empirical content beyond its actual limits was, according to him, the true hard core of the inductive process.

13-3-3 Cassirer and the "Independent Reality" Question

The strong, tight-built criticism Cassirer made of phenomenalism stands in striking contrast with the manner in which he dealt with the "thing-in-itself" question. True, his refusal of this notion was clearly expressed and radical (he went so far as to mention "the justified and necessary fight against ontology"). But he hardly took the trouble to motivate his refusal. Or, at least, the way he did so was not precise enough to satisfy whoever is not already convinced. In fact his reader has the impression that Cassirer considered his predecessors already settled the question and that he didn't see the point of stressing the matter further. In his book I could find but two rough drafts of a refutation and neither of them is convincing.

In the first (and more significant) one he started from the observation (quite generally acknowledged true) that naive realism in not tenable, in other words, that sense data do not reveal the things themselves. He implicitly assumed that this is also true when the data are gained by means of scientific instruments. In other terms, he took it for granted that, however sophisticated they may be, such data may in no way be viewed as indicating the existence of anything that might sensibly be interpreted to be an element of some reality-per-se. This is a Kantian thesis that indeed is endorsed by most philosophers, in spite of the fact that, as we saw, the arguments Kant gave in its favor are not any more fully convincing.

Being explicitly or implicitly aware of this loophole, most present day scientists seem to be rather skeptical concerning the thesis in question. As we know, a great many of them, implicitly judging that the no-miracle argument has at least some "background of truth," more or less instinctively rally realism of the accidents. They take it to be, if not strictly proved, at least most likely true. And to persuade them to give up this stand, it is necessary to gather together all the arguments and proofs reported in this book first part. However, what finally counts is that in the last analysis realism of the accidents *must* be given up. The present-day physical descriptions, making use of such notions as those of forces, electrons, virtual particles and so on, can strictly not be considered descriptions of reality-in-itself. We must therefore follow Cassirer in, if not his reasoning, at least his conclusion, when he branded, under the name of *transcendental realism,* the inconsistency of a realism that in fact, as he presented it, is just a metaphysical realism of the objects. In this respect we should however incidentally note that the objections he (apparently) had in mind—as well as the ones just mentioned—only hold good against the said metaphysical realism of the objects, that is against a conception that is diametrically opposed to the "veiled reality" one. They cannot therefore be directed against the latter.

Later we shall see that Cassirer nevertheless managed to build up a "realism of the object," but one that is not "metaphysical" in that it is free from transcendence or rather from any transcendence exceeding some given limits. Here let it merely be noted that his procedure for specifying the limitations in question indirectly provided him with the substance of an objection to the notion of mind-independent reality. To fix up the limits of "acceptable" transcendence it is—he claimed—necessary to compare the objects of experience to the objects of pure mathematics. Objects of mathematical knowledge, numbers, geometrical figures and so on, are not absolutely existing, separate things. On the contrary, they express specific, universally valid and formally necessary concatenations. Once we have firmly established this point, he went on, it is an easy matter to transfer it to physics. Objective physical statements are as a whole but the outcome and conclusion of a logical undertaking during which we progressively remodel experience according to what the mathematical concept demands. Everything that, within the realm of the "existence" concept, would go beyond such formal and logically necessary concatenations would thus be discredited and should be banned.

Now, does the foregoing really offers a basis for a decisive criticism of the mind-independent reality notion? We may be skeptical on this, for three points look worrying. One of them has to do with pure mathematics. It is the fact that, without offering any argument, Cassirer just denied the thesis of mathematical realism (although quite a number of pure mathematicians believe it is true). The second one is that, to constitute a true refutation of objectivist realism, the parallel drawn between conceptualization in mathematics and conceptualization in the natural sciences should be something more than just an analogy, for analogies prove nothing. It should be a genuine identity, that is, this Cassirer-like conceptualization should have, within the sciences of nature, the decisive, universal role that is has, as Cassirer pertinently noted, in mathematics. Now if, in Cassirer's time, this may have seemed to be the case, it has now become clear that in fact it is not. This point was already alluded to, apropos of Kant. Quantum physics does not—or at least does not primarily—develop through the method of first carefully constructing elaborated concepts—not even Cassirer-like "functions" potentially rich in infinitely many virtual realizations—and then applying them as they stand to "objects-for-us." It is true that, from time to time, it still builds up such concepts. But during the last century it had to admit that some of the most basic problems it got faced with simply could not be solved this way. Concerning them it therefore made use of quite a different method, which essentially consists in inventing (on the basis of analogies of any kind, viewed just as incentives for imagination) rules that, though they are expressed with words seemingly referring to concepts, are essentially observation predictive statements. This makes it possible to circumvent the difficulty, inherent in this realm, of precisely defining elaborate, descriptive concepts. In fact, the words made use of, "particle," "wave," "spin," etc., are mostly everyday ones, but they are not taken in their usual sense and, in practice, are not even imparted quite sharply defined ones. This is not an obstacle for, in the said rules, what finally counts (and correlatively, is made very precise) is the description of the way by which physical systems are prepared and observations are gathered. Equations then serve to connect the two together. Given a definite preparation they make it possible to calculate the probabilities we have of observing such and such outcomes. Of course, only the rules that successfully pass the tests are kept, and they are kept only so long as they keep on being successful with respect to new tests (this, for the time being, is the case of those constituting quantum mechanics). On the whole, therefore, even though it can not be denied that physical science is quite tightly linked to mathematics, in view of present-day physics the quite specific link Cassirer described and used here as an argument for discarding metaphysical realism has lost, by now, much of its persuasive power.

The third reason for considering that the Cassirer-like conceptualization procedure does not deprive the notion of "the Real" of interest consists in noting that the resemblance between physics and mathematics, even though significant is, nevertheless, but partial. That Cassirer was right when he claimed that the "transcendence" we attribute to the physical object is comparable to the one we attribute to the mathematical idea of "triangle" as opposed to the intuitive picture we have of material triangles, is indeed shown by what took place a few decades later. The turn physics then took does show that attributing to atoms or electrons a reality "deeper" or "more fundamental" than the one we assign to mathematical triangles would have been an error. That, on the other hand, the resemblance is but partial comes from its leaving out of account one of the main roles of the experimental procedure. A role well known to scientists and that Popper finally revealed to philosophers. I refer of course to the one, labeled point 2 in section 10-4-1, that consists in refuting a theory. Nothing, in mathematics, corresponds to such a role and it may be said that this difference sets mathematics quite apart from the empirical sciences. While all that is expected from mathematics is internal consistency, from the theories and concepts constituting the sciences in question we request something more. We demand that they should not be refuted by experimental results, that is, by elements that, even though we do contribute to them (the "theoretical admixture" within experimental data is well known) do not depend exclusively on us. In other words, by elements also depending on something external to us. This point should not be overlooked. Hence Cassirer's statement that what stands in opposition with the psychological immanence of impressions is, not some metaphysical transcendence of things but the universal, logical validity of the ultimate principles of knowledge, is not fully satisfactory. (Incidentally, we may remember that more recent philosophers gave reasons for doubting the validity of the "ultimate principles" in question). And we may feel the same is true concerning Cassirer's claim that Reason and the universal validity of its rules constitute the framework within which we sketch the very concept of Being. Both statements make me uneasy for I consider that the approach they describe just forgets about the "something that refutes and says 'no' " which we just observed must not be ignored.

REMARK 1

We may—and even should—wonder whether Cassirer would not have had an answer to that objection. And indeed once the question is asked the idea of a possible counterargument quickly comes to mind. A comparison with what takes place in dreams yields a raw notion of it. In a dream, what we call the external world normally does not intervene. Only our own representations come into play. But still, it is a fact that, even in dreams, difficulties are liable to appear. (In nightmares they even seem insuperable; but pleasant or half-pleasant dreams, in which the difficulties get overcome in the end, offer a better comparison with our problem). Hence a systematic idealist may consider that after all, since experimental outcomes are themselves just sense data (that is, in last resort, modifications of our mind), in physics just as in mathematics and in such dreams the whole process merely involves the "inner sense." On this basis a Cassirer adept might rather naturally claim that the above stressed difference between mathematics and empirical sciences is, after all, superficial. He would stress the fact that what, in physics, "says no" always is a set of sense perceptions, and that the latter are of the nature of ideas. He would therefore maintain that, there as in mathematics, the task finally consists in making various modifications of the mind fit together.

I personally think that the validity of this refutation of our objection is itself questionable and that between the two situations, the one of physics (and other empirical sciences) and the one of mathematics, a difference does remain. In pure mathematics, if we have to do with a consistent theory—Lobatchevsky's non-Euclidean geometry for example—this theory is ipso facto immune to any further discovery. It may become improved, enriched with further theorems, but never will it be refuted. It is self-consistent and will remain so. In physics, on the contrary, many people build up theoretical developments, take good care that they should remain consistent with well-verified preexisting knowledge, and still have the unpleasant surprise of being informed that new experimental data falsify them. In such a case, putting forward an idealistic, dreamlike explanation of the above described type would imply assuming that what takes place in the theorist's mind has to change for the sole reason (since there is no independent Reality) that the mind of the experimentalist underwent such and such an evolution. In itself such an idea seems strange indeed. Moreover, since only minds are at stake the question arises why it should always go that way and never the other way round. Why should we not keep experimentally disproved theories (not always, of course, but sometimes; when they are "exceptionally beautiful")? True, the argument from strangeness is not "logically watertight." What looks "strange" within one approach may look "natural" within another one, based on a deeper theory. But still, at least at the present stage in our knowledge the assumptions that have to be made in order to consistently rule out the "something that says no" notion seem to me to be distinctly more difficult to accept than the straightforward one that this something does exist.

REMARK 2

Unfortunately, philosophers with an idealist stand hardly ever tackled the problem of error. True, in the *Critique* first edition Kant did briefly mention it. Knowledge of objects, he explained, may be derived from perceptions through either the mere effect of imagination or experience. Then, misleading representations may be formed, to which the objects no more correspond and in which the delusion comes, sometimes from imagination (as in dreams) and sometimes from faulty judgement (as in illusions of the senses). However, the last parenthesis indicates that the type of error Kant was primarily concerned with was not the one in which a theory is erroneous because it is at variance with experiment. In fact, it is clear that what he considered here was just the converse process, the one of rectifying experience by theory. And indeed his additional comment that to avoid error we must follow the rule according to which "what fits with a perception according to empirical laws is real" corroborates this. Clearly, what Kant had in mind was what we do when we make use of known laws in order to rectify rough testimonies of our senses, as in the broken stick experiment.

Now, independently of what Kant himself explicitly stated or did not state on the subject, within Kant's theory *proper* the fact that theories are liable to be disproved by experiment still is, in principle, quite easily understandable. It is so in virtue of the fact that, as recalled above, Kant did evoke—under the name "transcendental object"—the notion of an external, purely intelligible cause of our representations. The circumstance that, according to him, we can strictly assert nothing at all concerning this object (which coincides with my "independent Reality") did not prevent him—and should not prevent us either!—from having it in mind. But then, why should it prevent us from seeing in it the true, mysterious "what says no" we are in need of? Of course, it is not by means of such considerations that a neo-Kantian such as Cassirer could have dealt with the problem in hand since neo-Kantians reject the very notion of a transcendental object. But nothing prevents us from, in this precise "transcendental object" issue, opting in favor of Kant's opinion.

As mentioned near the beginning of this subsection, Cassirer's book has in it an indication of another objection to the independent reality concept. Starting from the "fact" that we have access only to experience and logical necessities, this author stressed that, consequently, accepting the mind-independent-reality concept would imply accepting the view that thought is definitely unable to reach the "heart of the matter." Hence, he claimed, we feel prostrated under the weight of a vague idea of something the presence of which must forever remain indescribable and that we know we shall never reach. But does this constitute a valid argument?

This may be questioned for two reasons. One of them is just that (unfortunately!) the fact that something is regrettable does not necessarily imply it is not true. And the other one is that the frustration Cassirer thought we must feel at not having access to a genuine knowledge of Being itself is, in reality, unmotivated and, in fact, not actually present. On the contrary, what is exalting is the very deepening process. Fortunately, in the course of centuries—and millennia—this process never grew weaker and there are no indications it will. The sad thing would be the limit. What would be a pity would be if we got, one day, to an exhaustive, accurate knowledge of all that exists. The veiled reality conception guards us from such a danger.

13-3-4 Cassirer and the Construction of Empirical Reality

The foregoing outline and commentaries of a very rich book should suffice to form a proper idea of what its author had in mind. It seems that, after all, Cassirer aimed at scrutinizing, not so much the content as the process of knowledge. The difference is not inconsiderable. It is true that the fact of better understanding what takes place—of, say, grasping more clearly what "conceptualization" is—does shed light on the bearing and limits of knowledge and thereby, indirectly on its very *objects*. But this light is only partial. If, for example, somebody bluntly claimed that what marks the boundaries of discursive knowledge automatically marks the boundaries of whatever we are entitled to think, he/she would thereby put forward *as a postulate* the view that to evoke the notion of "existent but scientifically unknowable" is meaningless. So that if, in a second stage of her reasoning, she made use of this postulate to prove there is no other level of reality than empirical reality we could point out to her that she argues in a circle. In order to make her point what she would have to prove (instead of postulating it) is that there is no sense in having in mind the idea of something that cannot be discursively known. For this, she would first have to specify what sense she imparts to the word "sense" and give the reasons of her choice, which would not be a small affair. Most appropriately Cassirer avoided having to cope with such problems. For example, focusing as he did on just analyzing the process of gaining knowledge spared him the task of having to explicitly take position against Kant's above mentioned opinion that we may "think" the thing-in-itself, even though there can be no question of describing it.

Instead of tackling such subjects, Cassirer took up the, better defined, task of analyzing the ways in which human mind finally builds up *empirical* reality. Concerning this, it is worth noting that the empirical reality notion he had (at that time, in 1910) was not restricted to a specified class of objects or object attributes, such as "macroscopic objects" or "measurement outcomes." In this respect, twentieth century physics led, as we saw, to a conception of empirical reality (still taken in the sense of the set of all objects-for us and object-for-us attributes) that differs very much from the one he then had since (as we saw in chapters 4, 8, and 10) it is strictly conditioned by factual limitations in human aptitudes and, for this reason, may not be applied to the greater part of the "microscopic" realm. Within the whole realm of quantum physics the basic concepts are, to repeat, those of system preparation and measurement of observables, and the role of mathematics essentially is to link the two together. This role is more modest that the one of "mold of concepts" assigned to them by Cassirer as we just saw. (Of course, it is not within the book of Cassirer considered here, which was written quite a long time before the advent of quantum mechanics, that we could find appropriate developments on this. But still, we may be surprised that, in it, the notion of correct prediction of observational outcomes, which, in science, is nowadays seen to be a basic criterion not only of truth but even of meaningfulness, is not even mentioned.) With Cassirer, just as with Kant, it is at all scales that empirical reality is conceived of according to the classical pattern, that is, as made up of objects endowed with definite properties (known or unknown). Only, contrary to what is the case within physical realism, with him these objects are essentially constructs. When we analyze the "object" notion, he claimed, when we strive to become aware of what is implied by such a concept, we are inevitably referred to some logical necessities that thus show themselves to be the indispensable and constitutive "factors" of the said concept.

Interesting in this respect is the comment Cassirer made of Planck's views about realism. He first recalled the basic unity condition Plank expected any physical theory to fulfil. It must be a component of a universally valid consistent whole, be true at all places and for all times and hold good for all people and any culture. He then asserted that this triple condition yields the genuine meaning of the very concept "object," and claimed that Planck was therefore right in calling his conception realist, in contrast with phenomenalism, inasmuch as the latter remains exclusively tied down to impressions. But, he continued, far from being its converse, such realism is the correlate of a well-understood logical idealism. For the independence of the physicist's object with respect to all the particularizing processes of impressions sets in full light its dependence on logical principles endowed with universal validity. And being, as they are, at the source of the unity and continuity of knowledge, these principles and they alone guarantee the possibility of finding and ascertaining the content of the very concept of object. So, Cassirer's entire effort in this book ended at reconstructing and justifying, otherwise than by referring to metaphysical realism, the picture of a world basically composed, at all scales, of individual objects endowed with well-defined properties not depending on the choice of which one will be measured. As we see, Cassirer (as well as Kant) could easily have endorsed Berkeley's assertion that anybody desirous to keep the old realist language is quite entitled to do so provided only he should adopt a mental restriction void of practical consequences (the one of not attributing a reality-per-se to objects). Now, as we noted, quantum mechanics on the one hand and nonseparability on the other hand definitely ruled out this possibility. Which, incidentally, shows how false the opinion would be that, in the realm of basic philosophy, twentieth-century physics contributed nothing essential. More specifically it may be observed that the Einstein, Podolsky, and Rosen criterion (section 7-1)—whose validity nonseparability disproves—concerns, not, actually, a priori metaphysical notions but, rather, notions that more or less fall into the category of Cassirer's "logical necessities." In fact, it is to a kind of objectivist realism—only, cleared from its metaphysical veneer—that Cassirer's analyzes most naturally lead us, whereas we know that, even when thus purified, this realism is no more tenable. Hence we may well question the idea that these logical necessities constitute the basis on which our understanding of the ultimate reference of science must rest.

Even apart from these reservations, the foregoing clearly incites us to examine especially carefully what constitutes the very foundation of Cassirer's proposed solution to the reality problem. Much along the lines of Kant, Cassirer aimed at escaping from both realism (which, basically, he denied) and Berkeleyan idealism (which raises to the level of certainties the sense data of beings embedded in time whereas it takes things to be mere outward appearances). He judged he could do so by considering the object to be, not a substance but something of the kind of an *objectivity plan (or draft)*, which gains consistency as experience develops. Strictly speaking there is therefore, in his approach, no absolute Being. There merely is relative being but, far from entailing a physical dependency with respect to the thinking subjects, such relativity merely signifies a logical dependency with respect to the content of some universally acknowledged premises of knowledge in general. And Cassirer added that therefore, to claim Being is a "product" of thought implies no reference to any causal relationship whatsoever, be it physical or metaphysical. It is only a question of recording a purely functional relationship, a graduation in the validity of such and such judgements.

This point is significant. Through it we catch sight of—taking form in Cassirer's thought—the idea of a transpersonal element, represented by the content of some universally acknowledged premises of knowledge ("logical necessities" are undoubtedly part of it) and which, in a sense, appears to be prior to the subject-object splitting. Unquestionably this transpersonal element is "at the origin" of both the objects as Cassirer conceives them and our states of consciousness. Admittedly, Cassirer himself would perhaps have objected to this idea. Maybe he would have considered that it amounts to reintroducing the very transcendence he was so keen at eliminating, and that, to speak of an extraphenomenal cause is just simply heretical. But, after all, is it so sure, and, it if is, should we admit that he was right? Conceptually, the kind of "independent reality" evoked here has nothing in common with the physical objects (things, energy forces, acceleration) against the notion of the transcendence of which he appropriately protested. And as for the use of the word "cause," in the context where we are now to criticize it would just be quarreling on words. For, after all, between "to be a cause of" and "to be a premise of" the difference is hard to tell.

Admittedly however, it remains that to mention the "premises of knowledge in general" seems to imply some basic referring to human beings. If the said premises are indeed at the origin of objects and the human being is the "seat" of the said premises, then (transitivity) the human being is at the origin of objects. Not, of course, such and such individual human being but "the human being" in general, or more exactly, the human mind, since parts of bodies are themselves objects. To be sure we thus escape the notion of a transcendent existence of the objects (that is, the naive but natural view that objects exist out there as we see them, quite independently of us). But instead it seems that we have to face the necessity of imagining as an origin or "seat" of all things (perceived or not), a vague abstract entity, namely, just the said "human mind in general." What remains to be seen is why such a view should be more satisfactory than the one I suggest in which it is openly granted that existence is logically prior to knowledge since to think we must first exist. And in which it is therefore considered that the notion of an independent reality conceptually prior to the subject-object splitting can hardly be dispensed with.

Would Cassirer have bluntly discarded this view? This is not certain for, toward the end of his book, he came back to the mathematical realism conception and what he wrote should certainly not be considered a flat refusal of it. Indeed, at that place (chapter 7) he acknowledged the validity of Russell's standpoint that to know in a nonillusory way finally amounts to an act of recognizing. And that the meaning to be attributed to the discovery of arithmetic should precisely be the one the discovery of the West Indies had for Columbus. We do not create the numbers, Russell claimed, any more that he created the Indians. It is true that Cassirer then somewhat watered down his approval, but he refrained from any systematically negative conclusion. He even went so far as to grant that our knowledge is bound by two constraints, one coming from the nature of our sense impressions and another one determining our thinking in the field of logic and pure mathematics. For example—he noted—we shall state that the hundredth decimal of the number π is theoretically predetermined even if nobody actually ever calculated it. Similarly, he quoted in an apparently approbatory way a sentence from William James asserting that to avoid contradiction and illusion our ideas must somehow finally mach elements of reality. It is true that he did not consider the question whether or not the hundredth decimal of π would be predetermined even if no human being existed. But it seems clear that, between the three possible answers "yes," "no," and "the question is meaningless" he would not have chosen the second one.

As we saw, the views of the neo-Kantians nevertheless differ very much from those of Kant. And the differences have, in particular, some impact on the status of one of the arguments that may come to mind (I sometimes mentioned it myself) for the purpose of justifying the independent reality notion. This is the observation that Kant's ideality of space hypothesis was plausible at a time when space was taken to be Euclidean since it then matched our sensibility but is no more so now that physics showed space is curved. In this form, the argument cannot be brought against Cassirer's views. For with him space primarily is a mode of shaping up experience by means of *understanding*, that is, in this case, through geometry. And nowadays geometry nicely incorporates the notion of a curved space. But in a generalized version the argument still retains some significance since, as already noted, mathematics swarm with developments of various kinds. Why, therefore, the particular structure of space that we observe? Why this particular metrics, rather than any other one? The idea that there is, influencing this "choice," a reality distinct from us seems a plausible, natural one.

Besides, it seems to me that compared with the neo-Kantian theses my other arguments—arguments 1 and 2 at least (section 10-4-1)—remain valid as they stand. For, as we just saw concerning mathematics, the conception that imparts a prominent role to the universal logical necessities of understanding in fact fails to disprove the idea that the notion of existence is prior to that of knowledge. And, to repeat, the fact that our theories, though built up in strict observance of all conceivable logical necessities, sometimes break down under the weight of experimental disproof would be hardly comprehensible if even the experimental data stemmed exclusively from our mind. Obviously, neither the, nowadays acknowledged, nonexistence of really crucial experiments nor the, also recognized, importance of a theoretical admixture within observation invalidate this argument since, in spite of these two "effects," experimental refutation of theoretical hypotheses does, in fact, frequently occur and is unanimously viewed as implying their rejection.

13-4 Kant and Our Contemporaries

Nowadays many philosophers take an interest in Kantism and in post-Kantian empiricism. And all of them, of course, not only discard the notion of a *knowable* independent reality but also claim they can do so by exclusively making use of philosophical arguments. To what extent are they right on the last point? The present section is aimed at yielding some clues relative to this question.

First, let it be noted that, contrary to Cassirer, not all present-day philosophers with Kantian or empiricist inclinations radically reject any notion of a reality-per-se. Jean Petitot, for example, seemed (Petitot 1997) to unreservedly agree with Kant at the places where the latter stated that the thing-in-itself is a "ground of reality" (in German, just simply "Grund") for phenomena. This "ground" is nonsensible, he asserted, it is intelligible, but it still lies outside the realm of knowledge since genuine knowledge demands that causality be restricted to phenomena. Petitot therefore approved of Kant having considered that the usefulness this notion has is (and merely is) that it serves for understanding why a set of empirical laws can exist at all. Along these lines he did not seem opposed to a view—somewhat akin to structural realism—whose pertinence within the no-hidden-variable approach to quantum mechanics was discussed in section 9-5 above (and with more details in d'Espagnat [1995]). Schematically my point was, to repeat, that, within such an approach, for a realist interpretation of the mathematical quantum laws not to be straightaway ruled out, a necessary condition must be fulfilled. This condition (which is a very demanding one indeed) is that, contrary to what is the case in classical physics, the *contingent values* of the entities appearing in these laws should not be considered real.

In the United States, thinkers such as Hilary Putnam and Bas van Fraassen take up, concerning the independent reality question, an attitude that, reticent as it is, still is not tantamount to simple rejection. The first named one, for example, made the following concession: "But perhaps Kant is right: perhaps we can't help thinking that there is *somehow* a mind-independent 'ground' for our experience" (Putnam 1981). The second granted that: "there may or may not be a deeper level of analysis on which that concept of the real world is subjected to scrutiny" (van Fraassen 1980; quoted by Shimony 1993, I). On the other hand, Putnam made his concession only in brackets and as for van Fraassen his quotation above continues with: "and [that concept of a real world is] found to be . . . what? I leave to others the question whether we can consistently and coherently go further with such a line of thought. Philosophy of science can surely stay nearer the ground." In other words, both considered—as did Jean Petitot and many others—that the "ground" in question is not genuinely attainable by knowledge, a view that, on the whole, contemporary physics seems to corroborate, as we saw.

But, as we know, these two thinkers based the said view on exclusively philosophical arguments. And (this will be our second point) it is worth noting that, within this purely philosophical realm, they meet with opponents (not very numerous ones but still . . .) who part with strict neo-Kantism even more than they do. Indeed, nowadays there still are realist philosophers. Some of them rest content with just stating their opinion, as we saw Comte-Sponville did. But others undertake to refute the empiricists' and Kantians' arguments. Abner Shimony (1993, I) is one of them.

To properly understand the objections this author raised against Putnam's views we must start from the basic distinction Galileo made, as we saw, between the properties of bodies that he considered intrinsic (form, position, motion) and those such as color, taste, etc., that he viewed as being projected on them by the perceiving subject. The basic fact then to be noted is that classical physics built up a representation of the world exclusively formulated in terms of Galileo's *intrinsic* properties (with adjunction to them of such entities as electric charges and fields) whose most remarkable feature is that it is self-sufficient. The secondary qualities (color, taste, etc.) just simply don't appear in it.

The difficulty Putnam stressed consists in understanding how it is possible that we have access to this world. If, following Descartes, we postulate a genuine matter-mind dualism we have to build up the theory of their interaction, which neither Descartes nor his followers succeeded doing. Another conceivable possibility, the one of parallelism (Leibniz), is hardly convincing either. The only approach Putnam considered, for a while, to be promising is the one of "functionalism," which identifies sensations with a brain state, but with the proviso that the state in question is defined, not in physical terms but rather in terms of programs. Later however he considered that even this possibility, which he had himself put forward, meets with quite serious objections, so that any totalistic explanation (to use his terms [Putnam 1990]) of the nature of thought and its relationship with its object seemed impossible to him. The knowledge we have of nature cannot be accounted for without referring to intentionality and Putnam came to consider that the latter couldn't be adequately explained by a naturalistic theory of mind. Consequently, he built up a scheme he called "internal realism" that, as he pointed out himself, is not very far from Kant's views. According to it, the truth (of some world-description) consists in "some sort of (idealized) rational acceptability—some sort of ideal coherence of our beliefs with each other and with our experiences as those experiences are themselves represented in our belief system." In no way does it consist in a "correspondence with mind-independent or discourse-independent 'states of affairs' " (Putnam 1981).

If I understand well, what Abner Shimony criticized within Putnam's approach was that—faced with the just mentioned difficulties—Putnam, according to him, too hastily gave up trying to build up a theory aimed at what he, Shimony, called "closing the circle." He meant by this a theory intended at showing "how claims to human knowledge of the natural world can be justified and in turn how the resulting view of the world can account for the cognitive power of the knowing subject" (Shimony 1993, I). But most of the thinkers who have such an end in view are, he wrote, "naturalistic epistemologists, insisting upon the relevance of biological and psychological discoveries to the assessment of claims to knowledge." Shimony explained (along with other points) that "[his] own approach is atypical in maintaining that there is an irreducibly mentalistic component in nature." Very schematically summarized his thesis was that accepting views somewhat akin to Whitehead's—that is, grounded on the notion of some proto-mentality participating, in a way, in the very essence of things—should remove the difficulties met with by Putnam and others. In this connection he referred to Whitehead's view that the ultimate existing things are (mostly nonhuman) "occasions of experience"; a conception that, indeed, seems of such a nature as to remove the intentionality difficulty, even though (he granted) it is unclear how it might fit with those of classical physics. But he pointed out that, on the other hand, it has some formal similarity with quantum mechanical notions so that the idea of some protomentality inherent in nature at a fundamental level might be supported and refined by taking some quantum features into account (in this connection see section 18-4-2).

With Bas van Fraassen's *constructive empiricism* the question comes up differently. In his book *The Scientific Image* (van Fraassen 1980) this author took the view that science should keep clear from metaphysics and expressed skepticism about the no-miracle argument and inference to the best explanation. He therefore rejected scientific realism. Instead of stating, as scientific realism does, that through its theories science aims at providing us with a literally true description of the world, constructive empiricism states that it merely aims at yielding "empirically adequate" theories, that is, theories that save the phenomena. Van Fraassen did not deny that to consider structures and processes not directly accessible to observation may be useful, but he refused to endorse any positive claim concerning their intrinsic reality. However, at the same time he parted

from the classical empiricists in that he claimed that the distinction between what is and what is not observable is a scientific, not a philosophical one. He called this the "Grand Reversal." "Science itself"—he wrote—
"delineates, at least to some extent, the observable parts of the world
it describes. Measurement interactions are a special subclass of physical
interactions in general." And he added "It is in this way that science itself
distinguishes the observable which it postulates from the whole it postulates. The distinction, being in part a function of the limits science discloses on human observers, is an anthropocentric one. But since science
places human observers among the physical systems it means to describe,
it also gives itself the task of describing anthropocentric distinctions" (van
Fraassen 1980; cited by Shimony 1993, I). Shimony pointed out that these
quotations (I would say, particularly the first one, about measurements
just being physical interactions) show van Fraassen's Grand Reversal program comes close to the one of "closing the circle." Both aim at locating
the knowing subject within the natural world.

But is the van Fraassen procedure for doing so while keeping within
the borderlines of empiricism (as opposed to realism) entirely self-consistent? It is on this point that Shimony expressed some reservation. He
granted of course that science is entitled to describe the observer as an
object (to deal with him or her using the grammatical "third person," as
he wrote). This "object" may be treated by the methods of neurology
and described with the help of the usual concepts, receptors, transducers,
information processors, etc., but—Shimony stressed—as a matter of principle constructive empiricism is fundamentally *agnostic* about the ontological status of such entities. At the same time however, due to its very
status as an empiricist doctrine, constructive empiricism is grounded on
the notion of a knowing subject who exists in an exemplary way. Indeed,
far from being an element in the theory, this knowing subject has to be
"the auditor of the story and the judge who has a certain attitude towards
it, such as belief and commitment." Now, Shimony claimed, this sets us
on the horns of a dilemma. If we keep distinguishing this "observing subject in the first person" from the above described "subject in the third
person" we fall back on the traditional conception of an epistemology
antecedent to science that the Grand Reversal is meant to eliminate. But
if we take them to be identical we thereby identify an element of the theory
(the "third person" subject) to an entity (the observing subject in the first
person) that (to repeat) "exists in an exemplary way," as the primary
repository of the empirical evidence that the theory must fit. We are
thereby unfaithful to one of the constitutive principle of constructive empiricism, which, again, is a basic agnosticism implying that no assertion
about "what exists" may be claimed to be valid beyond the mere empirical adequacy level.

Admittedly the question thus raised, hinging as it does on the importance to be imparted to the "first person" in the very definition of empiricism, is, within pure philosophy, a subtle, debatable one, the analysis of which falls somewhat outside our subject proper and will not be pursued further. Let us merely note that, on this point also, Shimony expressed the view that his thesis of an irreducibly mentalistic component in nature opens the way towards a solution. Briefly summarized, the latter consists in letting ourselves be guided by commonsense as well as by "the massive evidence that a normally maturing child, without any philosophical sophistication, constructs a rough picture of the world and places himself or herself within it." Such considerations incite him to accept the above mentioned identification of the first and third persons while conserving the crucial features of each one of these two aspects. The implication is that empiricism has to be dropped in favor of, as Shimony puts it, "a modest but nontrivial version of realism" (nontrivial because, to repeat, nature has, according to it, an irreducibly mental component). Shimony stated he is aware that "the nature, constitution and boundaries of the knowing subject *qua* subject, the first-person aspect, are all problematic." But still, concerning the Cartesian *cogito* he judged that "the core of Descartes' argument can be salvaged." And as for the empiricist's concern that metaphysics be strictly avoided, clearly he did not raise it to the level of an absolute imperative. He noted that it may well entail agnosticism and that agnosticism is uninformative.

Concerning the theme of this book, what I think should be remembered from all this is that even though leading contemporary philosophers could cast most serious doubts on the validity of the independent reality notion, still, as long as they (purposely) kept to pure philosophy they succeeded no more than Cassirer and others in radically discrediting it, nor even in discrediting realism. Other philosophical approaches—the one of Shimony has been given as an example—point to the opposite direction.

On the other hand it should be noted that all these developments took place within the framework of the classical, philosophical matter-mind debate and that, within them, the discoveries of contemporary physics were, on methodological grounds, just ignored. It may be wondered whether this approach is entirely sound. In fact, in this connection Shimony—himself an expert in the matter; see, in appendix 1, the crucial improvement he brought to the Bell's inequalities—referred to a "dark cloud" showing that any possible realization of the "closing the circle" program will have to be quite intricate. The cloud in question is just simply the impossibility, noted in this book first part, of imparting an ontological interpretation to quantum states.[5] In virtue of this, the realist hypothesis concerning things is no more just "doubtful and groundless," as empiricists long took it to be. It is most highly questionable and even straightforwardly false concerning some of these "things." This generates an entirely new field of philosophical questioning, which, of course, is just the one that we here strive to explore.

[5] Incidentally, concerning what additional light quantum mechanics may cast on such matters, note that, obviously, one of the Grand Reversal claims, namely, the idea that science is what determines what is observable and what is not, is well supported by the quantum impossibility of simultaneously measuring incompatible observables. But also note that the correlative one, the claim that measurements are just ordinary physical interactions, does not—by far—fit so naturally into the quantum picture. We discussed in section 9-4 the pros and cons of an interesting proposal from van Fraassen, partly aimed at carrying out such a fit.

CHAPTER 14

CAUSALITY AND OBSERVATIONAL PREDICTABILITY

§℘

14-1 Introduction

ABOVE WE MAINLY dealt with problems concerning the notion of reality, and in what follows this will still be our main subject. However, in studying it we could not but note that investigations of this kind require making use of such ideas as explanation, cause, prediction of observations and so on, whose apparent clearness, as is well known, is deceptive. To remove possible uncertainties concerning already considered points and get adequate means for forthcoming studies we must therefore spend some time reflecting about these concepts. Of course, concerning such notions as, for example, the one of cause, on which deep thinkers pondered for centuries, there can be no question of systematically reviewing the relevant literature. But by restricting ourselves to what is of direct interest for our subject we may nevertheless hope to make a few points clearer. In this chapter we shall successively examine determinism, chaos, quantum indeterminacy, and the question of understanding why and to what extent predicting what will be observed is more reliable than trying to describe things as they are. The next chapter will deal with the notion of "explaining."

14-2 Causes and Laws

Again, this is not a proper place for summarizing the rich literature that, for two centuries and more, has been dedicated to the relationships and differences between the "why" and the "how" or, more or less equivalently, between the notions of "cause" and "law." From a historical viewpoint, let us just note that towards the beginning of modern times, pioneers such as Galileo and Descartes were confronted with a physical science inherited from Aristotle that, for explaining phenomena, made use of a notion of cause more or less modeled on human will. Final causes in particular (a stone falls so as to get nearer to its natural locus, as if some obscure yearning at reaching its normal rest-place inhabited it). And, more generally, causes inherent in the objects themselves. Even Galileo, at first, believed that the force governing the motion of an object, far from being external to it, lies inside it. Considering the case when some object is thrown upwards, he wrote: "the force gradually diminishes (*gradatim remittitur*) within the projectile when the latter is no more in contact with the *projiciens,* same as, within an iron object, the heat diminishes when the fire is no more present."[1] Long and difficult was the process thanks to which matter was finally cleared of the mysterious internal properties Aristotle had naively endowed it with.

Indeed, even now the word "cause" is said to refer, implicitly at least, to somewhat similar views, so that some authors take the notion in question to be crypto-anthropocentric and subtly metaphorical. As G. Toraldo di Francia noted (1981), we know we are able change a number of things around us in the way we want. Never mind whether or not this freedom is illusory. What counts is that every one among us psychologically feels quite sure he/she enjoys it, and also that, to all appearances, it lies at the very core of our notion of a cause. When, following a free decision we take, we change something in our environment we claim, by definition so to say, that we are the cause of the change. To take up Toraldo's example: if we want to break the glass of a window we move our arm, pick up a stone and throw it. We then know that the glass is broken in virtue of that free decision of ours, which entitles us to claim that *we* are *the cause* of the event. But still, for achieving this result we had to make use of an "intermediary," namely, the stone. It is therefore natural that, as soon as we try to reflect on our action, we should think: "Really, I was not the direct cause of the breaking. The stone was." And the point is that we have a natural tendency to think this way while keeping to our old instinctive definition of a cause, which tightly links it with the free decision of a mind. According to Toraldo *(loc. cit.)* this tendency may well have been inherited from our far ancestors. As he puts it: "humans have believed in animism for thousands of years and easily assumed that there is exactly the same relationship between the stone and the breaking of the glass as there is between our will and the movement of our hand." It is clear however that stones do not take such free decisions, so that finally the very notion of "cause," at least in its intuitive form, appears not to fit smoothly within the general structure of present-day knowledge. The fact that, in the mind of scientists, it is nowadays very much devalued compared to the one of "law" is therefore understandable.

On the other hand it must be granted that within our everyday experience of the physical world it is for us extremely difficult—not to say practi-

---

[1] Galileo *De Motu*, quoted by M. Clavelin (1968).

cally impossible—to entirely give up the use of the said notion. When we assert that, on the oceans, wind is the cause of waves we do not have the impression of taking an animist stand. We think we are just stating a trivial physical fact, and we feel that for doing so only a language involving the word "cause" is appropriate. Are we right? Yes, I would say, but this is because, within this framework, we impart to the word "cause" a meaning appreciably differing from the just analyzed one. We take it to be synonymous with the expression "initial condition" as it is made use of in connection with a physical law. Of course, there always are several, in fact, infinitely many, such conditions. But since most of them are normally satisfied we do not even think of them. So that finally the "cause" in this other acceptation of the word is just the one among the initial conditions that at the start (or "a priori") was the least probable one.

We meet here with the notion "law" and we immediately see how enlightening it is. Indeed, let us fancy a moment a strict follower of Aristotle trying to understand the (small) motion of a pendulum and bent on making use of the notion of cause, taken in its ancient and medieval sense. Imagine him pondering the importance of the "half-conflicting" relationships between two causes. The first one—gravitation—inherent in the pendulum, strives at bringing it to equilibrium. The second one—inertia—also inherent in the pendulum, strives, now at bringing it nearer to the said equilibrium, now at driving it away from it. None of the ideas this thinker would link together (by concatenations suggestive of the reasoning procedures made use of in economics!) would be demonstratively false. But all the same, a research work making use of so subtle a dialectic would be arduous, painstaking and (to say the least!) presumably hardly fruitful. Whereas combining the law of gravitation with the one of vector addition of forces (adding up a smattering of trigonometry) quantitatively clears up the matters in a few strokes and very well.

On the other hand it must be granted that also the notion of law involves definitional problems. It calls forth the idea of necessity and in the past, along the lines of the English empiricists, many philosophers of science considered the idea of physical necessity to be but a remainder of redundant metaphysical views (a misplaced evocation of some divine commandments). Consequently they claimed the concept of laws had to be restricted to that of purely *empirical* laws. If, they argued, we know that event A is always followed by event B, we may legitimately ground observational predictions on such a knowledge; and it is clear that replacing this empirical law by the statement "A is always *and necessarily* followed by B" would in no way improve the predictions in question. Such was more or less, Hume's standpoint. It is a self-consistent one. But, on the other hand, as soon as we try to define causality, as above suggested, by means of the notion of laws (Convention 1) and decide, following

Hume and others, to reduce laws to empirical laws (Convention 2) we encounter serious difficulties. The most obvious one is the following. If we observe (proposition *L1*) that *A* is always followed by *B* and (proposition *L2*) that *B* is always followed by *C*, according to Convention 2 we must say that both *L1* and *L2* are laws. Then, according to Convention 1, we shall say, "*A* is an initial condition, *L1* is a law, hence, by definition, *A* is a cause of *B*." And, for the same reason, we shall have to say, "*B* is a cause of *C*." Now it may very well happen that one or the other of these two assertions in no way corresponds to what we have in mind when we make use of the word "cause." More precisely, what often happens is that the statements that, in this respect, are deemed correct are "*A* is a cause of *B*" and "*A* is a cause of *C*." Such, for example, is normally the case when *A* is the arrival of an international flight at an airport, *B* the appearance of a waiting file at the passports booth and *C* the delivery of luggage in the luggage reception hall.

These and similar considerations induced, as it seems, most epistemologists to give up the idea of grounding causality on the notion of purely empirical laws. Moreover when studying the notion of explanation (section 15-2-2) we shall see that the "spiritual heirs" of the classical empiricists finally found themselves obliged to forgo reducing the very notion "law" to pure empiricism, and that therefore they had to grant that no genuine explanation—to the extent that one is required—can be grounded merely on the empirical law notion. Hence, when all is said and done it appears that causality can be reduced neither to just a naive anthropocentric delusion nor to the simple fact that phenomena undergo regularities.

14-3 Determinism and Causality

"An Intelligence capable of knowing, at a given instant, all the forces and the disposition of all the entities present in nature, and sufficiently profound to submit all these data to analysis, would grasp within the same formula the motions of the largest bodies in the universe and those of the lightest atoms; to her, nothing would be uncertain and the future as well as the past would be present to her eyes." This passage from Laplace (1814) is presumably the most glamorous definition of determinism ever put forward. Some elementary remarks are in order concerning it.

First it may be noted that it is in keeping with realism, nay, even with objectivist realism. However, it is clear that we could, without depriving it from meaning, substitute, in it, to the phrases "entities present in nature" and "motions of the bodies" the phrase "outcomes of nondisturbing measurements." In other words, in the abstract nothing, apparently, prevents us from fancying a weakly objective theory that would be determinist in an (almost) Laplace-like sense.

Next, let us note that, as Laplace's sentence makes it clear, determinism proper introduces no time arrow into the picture. In it, past and future are dealt with on equal footing. This distinguishes determinism from the causality notion, in which, quite on the contrary, the difference between past and future has the crucial role of distinguishing the cause (occurring first) from the effect (coming second). Incidentally this is just what makes so surprising the "backward causality" notion that Feynman introduced—on the basis of calculation procedures—in order to describe antiparticles, as we saw in section 9.6.

To prevent confusions that semantics might generate, let us finally note that, in connection with the advent of the theory of special relativity, the temporal distinction between cause and effect resulted in that progressively the word "causality" took up two different meanings. More precisely, we saw (section 3-2-1) that—according to the relativistic and "pre-Bell" conception—given some event B, the only events liable to have had an influence on B, and hence to rank among its causes, are those lying within its backward lightcone. For symmetrical reasons, the only events B is liable to have an influence on are those lying within its forward lightcone. This state of affairs was given the name "Einsteinian causality" but, for shortness sake, physicists often just simply say "causality." Since classical special relativity still was a deterministic theory, within it Einsteinian causality was obviously just an adjustment to it of the causality notion, meant in the sense defined above (determinism with adjunction of the cause-effect distinction). However, it turned out that the advent of quantum indeterminacy naturally led to another meaning being imparted to the expression and to its shortening as well. Today they are used even when no determinist link is assumed between the involved events. In other words, the expression "Einsteinian causality" nowadays merely refers to the thesis that all influences are propagated from the past to the future and that none of them travels faster than light. And by extension, the word "causality" is very often given the same meaning. We see that therefore this word has, in physics, two very different meanings. The fact, of course, is regrettable but fortunately the context removes the ambiguity in practically all the cases.

14-4 Determinism and Chaos

The laws of classical physics are determinist. This means that if the position of the system representative point in phase-space (i.e., the set of the position coordinates and velocity components of the material points composing it, or equivalent data) is known at a given time, applying these laws yields the trajectory of the said point (that is, it yields the values of the said positions and velocities at any earlier or later time). Today however, most authors stress the impossibility of knowing the position of such a representative point with an infinite accuracy (and even of imparting a precise meaning to such a notion) so that questions relative to the stability or instability of the trajectories take up, in their approach, a considerable importance. This induces them to introduce a distinction between the determinism defined above, which they sometimes call "mathematical determinism," and something they call "physical determinism," which is a property that, even in classical physics, only some systems possess. Roughly speaking the latter are those the trajectories of which are sufficiently stable with respect to space and time, that is, are such that if they proceed from representative points that are close to one another they remain close to one another for an arbitrarily long time. In fact, it is now known that even within Newtonian mechanics "physical determinism" is obeyed only by some very simple systems such as the two-body ones. Henri Poincaré's works relative to the three-body systems showed that even such simple systems do not enjoy the stability in question (they are said to be unstable). We all know that this discovery gave rise to very important theoretical developments, known under the name "chaos theory."

Is this violation of "physical determinism" within classical mechanics a matter of *strong* or of *weak* objectivity? Though never asked, the question is highly significant. It comes all the more naturally to mind as, as we just saw, Poincaré's works were at the origin of the discovery in hand. For in epistemology Poincaré developed a conception—"conventionalism"— according to which physics is aimed, not at unveiling Reality but merely at discovering relationships between phenomena and providing us with the most *convenient* accounts of the latter. Within such an (obviously "weakly objective") philosophical approach "mathematical determinism" is of course devoid of any physical meaning so that within it there exists no sense of the word "determinism" in which an *unstable* physical system could nevertheless be termed "determinist." But conversely if, contrary to Poincaré, we opted in favor of objectivist realism, the question would arise whether—still within the classical realm—the concepts of chaos and "merely physical" determinism would, for us, retain a genuine physical meaning. For indeed, within objectivist realism it does not seem consistent to impart a basic, physical signification to notions defined using expressions such as "the impossibility of knowing" or similar ones, which, clearly, refer to human possibilities. And, correlatively, still within the same conception it seems difficult to deny that the Laplacian definition of determinism remains meaningful. True, it too refers to knowledge. But contrary to "physical determinism" it does not use knowledge or lack of knowledge as an ingredient in an attempt at explaining or classifying physical facts. True also, this definition refers to data (the positions and velocities of all material points) to which we have no real access. But what counts here is the fact that—according to the views that lie at the very basis of objectivist realism—these physical quantities do exist. These observations should make us suspect that claiming at the same time (as is sometimes done) that objectivist realism is right *and* that the instability phenomena found in classical chaos violate determinism is a logically inconsistent stand.

In fact however, before being able to claim that this conclusion is a firmly established one we must examine the question somewhat further. This is because, today, authors who deal with chaos sometimes refer to Poincaré's following passage. "When the same antecedent occurs again, the same consequent must also occur; such is the usual statement. But, such a general statement expressed this way is of no use. For, for us to be able to state that the same antecedent occurred again it would be necessary that *all* the circumstances occurred again . . . and that they got *exactly* reproduced. This never being the case, the principle could never be applied (Poincaré 1905)." Poincaré inferred from this that we must bring in the stability concept and that determinism finally boils down to the above-defined "physical determinism." People who uphold the view that in chaotic phenomena determinism *is* violated and, at the same time, are reluctant at giving up objectivist realism might feel inclined to point out that actually this passage does not bring human abilities into the picture since it just refers to events that "occur." In particular, it makes no reference to the human-centered notions of convenience, etc., that characterize conventionalism. Hence these people might feel inclined to see in it a possible starting point for a refutation of determinism carried out within an objectivist realism framework. However, it seems to me that, in fact, such a refutation would be unconvincing since, as we noted in chapter 1, objectivist realism goes in par with counterfactuality. Admittedly, Poincaré was right in stressing it never happens (that all the circumstances occur again exactly as they once did); and since he did not believe in objectivist realism this remark was, for him, a pertinent one. But for somebody who accepts objectivist realism (and hence also counterfactuality) it is not, since this person must take into account the fact that the same circumstances *might* very well occur again.

Incidentally, there is another argument that goes the same way. For we may note that, contrary to the definition of determinism that refers to an "Intelligence" (also called the "Laplace demon"), the one we just read, "When the same antecedent occurs again, the same consequent must also occur," refers to *several* sequences of events, all of them really taking place. It follows that Poincaré's reasoning finally rests on the fact that the world is not large or not old (or not, etc.) enough for it to be the case that, spontaneously, the initial conditions should all, simultaneously, occur again. And we may consider, somewhat as we did in section 9-7 concerning the very notion of reality, that within realism it is awkward to have to ground the definition of such a basic notion as determinism on contingent circumstances such as this one.[2]

Hence, to sum up, contrary to a view that seems to be widespread it appears that whoever considers himself to be an adept of objectivist realism cannot logically claim that the phenomena related to chaos violate determinism just because they are "chaotic." Let it however immediately be added that, anyhow, such considerations are, to a large extent, academic since they bear on classical physics and its laws. It is well known that, at the level of anything that might deserve the name "ultimate reality" or even just that of "microscopic reality," these laws are violated (we know that, at the microscopic level, only the quantum ones are correct). Elementary as it is, this observation severely diminishes the pertinence of the, sometimes uttered, statement according to which the advent of the theory of chaos constitutes one of the most important conceptual upheavals that ever took place in physics. But of course this is not to say that the theory in question is uninteresting as far as basic ideas are concerned. Quite on the contrary it has the considerable interest of showing that within classical physics there are phenomena that imitate intrinsically random ones to such an extent that they are operationally indistinguishable from the latter.

Finally, the notion of chaos was sometimes linked with a thesis to which some writers gave the impressive name: "end of certainties." Let us stress once more the extent to which such a view is questionable. In fact deriving it from chaos theory is consistent only if "certainty" is identified with "predictability" but this is much too strict a limitation. We may quite well be convinced that such and such physical laws are strictly true and grant that, in the complex situations that occur in practice, they make reliable predictions possible only for limited lapses of time.

14-5 Quantum Indeterminacy

Most people with a smattering in physics are inclined to think that what mainly distinguishes quantum physics from classical physics is the intrin-

---

[2] This is also what sounds unpleasant within the definition of cosmic time based on the (so-called "cosmological") principle of isotropy of the Universe. Incidentally, note also that the Laplace definition of determinism, making use of counterfactuality as it does, is thereby grounded—as any realist definition must be—on modal logic. This is not—or at least not to the same degree—the case concerning the Poincaré one.

sic indeterminacy of the former, contrasting with the basic determinism of the latter. And most of them feel confirmed in this view by the famous sally: "God does not play dice." Einstein once made use of in order to express the aversion he felt for the quantum mechanical approach. Consequently, some readers may have been surprised that, while this book first part abundantly dealt with quantum mechanics, in it the intrinsic indeterminacy of this theory was not stressed. In fact, this postponing to the present second part of an obviously quite important subject proceeded from two, tightly connected reasons. One of them is that the indeterminacy in question is conceptually quite different from what it is most often believed to be, so that, to be able to analyze it correctly, we first had to gather the relevant information. The other one is a point Einstein was already (as some texts from him show) quite aware of, even though his quoted sally does not reveal he was. It is the fact that, contrary to what is usually thought, the most important, distinctive quantum mechanical feature is not indeterminacy. It is weak objectivity, as analyzed and discussed in chapter 4.

Let it once more be granted that the fact quantum mechanics is merely weakly objective is far from being immediately obvious. For instance, it has no role in determining the energy levels of a system. The values these levels have are found by solving an "eigenvalue problem" of quite a classical type, conceptually similar to the problem of finding the possible modes of oscillation of vibrating strings. And even within the study of the quantum measurement problem, in which, we noted, weak objectivity plays a crucial role, the latter is, as we shall shortly see, somehow concealed by a wrong interpretation of quantum indeterminacy, which, at first sight, erroneously supports the view that strong objectivity is correct. Presumably this is why the weak objectivity of quantum mechanics is considerably less well known than its intrinsic indeterminacy, and correlatively it is also the reason why the latter is commonly taken to be the main characteristic of the theory in question. Sections 14-5-2 to 14-5-4 will show why, on the contrary, weak objectivity is its foremost distinctive feature, indeterminacy being just a relatively secondary aspect of the way the said weak objectivity appears. But, before that, it is appropriate that we should take a look at the historical development of ideas in this field, just in order to grasp how indeterminacy, after having been negated for a extremely long time, suddenly got to be accepted.

14-5-1 What about "Sufficient Reason"?

Today indeterminacy is accepted. And, apparently, nothing, in it, goes against rationality. However, strangely enough it was long considered unthinkable. Indeed, in Leibniz's world view contingent facts were ruled by a *Principle of Sufficient Reason* "according to which we consider that no fact can turn out to be true, or existing, no statement can be veritable without there being a sufficient reason for it to be so and not otherwise. Although the said reasons cannot, most of the time, be known to us . . ." (*Monadology*). Or, to take another quote: "Nothing takes place without a sufficient reason. Which means that nothing takes place without it be possible to he who would sufficiently know the things, to give a reason that would suffice for determining why it is so instead of otherwise" (*Principles of Nature and Grace*). And, similarly, it was in the sense of strict determinism that Kant meant the *causality* he took to be one of the categories of understanding.

It is most surprising indeed that on such a basic question the firm beliefs of the community of thinkers should have changed so drastically. It is therefore interesting to inquire somewhat about the reasons—apart from possible metaphysical or theological ones—that induced past philosophers to believe in causality in the sense of determinism. The question of course directly leads to examining Kant's arguments. Now when we study the passages in Kant's works that deal with the causality notion we observe that in fact they are centered on that of time. When he dealt with the apparent succession of phenomena Kant pointed out that, although time is an a priori form of our sensibility, time itself is not perceived. That all that we are aware of is that our imagination sets one phenomenon before another one but not that, within the object, one of them takes place before the other one. And that what is needed is therefore that the relationship between the two be conceived of in such a way that the order in which they are to be set—this one first, the other one second—should be made necessary. He noted that, concerning some, but not all, sequences of perceptions (e.g., those of a boat sailing down a stream but not those we get of the various parts of a house when we look at them successively) we have within ourselves a principle that makes us consider their order necessary. And this principle (a "pure concept of understanding" in Kant's vocabulary) is just that of the cause-effect relationship. According to him, this is the way in which the causality notion forces itself upon us. Within us, to repeat, there always are successions of apprehensions, but most of them we do not interpret as referring to events taking place. While on the contrary as soon as we perceive, or assume, that a succession follows a rule, we picture to ourselves something like an event actually occurring, which means that we recognize something that we must situate at some definite place in time. And finally Kant added that therefore something can get its well-defined place in time (be an "event") only in virtue of the fact that, in the previous state of affairs something else is assumed that it always follows, that, in other words, it follows *according to a rule*.

Relatively to the question in hand these last words are especially significant. Kant made them more precise by specifying, first that we cannot invert the time sequence and second that, given the anterior state, what follows it is strictly determined. He added, "hence the principle of sufficient reason is the ground of any possible experience." And he commented by asserting that if, given the antecedent, the consequent did not necessarily follow we should take it to be a subjective fancy and should consider what seems objective in it to be just a dream.

This mention of a dream may help us somewhat in grasping the substance of Kant's thinking. The question of knowing what differentiates reality from dream always puzzled philosophers. And, from an operational point of view, they hardly found any other answer than just "regularity," which means "a rule." When apropos of some remembrance or other, I ask myself, "was it real or did I dream it?" and finally judge it was a dream, it practically always is the incoherence of the whole story that makes me decide this way. It is the fact that the events follow one another in a "really absurd" manner. And this yields a possible criterion, namely, "are to be called (empirically) real only the sequences of apprehensions that conform to some regularity criteria called by us 'physical laws.' "

Is this criterion a good one? I think it is, and, moreover, I consider that, in spite of the fact that quantum physics is indeterminist it also applies to it. For let us think of phenomena taking place at such a small scale that we may easily build up statistical ensembles of them. If, concerning some such ensembles we were to observe none of the statistical regularities that have been observed in an extremely large number of similar instances I think we would infer from this that the involved events are dreamlike. We would not take them to be real. In other words, I note, following Kant, that regularity in time is a criterion on which we may reasonably ground our decision of considering phenomena to be (empirically) real. But if we are intent on calling "determinism" the regularity in question, we should immediately add the epithet "statistical." Kant's mistake was, I think, merely a question of omission. The idea of widening the determinism notion by extending it to *statistical* determinism simply did not occur to him. He did not realize that it might well constitute a criterion of empirical reality just as valid, for such small-scale phenomena, as the strict causal succession one is concerning the phenomena at a human scale. Considering that, in his time, the study of such very small-scale phenomena could not even be contemplated, his fault is, of course, infinitely pardonable. Still, it follows from the above that, while Kant's reasoning does show the necessity of something like statistical determinism it does not prove that *strict* determinism holds true. Moreover, since such statistical regularity notions took shape much after Kant's time, the foregoing considerations should, I think, help us understand why so dramatic a change as the giving up of the principle of sufficient reason—at least in its Leibnizian-Kantian form—could occur at all.

14-5-2 On the "Ensemble Theory"

But then, what is the real nature of this "statistical determinism"? The question is tricky and was given various answers. The one that comes first to mind, since, at first sight it looks the simplest, is the "ensemble theory," also called the "statistical" or "stochastic" approach. It rests on the view that the quantum formalism (grounded as we know on state vectors and Hilbert-space operators) describes the behavior, not of individual systems but just of ensembles of systems. According to this theory the probabilities the formalism yields do correspond, as stated in text-books, to the relative frequencies with which—upon measurements separately performed on every element of the ensemble—such and such outcome is obtained. But it is specified that the formalism does not yield more and, in particular, does not inform us about what happens to each individual system. More precisely, the adepts of this theory deny that the state vector representing the ensemble should necessarily represent the state of each individual system considered separately.

Up to this stage the theory hardly lays itself open to criticism. But it is a fact that many experiments do involve individual systems so that the question of knowing what may be said about the behavior of the latter cannot be evaded. For example, in the case considered above of measurements performed on an ensemble of systems we may well ask what the meaning is, with respect to an individual system, of the "relative frequency of the outcome." To this question Ballentine answered (1970) that the said frequency is identical to the probability that, just before the system-instrument interaction took place, the quantity measured on the system *had* the value in question.

Obviously, whenever it is not the case that all measurement outcomes are identical this interpretation implies that, before the measurement process took place, the values of the measured quantity were not the same on all systems and that therefore these systems did not all lie in one and the same objective state. This must hold true even though the ensemble of these systems was (by assumption) described by a state vector (alias, a wave function). It is therefore clear that the theory implies the existence of hidden variables. Now, Bell's theorem shows, as we know, that any local hidden variable theory is at variance with both quantum predictions and experiment. To avoid these contradictions any ensemble theory must therefore be identified with a nonlocal hidden variable one.[3] Clearly, the startling simplicity of the hypothesis on which this theory rests is thereby irretrievably lost. We observed above (chapter 9) that several aspects of such hidden variable theories render them most unattractive. It follows from what we just noted that the same is true concerning the ensemble theory.

14-5-3 On the "Intrinsic Probability" Notion

Classical mechanics—Newtonian or relativist—is embedded, as we saw, in the objectivist realism conceptual framework, and is, moreover, a determinist theory. A pencil resting on its point (an example Poincaré used) falls in a direction that is but apparently undetermined. According to the laws of the mechanics in question this direction is, in fact, determined by a myriad small parameters. True, the latter (as Poincaré stressed) are on the whole unattainable and irreproducible, but they do exist all the same. We may also imagine nails fixed on a wall and balls falling on them. When they encounter a nail the balls have their trajectories deviated, some to the left, the others to the right and, for a ball falling on a nail from straight above it, this also depends on structure parameters that are unattainable, irreproducible and so on. But still, they exist.

We already noted that, in the abstract, a pure thinker might conceive of a theory that would be deterministic but merely weakly objective. Still in the abstract (that is, independently, at the start, of anything we know), is it possible to conceive the opposite, that is, a "mechanicism" that would be strongly objective and thus in line with objectivist realism, but still would not be deterministic? Yes it is. Within objectivist realism we easily imagine a word in which the direction in which the pencil falls, rather than being determined by small parameters, would, even in the eyes of a Laplace demon, be strictly a matter of chance. And in which the same would be the case concerning the side on which the ball falls. In such an approach we may speak of an *intrinsic* probability that the ball should fall, say, to the right.

For a while some philosophers, and, following them, some physicists as well, believed quantum mechanics to be essentially a theory of such a type. And, in fact, it would seem that a few of them still believe it is. More precisely, what they claim—still within the framework of objectivist

---

[3] It is therefore surprising that L. E. Ballentine, one of the main supporters of the ensemble theory, wrote (*loc. cit.*) that in view of Bell's theorem any hidden variable theory is physically unreasonable. On these questions see, e.g., D. Home and M.A.B. Whitaker's detailed report (1992) and *Veiled Reality*, section 13-6, which reach the same conclusion.

realism—is that the state of a quantum system is characterized by, along with the properties we well know about, an additional one. This additional property the system has consists in that, in such and such circumstances, there is a certain intrinsic probability—sometimes called *propensity*—that it should behave in such and such a way. As we see, in this theory—call it "Theory *T*"—the basic idea is that strong objectivity should be preserved at all the stages of a system evolution, including those when it interacts with another system, be the latter a measuring instrument or any other, micro or macro, object. In Theory *T* everything takes place as in the example of a ball falling straight on a nail, when it is assumed that the ball has an intrinsic probability of being deviated, say, to the right while still being, at all times, a well-localized "real ball."[4] Supporters of Theory *T* could point out that since, in it, the state of a physical system is (or may be) endowed with the rather peculiar property we called "intrinsic probability" it is not—or at least not exclusively—descriptive. That it comprises a predictive element since this probability the system (or rather the global system made up of the considered system and the instrument) presently has will manifest itself only in the future, when and if some interaction takes place.

It is important to realize that, however attractive Theory *T* may look, it does not stand close examination. To see this, the best is to make use of the example considered in section 8-2—the one of an electron entering into interaction with an atom—and in assuming once more, just to make things simple, that the electron gets absorbed by the atom. When the electron initially is in the quantum state *a* + *b* there is, we said, a probability 1/2 that the final system *F* (the atom including the absorbed electron) be observed in state *A*, and a probability 1/2 that it should be observed in state *B*. (These observations being made with appropriate instruments that need not be described here.) Theory *T* interprets this by means of an implicit assumption of a realist type. For it amounts to state that, in an ensemble of systems *F* prepared this way, about half of them are, after the interaction but before anything else (such as being observed) happens to them, in state *A*, the other half being in state *B*. In other words, we should, at such a time, have to do with a *proper mixture* of systems *F*, half of which are in state *A* and the other half in state *B*. Now, we saw in section 8-2 that such a description entails, concerning the outcomes of possible later measurements of observable quantities pertaining to the elements *F* of this mixture, implications that are at variance with the quantum mechanical predictions, and hence erroneous. Finally therefore, as a gen-

---

[4] Of course, strictly speaking this property pertains to the ball-nail system rather than to the ball alone. Similarly, in the electron-atom example to be considered below the probabilities pertain to the composite electron-atom systems.

eral interpretation of quantum mechanics Theory T may not be kept. It is true, to repeat, that this theory does comprise a predictive element. In this respect, compared to the standpoint of classical physics, it is an advance in the appropriate direction. But this advance is but one step forward. It is clear that a further step is necessary and that we are forced to take weak objectivity into account. In other words, asserting that quantum mechanics is a predictive theory is not enough. As long as it is considered to be a general theory, applying to physical systems of any scale and complexity, it must be specified that it is predictive of *observational results*. Indeed, in view of the foregoing and the more detailed considerations developed in section 9-7, we see that to interpret the quantum mechanical probabilities as done in Theory T might, at most, be coherent only within the framework of a "realism of signification." And moreover, as we observed in that section, a realism of signification drastically parting with objectivist realism.

14-5-4 What then is the Real Sense of Quantum Probabilities?

To this question the foregoing yields the answer. Contrary to those of classical physics, they are not ignorance probabilities. Nor are they intrinsic probabilities of the type of those that, within Theory T, a ball falling on a nail would have of being deviated to the right or to the left. Quantum probabilities are both intrinsic and "of appearing," which mean, they are "probabilities to appear to observers." That the necessity of bringing in the notion of "observers" raises conceptual problems is not to be denied. The very existence of the present book is, in a way, a proof it does! But the existence of these problems should not bring us back to the view that the said probabilities just refer to our ignorance, or are intrinsic in the sense of Theory T, since facts rule out these assumptions.

14-6 Predictability and Reliability Revisited

The fact that the quantum laws are mainly predictive—and predictive of observational results—has very positive implications concerning their reliability.

This point we already noted, in section 7-2-1 in particular. As observed there, history of science shows that when a theory got replaced by a more general one, most often the descriptive concepts that had served as a basis for the first one were radically set aside and replaced by other, quite different, ones. Consequently, the formulations—in descriptive terms—of the laws were bound to radically change as well. Undoubtedly such considerations rank among those that led authors such as Michel Serres (1997) and Edgar Morin (1982) to announce the decline of the universal and the comeback of the event. However, the authors who stress these aspects of the evolution of sciences often tend to forget or ignore the fact that in all such upheavals, the predictive content of the former theory practically always remained valid, at least to a very good approximation. True, the remark is quite trivial. But it is important nevertheless. In general a "recipe" is suggested by some pieces of knowledge, and it may well happen that, in the course of time, the latter prove to have been misconceptions. But if the recipe worked it will go on working, at least within the realm of its former field of application. Who wonders that in practice Newtonian mechanics continues to be used, rather than Einsteinian general relativity theory, for calculating spaceship trajectories?

This being the case, a question comes to mind. Why is it that the predictive rather than strictly descriptive character of the laws of modern physics never gets mentioned by the people—scientists mostly—who, against Kuhn or Feyerabend followers (or against thinkers haunted by the ghosts of complexity and contingency), maintain that science is cumulative and pertinent? The answer seems to be that these people did not analyze lucidly enough the very nature of the predictive character in question. That they kept to a notion of "predicting what will be"—in other words to one pertaining to the strong objectivity realm—instead of realizing that what is finally involved is the notion of "predicting what will be observed," that is, a weakly objective formulation of the basic laws. For it is clear that, as long as one keeps to prediction of what will be, the predictive nature of the laws in question hardly increases their reliability. Since we remain inside the descriptive realm, we still run the risk that the concepts by means of which the descriptions are formulated may be made invalid by advances in human knowledge. A true gain in terms of certainty, and hence cumulativity, only appears when we take into account the fact that the said laws predict *observation outcomes*. Incidentally this shows that the weak objectivity of quantum mechanics is not to be brought to the discredit of the latter. Note in this connection that, even though classical physics was strongly objective, it is only because it was also weakly so (strong implying weak objectivity) that it remains significant even now (by correctly predicting observational effects).

Admittedly however, in a more theoretical mood we may wonder what it is that makes scientific prediction of observations more reliable than scientific description of facts or events. To investigate this question, the best is to try and imagine situations in which such scientific observational predictions might turn out wrong.

It is clear that this might happen if (in a realist approach) the laws of nature happened to change, if the sensorial abilities of human beings changed or if the conditions of application of our predictive laws were modified. The two first assumptions look rather unrealistic and the feeling of improbability the first one arouses is especially worth noting. Indeed, it reveals the extent to which the notion of a—knowable or unknowable but stable—reality underlying our experience is engrained in the mind of every one of us (including even radical idealists although it is at variance with their views!). The third assumption is therefore the only one concerning which we may rather easily imagine situations in which it would materialize. For example, if, during the nineteenth century, the Earth had come within the vicinity of a black hole the scientists of the time would indeed have observed that some of their observational predictive laws were progressively becoming less and less accurate. And not having the spatial curvature notion at their disposal, they would have been unable to appropriately modify them. Effects of similar types remain of course possible. And, within a less dramatic outlook, we know of Duhem's and others' approaches showing that a theory that steadily yielded well-verified observational predictions may suddenly start yielding erroneous ones, just because initial conditions imperceptibly changed. As a consequence of all this, correct prediction of observations should not be taken to be the unfailing touchstone of certainty. However, imperfect as it is, the touchstone in question still is, as compared to others, extremely good. Within the realm of normal scientific practice it is considerably more reliable than the sets of descriptive concepts that theorists use to build up.

REMARK 1: OBSERVATIONAL PREDICTION AND CHAOS

It should of course be clear that the claim "observational prediction is reliable" does not in any way conflict with the view that "the Real" is veiled (or totally hidden). What has been shown to be reliable is a certain human ability at predicting what human beings will apprehend. Basic reality ("the Real") is obviously something else. But the question as to how the said claim matches with such recent notions as those of unpredictability, undecidability, chaos, etc; is different. And it is one that may quite legitimately be asked.

To analyze it we must first of all set aside grossly incorrect interpretations of the just mentioned notions. Nowadays quite a number of people think they imply that no scientific prediction is reliable. This, obviously, is a trivial mistake, and one that experts in calculating tide schedules, space-missile trajectories, and so on run strictly no risk of making! Next, we have to realize that what these notions truly mark the end of is just a delusion we long labored under. Delusory indeed was the idea—inspired by objectivist realism combined with Laplace determinism—that everything that holds good concerning observational prediction could be extended without ado to predictions of imagined observations no human being will ever be able to perform. Clearly, we shall never observe such things as a point, nor the exact value of an observable having a continuous spectrum. Giving up the vague illusion that somehow we could in no way affects the reliability of genuine scientific prediction. Seen under this light, even quantum indeterminacy, far from jeopardizing the reliability of scientific prediction in fact corroborates it since, as noted above, statistical determinism normally goes with it.

From these observations—supplementing the foregoing ones—the conclusion, however, should not be drawn that, in the field of great conceptual ideas, the discovery of instability and chaos brought nothing new. True, as we saw above, the phenomena in question do not, by themselves, suffice for disproving "scientific realism" à la Laplace. But this hardly matters since quantum physics already convinced us that such a position in untenable. What counts is that the said quantum physics now yields most persuasive reasons to judge that, on this basic issue, Poincaré's philosophical standpoint was closer to the truth than Laplace's one. We thus have strong reasons—which the findings of chaos theory can of course but confirm—of considering the standpoint in question (I mean, Poincaré's) to be the right one. Now, accepting it somehow implies that we unreservedly endorse the limitations Poincaré brought to determinism, "unreservedly" implying that they have to be extended to quantum statistical determinism. Hence we see that, once thus replaced within a general philosophical conception, the recent developments in the field of instability and chaos do present a considerable interest. For indeed they show that, even though statistical determinism restores the predictability that counts in practice—in the macro field in particular—this predictability is not the conceptual straightjacket that it once appeared to be. We now have reasons to consider that even within our macroworld it leave room for such inspiring—though vague—notions as those of contingency and unpredictability, nay even—maybe—creativity.

REMARK 2: ON LAWS AND FACTS

Above, the hypothesis that the very laws of nature might change was set aside as "unrealistic." We therefore considered that if, by any chance, our laws came to lose some of their predictive power the only possible explanation would be that the initial conditions dramatically changed. (As we know, the expression "initial conditions" is a convenient one for designating "contingent conditions of application" of any kind.) Clearly, such a view is meaningful only if a strict distinction is made between the two involved notions, the one of laws of nature and the one of initial conditions; and the said distinction in turn implies that a similarly strict one should be made between laws and facts. Now, can such a distinction—which is attributed by some to the Ionic pre-Socratic thinkers and lay at the basis of the whole development of science—still be considered absolute today?

In view of present-day cosmology the question may indeed be asked. In fact it may be raised under different forms, some more pertinent than others. Among the former we find the one Dirac adopted, consisting in wondering whether or not the so-called "universal constants" (light velocity, Planck's constant, gravitation constant) really deserve to be called "constants." Couldn't they have varied in time? Particularized in this way the question, while difficult, hardly raises true conceptual difficulties for it cast doubts, neither on the notion of "law" nor on in its difference with the one of "fact." But other authors go farther. Some—who are not scientists but either epistemologists or sociologists or just pure philosophers—even seem, if not totally unaware of the difference between (unchanging) laws and (varying) facts, at least unaware of its significance and importance. Some, for example, seem to identify the notion of laws of nature with just the fact that events get repeated. So that the facts that the Universe has a history and that its constitutive matter itself has a history make them quite naturally consider that, finally, nothing at all is stable. I, for one, take this to constitute a regrettable shift toward a phraseology that may lead to a variant of nihilism. For indeed, if there are just scattered events without any fixed link between them can some meaning still be imparted to the word "knowledge"? Criticizing scientism is justified in many respects but should not induce us to throw overboard all the ideas—including the one of a difference between contingent and noncontingent—that helped building up knowledge. Just as "letterism" was not the acme of poetry, rejecting universality is not the climax of science. And whoever studied complexity and chaos theories knows they are not built up on a negation of laws.

14-7 The Influence Notion Revisited

In everyday language we often speak of "the" cause of an event as if it were the only one (which is not justified, as we saw), whereas the word "influence" naturally conveys an idea of plurality. We normally speak of the influences, or of one of the influences, that acted on the course of the events. This is why, in chapter 3, the word "influence" was used.

It is clear that such a difference is purely conventional and raises no conceptual problem. This is not the case concerning the point now to be studied. Above, we saw that, in order to avoid anthropocentrism the best is to identify the notion of cause to the one of initial condition. And of course this also holds true with respect to the "influence" concept. But within a definition of "cause"—or "influence"—constructed this way the temporal order obviously has a primordial role since it, and it alone, distinguishes between the cause and the effect. Is it possible to reconcile this with the fact that the "Aspect-type" experiments suggest taking into consideration the notion of an influence exerted between two events the temporal order of which depends on which reference frame we arbitrarily choose to make use of for thinking about the process?

One first element of the answer is of course that we must give up referring to this temporal order in the definition of the concept to be constructed. Instead of speaking of causes and effects we should just speak of causal relationships. To the notion of an influence of A on B we must substitute the one of influences taking place *between* A and B (unless, unafraid of neologisms, we venture to speak of "influential relationships"). But we must be careful. By generalizing this way, do we not run a risk of introducing meaningless expressions into the discourse? In other words, within the framework of these Aspect-like experiments is it still justified to speak of influences rather than of, just, correlation-at-a-distance? As we noted in section 3-2-3 the "supplementary theorem" shows that such a justification can in no way be grounded on operational considerations since these "influential relationships" cannot be made use of to send signals. Under these conditions the question indeed does arise whether or not (within physical realism, of course) it is at all possible to build up a definition of the influence-at-a-distance notion that would obey two conditions. First it should be a natural extension of the one to which the word "influence" normally refers and second it should be such that the outcome of the Aspect-type experiments should appear as expressing the existence of superluminal such influences. Unfortunately the problem of actually building up such a general definition proves difficult, if not impossible. So that is seems reasonable to fall back on a less ambitious purpose, the one of defining *cases in which* we shall say that such influences take place. This second purpose is definitely less ambitious than the first one since the cases in question may well correspond to rather special situations.

Here these special situations will be those that were considered at the place where the locality principle was stated (section 3-2-1). To take advantage of them let us consider the proposition—call it *Proposition A*—stating that, according to the precise definition of "influence" (a definition we do not dispose of but we imagine existing and not being at variance with our common use of the word), no influence is propagated faster than light. Then, whatever this definition may be it is clear that the events taking place in R' cannot influence those taking place in R (notations as in section 3-2-1). However, as we already noted, this does not imply that events in R and R' *cannot be correlated, for it may well be that both parties undergo influences from events situated in their common past.* More precisely, *call E* one given event in R and call K the events situated in region V (section 3-2-1, fig. 3.2). For the just stated reason, if a Laplace daemon is anxious to gather the best possible information as to whether or not she will see E happen but is informed of the nature of only some of the events K, it is in her interest to supplement her information with data relative to region R'. But, conversely, if she is precisely informed of the nature of all the events K in V and takes them all into account, then, in view of proposition A and our conventional acceptation of the "influence" notion, we may consider it obvious that for her such a supplementary information is redundant. In other words, in view of the said proposition we may safely assert that the information in question will not modify the probability of seeing E happen that she evaluated without making use of it.[5]

This shows that if the said assertion turns out wrong—that is, if the locality principle (section 3-2-1) is found to be violated, it will be very difficult indeed not to interpret this effect as being a violation of Proposition A, that is, not to attribute it to faster-than-light influences. "But, we might feel tempted to observe, the demon does not exist! How could the locality principle—which refers to hidden variable values that only the demon could know—ever be tested?" The answer to this is that arguing this way is tantamount to forgetting Bell's theorem and its surprising power. As we saw in chapter 3, thanks to this theorem the locality hypothesis can indeed be tested, *was* tested and was found violated. So, as precisely defined above proposition A must be declared wrong. If existing things are considered imbedded within an Euclidean, of near-Euclidean, space, it seems about impossible to deny that, in a sense akin to the common one, faster-than-light influences (or "influential relationships") do indeed exist.

---

[5] In the dart case (section 3-1) this is indeed what happens. If we are not informed of their initial orientation, hence, in particular, of the orientation of the left dart during its flight, then, for predicting as best as we can what will be observed on the left-hand side it is our interest to take what was observed on the right-hand side into account. But if we know precisely the initial orientation of the left dart such an additional information is strictly of no use to us.

CHAPTER 15

EXPLANATION AND PHENOMENA

15-1 Introduction

IN THIS CHAPTER, which is a logical continuation of the former, we study—still, of course, in the light of contemporary physics—several approaches to the "explanation" notion. We also inquire whether or not (and in what sense) empirical reality may be considered the carrier of causes.

15-2 The Notion of Explanation

To fully explain an event such as the falling of a stone we need to know two sets of data, a general law (e.g., gravity) on the one hand and some initial conditions (e.g., the angle and velocity with which the stone is thrown) on the other hand. In everyday language these conditions are called "causes." However, when, still within everyday language, we claim we know the explanation of something we usually have but one of these two elements in mind. In the example in hand—and, generally, when we are interested in some individual event—this privileged element is the second one. We assert that the explanation is the throwing of the stone because in our mind the law—gravity—goes without saying. On the contrary, when we have to do with repetitive phenomena no individual cause or effect is singled out and what is normally called "explanation" is knowledge of the law. It is the law of gravity that "explains" the uniformly accelerated motion of objects falling in vacuum. Theoretically this might induce some ambiguity in the use of the word "explanation" but fortunately this ambiguity is practically always removed by the context.

This being said, still it is true that the terms "explanation" and "to explain" appeared at several places in this book, that using them always seemed natural and unproblematic, but that they did not take on quite the same meaning everywhere. Some questioning followed. In section 2-5, concerning bubble chamber tracks, it seemed appropriate to consider that the quantum mechanical laws yield a genuine "explanation" of the said tracks. And that they do, even when conceived of to be just observational predictive rules synthesizing our experience in this domain. But on the other hand, universal as they are, still, in view of their very nature these predictive rules do predict something only because, implicitly, induction has been postulated. Now, at the beginning of section 3-1 it seemed clear—on the example of imagined correlation between phone rings—that induction grounded on such predictive past experience does *not* constitute an explanation.

We have here to do with an—at least apparent—inconsistency that we cannot truly sweep away just by relying on trivial observations (such as stressing the contrast there is between the extremely general nature of the quantum rules and the very special one of the phone-ring correlation). In fact, as we know (section 2-8), it can only be removed by considering that indeed there is a difference between explanation as conceived by an objectivist realist (the realist each of us intuitively is) and explanation as conceived by an adept of the view that physics is merely a description of experience. The latter rests content with an explanation by conformity with predictive rules while the former demands much more.

On the other hand, this problem may also be turned upside down. I mean: given the existence of these two explanation modes and the fact that each one corresponds to a set of situations in which it appears to be genuinely valid and appropriate, we may inquire in a somewhat philosophical manner about the conditions under which these judgments are legitimate. This is what now will be attempted.

To this end, the best is to go back to the root of the matter. Strict or nonstrict correlation-at-a-distance effects are, as we saw, especially striking instances of phenomena that demand an explanation. Of course, when such a statistical correlation is observed, the first idea is to try and find, among the known phenomena, a cause of it expressible in strongly objective terms that is, according to the above analysis, a known descriptive law *and* a plausible set of initial conditions. But suppose that no such cause is to be found. Should we invent one at all costs? Or shall we rest content with an "explanation" couched in weakly objective terms involving the observer, that is, finally, with some more or less direct reference to some efficient observation-predicting recipe? These seem to be the only two possibilities but each one, as we know, is open to some criticism. The first one may be dubbed an appeal to metaphysics. As already noted, to build up an explanation by assuming fields, particles, an "ether" or, more generally, parameters assumed to be "hidden" seems to be justified only if this assumption has verifiable implications, the verification of which should eventually yield some sort of an unveiling of the parameters in question. It has been rightly claimed that if it lacks such consequences this sets it on a level with mythology; that then, its rationality does not rise above that of the Iliad. But the other possibility also has its drawbacks. If the "recipe" to which it refers is special to a narrowly defined class of phenomena, the explanation has practically no substance. At the limit it becomes of the type "it takes place this way because it takes place this way." It becomes worth considering only when the involved "recipe" is a truly general, or, better, universal one. In correlation-at-a-distance phenomena between outcomes of measurements carried on on particle pairs this is typically the case. The theory—quantum mechanics—is, to repeat, an extremely general one. It is even likely to be universal. And it *is* of the "recipe" kind. I mean, among the pieces of information it yields, those that are absolutely certain bear neither on the deep nature of the involved systems nor on their attributes at given times. As we saw, these are matters that, in quantum physics, are interpretation dependent and the physicists do not all interpret them in the same way, which makes it difficult to fully believe in the resulting descriptions. The "sure" information the theory yields bears on the results to be obtained upon measurements performed on systems prepared in such and such specified way (or on their mean values upon repeated such measurements). And in the considered experiments it does predict what is observed.

Now, does such a well-verified observational prediction rule constitute, all by itself, an "explanation"? The "yes" answer seems to be—or have been—in favor with many thinkers and in particular those of the so-called "positivist" school. For them (see, for example, Reichenbach [1951]) "to explain" is "to reduce to some general law." And the fact that the said law is—as is here the case—stated in weakly objective terms is not liable to disturb them in any way since their ultimate referent always is of a perceptual nature. On the other hand, anybody who strictly keeps to the thus defined framework must be especially sensitive to the difficulty Hume mentioned about justifying induction.[1] For indeed, since this person decided to reason without the help of a realist presupposition grounding the notion of permanence, each time she will purpose to predict some observational outcome (by making use of the quantum predictive rules) she will have to make a new bet. She will have to say to herself: "I bet these predictive rules that worked up to now will work once more." Much as if, faced with the enigma of telephones ringing simultaneously at home and at my neighbor's, I thought: "As a matter of principle I refrain from attributing this to the existence of any agent unknown to both me and the telephone service. But, soberly, and out of habit, I bet that next time we'll again observe simultaneity." Would I take this for an explanation? Would I think: "Now everything is clear, I know the explanation"? We

---

[1] In section 14-5-1 I mentioned Kant's solution to the causality problem Hume raised. Is this solution transposable to the form the induction takes in this framework? It is far from clear that it is.

already answered this question in the negative. True, in matters of philosophical interpretation of physics such parallelisms with everyday life should be viewed with care. Again, the difference between the quantum case and this example is indeed considerable since the observational predictive quantum rules were found at the price of enormous efforts and are apparently universal. This makes us consider them valuable and naturally incites us to view their being successful in predicting an observed effect as an explanation of the latter. But still, let us not pay ourselves with words. Quite as in the imaginary case of simultaneous telephone rings, it seems fair to grant that the explanation—I mean the "true" one!—escapes us, and that, had we it, we should, intellectually, feel a little more comfortable. For we then should not be compelled, in every particular instance, to postulate anew the validity of induction.[2]

I just wrote, "had we it." In fact we do not have it but even so we may think about it. We may conceive that it exists and lies beyond mankind's intellectual abilities. Such an idea—some will say—is "shallow," "disappointing," "ad hoc." Maybe. This is largely a matter of opinion.[3] But I fail to make out in it any logical inconsistency. And the assumption is enough to remove my qualms relatively to the pertinence of induction. To think that a "true" explanation "exists" is for me "almost as good" as to actually know it. Now it seems inordinately difficult—not to say impossible—to conceive of an explanation having the here requested strength and that would not refer to some notion of "the Real" in the widest sense of the term. We see from this that the hypothesis of a Reality that would be veiled or even totally unknowable escapes the charge, most commonly brought against it, of being *fully* arbitrary. To me this, at least partial, elucidation of the induction problem seems preferable to the one Cassirer (1910) put forward, as we saw in section 13-3-2. In a case—in atomic physics, say—of strict correlation between two distant measurements, each one of the two outcomes is random (it is sometimes "yes" and sometimes "no," without anything determining beforehand what will be the case) but a strict correlation is nevertheless observed between them. For example, if one of them is "yes" the other one is "no" and conversely. It

---

[2] Epistemologists use to stress that due to the problem induction raises it is impossible to empirically demonstrate the validity of universal laws, and some consider that this is a reason for casting doubts on realism. Some of them even seem to think that this fact makes the idea of justifying induction through realism unacceptable. To agree with this view is not a logical necessity. If once and for all we postulate realism—even just *open* realism—we thereby explain the existence of universal laws and we are no more in the necessity of postulating induction again and again, every time we do anything.

[3] And of course I fully agree that within the framework of our usual research activities, in which we look for the specific causes of well-specified phenomena, the idea would be defective. But clearly we are concerned here with a problem of quite a different nature.

is not clear how "logical necessities" bearing (as claimed by Cassirer) on one particular event in the sequence might explain this (except by assuming source-correlated hidden variables, a hypothesis much too "metaphysical" to be accepted by Cassirer and which, moreover, the Aspect-type experiments refute). True, the somewhat vague expression "logical necessities" might still, perhaps, be understood to refer to some *logos* preexisting, in a way, everything. But this would amount to tacitly reintroducing as a primary notion the notion of absolute existence that, precisely, neo-Kantians mean to discard. Besides, and to repeat, my notion of existence is a broad one. There would be no inconsistency in considering that the "Real" it refers to essentially consist of the said logos.

15-2-1 Objection

It consists in questioning the very need for an explanation notion. To ask for an explanation of some event is to inquire about its cause (or causes). In other words, it is to ask about its "why." The here considered criticism focuses on the idea that science owes its successes to the fact that it got to be more modest. It now rests content with trying to find "how"—in other words, according to what *laws*—events get produced, a law merely being knowledge of regularities in the concatenation of phenomena. Asking oneself questions other than such ones is running the risk of overstepping the bounds of general rules of understanding. It is therefore appropriate to either set aside the very notion of explanation or unreservedly subordinate it to that of law. In *Philosophical Foundations of Physics* (1966) Carnap, setting himself within the framework of pure classical physics, clearly explained this standpoint by means of an example, the one of the "entelechy" notion. This word was introduced by the biologist Driesch and was meant by him to refer to some assumed nonphysical force making living beings behave as they do (each type of organism or organ having its particular entelechy). Driesch saw entelechy as a philosophical explanation of phenomena, such as tissue regeneration, that, in his time, had no scientific one. In order to justify his approach Driesch pointed out that also the physicist, for explaining processes such as objects being attracted by one another, assumes the existence of forces, magnetism for example, that are not seen. And the criticism Carnap addressed to him consisted in observing that the physicist does indeed call magnetism in, but that this is, for him, merely the starting point of the explanation. That the actual explanation lies in stating the *general laws* of magnetism such as, for example, the simple one that magnets attract small iron objects. In other words, Carnap stressed that, in order to produce an explanation, or a prediction, what counts is not to introduce a new entity, it is to make use of laws. Now Driesch stated no entelechy-involving law. He merely referred to most commonplace empirical laws. And Carnap felt therefore entitled to ask him: "What do you add to the explanation yielded by these laws and the observational predictions grounded on them by claiming that the phenomena they govern are due to the specific entelechy of the living being under study?" The here considered objection to the notion of a veiled (or unknowable) reality may equivalently be expressed as "the predictive laws being known, what does the idea of some—veiled or not veiled—underlying Reality add to the explanation they yield?"

15-2-2 Answering This Objection

Before stating possible replies to this objection let us note that Carnap's criticism of Driesch merely stressed the fact that any valid explanation has to refer to one or several empirical laws. It did not prove that such a reference is sufficient, and indeed, as we shall see below, Carnap himself was progressively led to consider that it is not *quite* sufficient. That much being noted, there are at least two possible replies to the objection.

First Reply

It is grounded on the evolution of the notion of physical knowledge. As we noted, in classical physics time, even without explicitly committing themselves to realism scientists could implicitly entertain the view that somehow physical knowledge consists in a *description* of events actually taking place "out there." Now, for people who thought with such an idea at the back of their mind it, admittedly, was rather natural to consider that the general laws governing sequences of events of a given type yield, ipso facto, an explanation of such and such particular event of the said type. The reply proposed here lies in conjecturing that the fact Carnap and many others came to relate the notion of explanation to, exclusively, that of law was just due to this circumstance. It consists in claiming that such a way of thinking was then so intuitive and familiar to everybody that the idea of taking *explicitly* into account the concept of some underlying reality came to be considered superfluous, "metaphysical" and so on. In these people's mode of thought, the notion of "something that is described" thus became just a sort of an uninteresting, "borderline" concept, while still a vague notion of description (of events, considered as phenomena) was kept. However, as we have seen, with the advent of quantum physics the very possibility of such an ambiguity vanishes. Knowledge clearly is not identifiable with description. Nowadays, instead of being descriptive, the basic laws are, in fact, just rules enabling us to correctly predict observations. Now, are such laws, by themselves, still truly explanatory? As we just noted, this may well be questioned. And indeed we already observed that among the present-day thinkers who are critical concerning (traditional) realism some do not categorically reject the notion of a "ground," unattainable through discursive knowledge. What the present reply suggests is that in fact the said notion, although it does not constitute a piece of knowledge, is valid as an (ultimate) explanation.

SECOND REPLY

It is grounded in the remark that if we were bent on limiting ourselves to some totally strict empiricism this would restrict our prediction abilities to actually observable facts, which implies that in our analyses we could take into account only the empirical laws that are truly—i.e., presently—available. Now, as Carnap himself granted (1966), a most unpleasant consequence of this is that strictly speaking it would make it impossible to ground the notion of causal relationship, hence also the one of explanation, on that of law. For—he pointed out—if we tried to do so we should say that today such and such causal relationship is true, since there is a law that accounts for it, but that yesterday it was not, since that law was still unknown. Which sounds absurd.

To overcome this difficulty, Carnap was led to identify the meaning of the statement "event B is caused by event A" with the one of the assertion "basic laws exist in nature from which, when associated with an exhaustive description of event A, event B is inferred logically," it being specified that this definition is valid whether or not the laws in question are actually known. In other words he considered the assertion that a causal relationship exists to be meaningful as soon as the existence of such laws is postulated, although of course its *truth* cannot be ascertained as long as the said laws remain unknown.

What remained then to be done was to specify what features distinguish such "basic natural laws" from accidentally universal statements. This Carnap did by stating that only the first named ones make it possible to corroborate counterfactual statements (such as: "if this iron rod were heated it would extend"). In view of the fact that counterfactuality and realism are linked as we saw, this may be taken to be, by itself, a step towards a recognition of the role of the notion of reality. But what is perhaps even more significant in this respect is just the above noted fact that Carnap—however anxious he was to keep away from metaphysics—had to impart a meaning to the notion of natural laws assumed to exist even though unknown. If such a view is not identical with open realism, at least it stands quite close to it. For indeed, whoever says "there are in nature laws that . . ." postulates thereby that the notion of a "nature" endowed with "structures" is meaningful. And if the said "laws" or "structures" are assumed to exist even when unknown, this implies that the "nature" that carries them is somehow external to us. Hence, to express oneself as Carnap did there finally amounts to grant that the notion of a human-independent nature—or "Reality"—is meaningful and useful, independently of specific assumptions as to the possibility of getting to know its structures.

The basic quantum mechanical laws are, as we know, merely predictive of observations. They therefore cannot be identified, as such, with the said "structures of the Real." And if they constitute the most basic elements of physical knowledge, as I believe is the case, this implies that these structures are not knowable as they truly are. Still the foregoing argument indicates that we should take them to really exist. This therefore is another path leading to the conclusion that the notion of an either unknowable or, at least, Veiled Reality does have a meaning.

15-2-3 Explanation, Linguistic Framework, and Relative Ontologies

To the reasoning described above it could be objected that Carnap himself seems to have had in mind another solution, grounded on his notion (summarily reviewed in section 5-4) of a "linguistic framework." If we suppose that when (1966) he wrote the cited passage about laws existing in nature he had the just mentioned notion in mind, we have to interpret the said passage otherwise than we did above. We have to consider that, if—at the risk of being misunderstood—he introduced in it the concepts "nature" and "existence" without stating that he meant them in a sense differing from the usual one this was in order not to weary his reader with too highbrow philosophical analyses. Had he been questioned on this, maybe he would have answered that the said passage implicitly presupposed the choice of an appropriate linguistic framework. either the one of "things," or the one of "sense data," or the one of "propositions," or etc. And maybe he would have explained that the natural laws he mentioned deal with concepts belonging to the said framework rather than with the structures of some "independent nature" external to it. At a time when it seemed classical physics yielded the keys of all phenomena, the appropriate framework was the one of things and their attributes. In the present quantum era, it may be maintained that it is the one of the sense data. But in either case we remain within a purely scientific realm, free from any Platonic ontology or the like.

Is the answer I thus attribute to Carnap a really satisfactory one? As for me, I think it is, but only in part. From the angle of science understood in quite a strict sense, I think it is. I mean, if our purpose just is—as apparently Carnap's was—to provide science with a fully consistent framework, within which statements have clear, unambiguous meanings, then indeed the notions of linguistic framework and our choosing the most suitable one are appropriate. Within classical physics, choosing the framework of things and their attributes enables us to state—or even imagine the existence of—basic laws that, within this framework, do "explain" in detail the unfolding of the phenomena. Within quantum physics, choosing the sense-data framework in principle enables us to do the same. But if we aim at philosophical understanding I consider that in both cases the thus provided "explanation" is, when all is said, not basically satisfactory. I mean, it comes too near to just being a way of making everything explicit and does not really quench our thirst for genuine explanation. In the first case, this deficiency is masked by the fact that the idea of the "things" composing the chosen framework is isomorphic to the one we intuitively build up of things composing some, completely independent of ourselves, "outside world." But "to mask" is not tantamount to "to eliminate." And when we remember that the said "framework of things" is, in Carnap's very words, essentially "linguistic" and "chosen by us" we come to feel that the explanation in question could be a fake. In the second case, no faking is even at work, so that the deficiency is more manifest. For indeed, in that case, the involved basic laws are just universal observation-predicting rules, and the notion "observation" has no correspondent within our intuitive conception of an external world independent of us. For example, if the said predictions are about correlation effects between outcomes of measurements performed at different places, the fact of disposing of a recipe enabling us to predict these effects, though intellectually satisfactory, does not, as we repeatedly noted, give us the impression of constituting a genuine explanation. On the whole it therefore appears that taking the linguistic framework notion into account does not, by itself, suffice to modify the general conclusions reached at in the foregoing section.

I feel inclined to say the same apropos of Quine's relative ontology notion (section 5-4). With the difference however that, contrary to Carnap, Quine did not stress the role of choice. Indeed, as we saw, he mentioned "the ontology to which one's use of a language *commits* him" (emphasis mine). Since, in view of the context, we here obviously have to do with the (relative) ontology of the "world of things" this may suggest transferring somehow to the there considered "things" the explanatory power we, above, acknowledged "Nature," or "Reality" has. Now, it is true that, in ordinary life, we unhesitatingly attribute such an explanatory power to *empirical* reality—as defined at the end of section 4-2-3—just because we conceptually mix it up with human-independent Reality. But on this we are wrong. From contemporary physics and the developments of this book first part it clearly follows that empirical reality—the "world of things"—cannot be identified to reality-per-se. On the whole, it would be much less misleading to call it an "appearance." Now, as a rule, appearances (a ghost, the television image of a gunshot) are not susceptible of constituting genuine explanations of anything. Still, the idea here in germ might well be that in the case in hand the appearance is so "robust," involves ourselves to such an extent, that we are entitled to transfer to it the explanatory power that, in the eyes of the realists, only "the Real" has. (On this, see section 15-4.)

15-3 Back to the "Explanatory Power of Predictive Rules" Question

When a theoretical physicist, having heard of the discovery of some new physical effect, succeeds in showing that, according to the quantum mechanical laws, this effect was indeed to be observed, he is pleased. The fact of knowing that the laws in question are merely observation predicting ones—that they are "weakly objective" only—in no way diminishes his legitimate satisfaction. And his colleagues just as himself consider, with no reservation whatsoever, that he truly "explained" the phenomenon in question.

On the other hand, this notion of explanation, is not a one-stage rocket. Whether what is to be explained is some particular event or a general phenomenon, there is, first, the quest of explanation through a law and then, once the explanation—that is, the law—is found a desire to explain that very law. As we just observed, this desire—or, better to say, the feeling of a task remaining to be done—is particularly acute in the case of laws that are just predictive of observations. Our mind then somehow feels compelled to attribute to them some support in "the Real." And this is indeed the idea that, up to this point, has been steadily upheld in this work.

In a way, this yearning for explaining the law, then explaining the explanation of the law and so on involves some risk of infinite regress. Obviously we must stop somewhere, and the question is, where? Which stage is to be the last one? Concerning the, most enlightening, example of quantum correlation-at-a-distance effects my own choice up to here was, on the whole, to hold the following discourse. "Our observational predictive rules are fully general. In cases of strict correlation they do predict that two distant experimentalists who, after a measurement run, meet and compare their respective outcomes will find that, on each individual pair, both outcomes were always the same. From a strictly scientific point of view I take this to be sufficient ground for considering the correlation effect to be *explained*. And then, in a second stage, which however is a purely philosophical one, I account for these very rules by assuming, without any attempt at preciseness, that they are partly isomorphic to the deep structures of 'the Real.' " But, some will ask, should I not have shifted—or at least tried to shift—the borderline between scientific analysis and philosophical speculation somewhat further?

What makes the question susceptible of being investigated is the fact that there indeed are scientific theories—or at least "models"—that, supposing we trust them (they are highly speculative), do render such a shift possible. One of them is the Broglie-Bohm model. As we saw (section 9-3-2) it does account for the observed correlation, and, being ontologically interpretable, it does so without referring to the observational prediction notion. However, the price it makes us pay for this is very high. In fact, let us consider the outcomes of the two measurements made on a given pair (say, of photons, as in the section 3-1 example) characterized by the values of all of its defining parameters, the "hidden" ones included. From a detailed analysis of the problem (see appendix 3) it follows that, considered individually, the said outcomes may well drastically change, for the *same values* of all these parameters, merely according to the order in which the two measurements take place in time. This, as shown in section 5-2-4, just follows from the "Bell calculation" reported there. Under such conditions we may easily imagine the difficulties that would be raised, in this example, by any attempt at making the model relativist. But then, may we really consider that the Broglie-Bohm model yields a genuine *explanation* of the correlation under study? This is, to say the least, questionable. Better take it to yield (at least at the nonrelativistic level) a "glimmer of explanation" concerning the validity of the quantum predictive rule, without aiming at being more precise than that.

The same standpoint may be taken concerning other proposals, also aimed at yielding a "second degree explanation" of the "first degree" one provided by the quantum observational predictive rule. Consider, for instance, the, openly dualistic, theory that I once put forward as an example, consisting in explicitly introducing the concepts of nonphysical minds and corresponding states of mind. As we know (section 5-2-5) the idea is that such a mind (for example the Schrödinger cat's one) chooses one of the branches of a quantum superposition. When, as in the correlation-at-a-distance experiments, two observers, and hence two minds, are involved, it may happen in this model that they do not both choose the same branch. And it might therefore be feared that the said model be incompatible with the correlation predicted by the quantum predictive rules and checked by the observers when they compare their notebooks.

But in fact, as we saw, the mismatch between the two outcomes, even if it exists "in reality," will never be discovered, due to the fact that, if one observer wants to know what the other one observed, she can never directly ask the other one's *mind*. She has to make use of the physical channels constituted by sound waves, paper, auditory or optical nerves etc. In fact she thus performs a "measurement" bearing on the physical state of her partner. And under such conditions the quantum rules predict that the outcome of the latter will coincide with the one of the direct measurement she made. This model, therefore, is in agreement with experiment, but, as we see, this agreement is just an appearance. In fact, each subject is confined within her own self and the—misleading—answers she gets from other people merely reflect her own ego. May we call this a "realist" explanation of the observed correlation? In a way, yes, since we have here to do with a "realist" description of the physical world (via the "state vector of the Universe") along with an equally "realist" (albeit nonphysical) description of the involved minds. But who would say "yes, I am convinced"? In fact, a disciple of Putnam confronted with either one of these two attempts (this one and the one stemming from the Broglie-Bohm model) might well comment that by engaging in them we, in fact, tried to take a "God's eye view." And she might well add that this was tantamount to committing the sin of pride, and that the disconcerting character (to say the least!) of the "explanations" we managed to elaborate just serves us right. In more conventional terms: above we purposed to, at least, *try* and push the borderline between scientific analysis and philosophical speculation beyond the observational predictive rules level. The attempt was honestly made, and it was hardly successful. The conclusion thus seems to be imperative: what lies beyond the said rules is not the scientist's job. It belongs to the realm of philosophical conjectures (or motivated opinions).

15-4 Empirical Reality and Abstractions, Explanation, and Empirical Causality

In section 4-2-3 when, apropos of the measurement problem, we introduced the empirical reality notion we did this by means of what may be called a "mental process of disregarding" (or "ignoring"). We had to do so because, due to the peculiar structure of the quantum mechanical rules, we had to *justify* the "commonsense" assertion that the pointer always lies in some well-defined graduation interval rather than in some unthinkable quantum superposition. And we found we could do so only by ignoring a number of "facts" along with the terms in the formulas that are responsible for such "facts." More precisely (see chapter 8 for details), we had to "forget" that, according to quantum mechanics, there must exist physical quantities that are theoretically observable—at least by some Laplace demon—and whose measurements, if performed, would yield outcomes belying the assertion in question. In section 8-2 we then found out that, in practice, decoherence renders such measurements unfeasible. But, on the other hand, the words "in practice" are obviously meaningful only relatively to the abilities of some observers or experimentalists. Which implies that the statement about the pointer having a definite position on the dial is itself only meaningful relatively to the abilities in question. This is why the said statement merely refers to *empirical* reality (see in section 9-7 a more elaborate discussion of this point).

Of course, such considerations are not limited to pointer positions on dials. For example, in quantum mechanics it is only by disregarding some terms directly or indirectly corresponding to possible correlation relationships that we may consistently attribute shapes to molecules. Finally it is therefore the whole set of the pieces of reality that—directly or by means of instruments—we apprehend through our senses that must be called "empirical."

It cannot be denied that the necessity we thus are in of just ignoring some terms within formulas, together with the implications they would entail, has something disquieting in it. And indeed, in this book first part we extensively pondered on the conceptual problems it raises. It is true that the expression "just ignoring" has at first sight a somewhat reassuring look. After all, don't we "just ignore" or "disregard" an immense lot of data at any moment? Don't we, in any domain, for brevity sake, pass over a mass of details that might have been mentioned without damage but, simply, are not relevant for what we purpose to say? Maybe this feature of everyday talk is what explains that, apparently, some commentators, having it in mind, consider that decoherence allows us to speak of pointer positions, molecule shapes, etc. with as much "ontological assurance" as if the involved objects were intrinsically classical. But in fact, as we know, the situation is quite different. The disregarding process here at work differs very much, at least in principle, from the just described trivial one since the details it ignores would, if they were kept, prevent our description from being correct. The difficulty may be compared—with some differences of course—to the one encountered by physics teachers who, for pedagogical reasons, limit their teaching of atomic theory to a description of Bohr's planetary model and implicitly make their pupils believe that atoms are *really* constituted this way. For, as a consequence, these teachers have to disregard—in the nontrivial sense of the word—many data that happen to be flatly inconsistent with the said model (as, first of all, the energy spectrum of helium, the simplest atom after hydrogen).

Now then, is such a "disregarding process" susceptible to serve as a basis for an explanation or is it not? We are here facing a question akin in nature to the one met with at the end of section 15-2-3. Is it possible to ground an explanation merely on the *empirical* reality concept? One of the main reasons that make us intuitively prefer realism to operationalism seems to be the feeling we have that operationalism merely predicts, whereas realism both predicts and explains. In classical physics, when, say, a billiard ball hits another one we feel we understand why the latter moves. It does because it was hit by a real ball, that is, by a massive object endowed at any time with a well-defined position, a well-defined velocity, etc. And of course our feeling of understanding this is not lessened by the fact that, in describing the incident ball in this way, we ignore many other properties it has, such as its substance, its color and so on. But this is because we have here to do with a disregarding process of the first of the two above-mentioned kinds. Within the quantum realm—assuming quantum mechanics is universal—our approach is, of necessity, quite different. Neither one of the two balls exists as such, since both are entangled with the rest of the Universe. It is we who, by ignoring some terms in some formulas—that is, through a purely mental operation—separate them from the latter. Under these conditions the colliding event also is largely a mental construction of ours, and it is therefore very difficult to consider that it yields a genuine explanation.

It is instructive to compare the process we just discussed, and in which we are not involved, with one in which we *are* involved. For example, I could imagine that I am the ball that gets hit. Alternatively we may imagine the following setup. Let us think of a particle the quantum state of which is (in the sense defined in section 4-2-2) a superposition of two totally distinct states, such as, for example, two states of quasi-rectilinear, uniform motions along different directions. We may then consider that the regions where there is an appreciable chance that the particle be detected are composed of two tubes—or "beams"—lying along either one of these two directions.[4] Imagine that, at some place in one of these tubes, a detecting apparatus A is set, which acts on a relay R, which in turn acts on a flasher F that I look at. For a theoretical physicist assumed to be entirely external to the whole setup the latter is conceptually very similar

---

[4] In the laboratory such "superposed states" may be obtained by means of just an inhomogeneous magnetic field (Stern-Gerlach experiment). If a particle endowed with a magnetic moment (having a "spin") goes through such a field with its spin pointing in the general direction of the field its trajectory is deviated in a definite direction, call it "up," and takes up thereby a definite "state." If its spin points in the opposite direction its trajectory is deviated in the opposite direction, "down," and takes up another "state." Finally, if its spin points in any other direction the particle, after having gone through the field, is in a state that is a quantum superposition of the two aforesaid "states."

to the above two-ball system considered quantum mechanically. She considers that after the process has come to an end I am in a quantum state entangled with the ones of the particle, apparatus A, relay R, and flasher F. She knows the mathematical structure of the corresponding wave function, and from this knowledge she derives the probability she has of getting from me such or such answer should she ask me what I have seen. Within this inference process she nowhere has to make use of the notions of cause and explanation. But we clearly see that this is just because in fact, from her "external" point of view, no event whatsoever actually occurred.[5] The event, in this scheme, is a private affair. Only I live it through.[6] It is I who saw either that the flasher flashed or that it did not. And in either case I am entitled to claim—in my own "private" language—that there does exist an explanation to the perception I had, namely, a certain concatenation of causes consisting in that apparatus A was (or wasn't) triggered, that it did (or, respectively, didn't) trigger relay R, and so on.[7]

In a way, it seems we spot here the very origin of the notion of empirical cause: I mean "cause" understood in a Kantian sense, that is, bearing on empirical reality. For indeed, in a less schematic approach the foregoing, individual "I" should be replaced by the whole collectivity of human beings and "my" private language by our common one, the above mentioned "theoretical physicist" being confined to the Empyrean of pure theory. Hence what this analysis seems to show is that the notion of "explanation of an event" is a relative one but no more so than the notion of "event" itself. Both are meaningful only within the empirical reality framework, but within *this* framework the first named one is just as valid as the second one.

15-5 The Rainbow Analogy

As is well known, analogies are deceptive. They all are imperfect, and taking them too seriously might well, therefore, lead us astray. Already in chapter 1, quantum mechanically viewed micro or macro objects were compared with rainbows.[8] The flaw in the analogy—let this be noted right

---

[5] See note 10, chapter 4.

[6] This of course is the Wigner friend problem again.

[7] Even within a quite definitely quantum mechanical realm—that of subatomic particles—an embryo of this "dialectic of the constitution of the notion of cause" may be discerned in the fact that, in a Stern-Gerlach experiment, when a particle manifests itself in the upward beam, if we are not keen on thinking quantum mechanically we naturally tend to say, "oh well, this is because its spin *was* up."

[8] Nick Herbert (1985) already put such a comparison forward.

at the start—comes from the fact that while we are able to manipulate objects we cannot do so with rainbows. But the analogy is instructive nevertheless. For indeed rainbow theory belongs to classical physics and raises therefore no basic conceptual problem. The similarity between rainbows and quantum objects may therefore be found helpful when we try to apprehend somewhat better the true nature of the difficulties raised by the study of the latter.

The analogy itself is clear. Even within a classical, mechanistic, approach a rainbow, obviously, may not be considered an object-per-se. For indeed, if we move, it moves. Two differently located persons do not see it having its bases at the same places. It is therefore manifest that it depends, in part, on us. It is worth noting that it depends *collectively* on us, at least if "us" stands for a collectivity of people all assembled at the same place; for they then all agree as to the rainbow position. But still, even though the rainbow depends on us, it does not depend exclusively on us. For it to appear it is necessary that the Sun should shine and that raindrops should be there. Now similar features also characterize quantum mechanically described objects, that is, after all—assuming quantum mechanics[9] to be universal—any object whatsoever. For *they* also are not "objects-per-se." The attributes, or "dynamical properties," we see them to possess depend in fact on our "look" at them (on the instruments we make use of and on how we arrange them). And still the theory predicts that we, observers, will intersubjectively agree concerning the results of the measurements (or other observations) that we may make on these objects. And lastly, at least according to the veiled reality conception, even though these micro or macro objects depend on us they (just as rainbows) do not depend exclusively on us. Their existence (as ours) proceeds from that of "the Real."

This being the case, concerning the objects in question quantum mechanics raises conceptual problems for the analysis of which the analogy in hand may prove useful. The simplest of these has to do with the measurement process. Let it be restated as follows. "In the case of a particle the position of which is being measured, quantum mechanics denies us the right to consider that, before the measurement, the particle already had a definite position. Somewhat similarly, in the case of a 'generalized measurement' of the type studied in chapters 4 and 8 (the cat problem) the said mechanics denies us the right to believe that, after the process was over but before we looked, the pointer, thought of as a 'being-per-se,' lay 'per-se' at the place where it is found. Apparently we are thus forced to consider that *we* are the ones who, by looking (or measuring), generate the said, definite position (of the particle or the pointer). Now,

[9] Assumed free from "ad hoc" terms; see section 9-8.

this view is so strange that it is not easy to endorse it. Is it truly unavoidable?" A variant of the same idea would be to assume the observation is done by means of an automatic registering device the records of which we read only later. The question then becomes: "Does quantum mechanics force us to say that it is the registering device that generated the observed value?"

To try and answer such questions let us consider the rainbow case. And, first of all, let us wonder whether the analogy we have in view is valid. In other words, let us ask ourselves the following, preliminary question: "Is it really true that—as we claim is the case concerning the particle (and the pointer) position—if no onlooker were there and no registering device operated the rainbow would, just simply, be nonexistent?" At first sight we feel tempted to answer by the negative for, clearly, light reflection-refraction processes within raindrops owe nothing to our being present. But still, if two observers are not standing quite close together they do not see the rainbow quite at the same place.[10] When N observers are scattered in the fields, each one of them sees the rainbow at a specific place, different from the ones where the others see it. In fact, under these conditions speaking of one and the same rainbow seems improper. It is quite definitely more correct to state that there are N of them, and that each observer sees his own "private" rainbow. But then, if $N = 0$ there is no rainbow. Otherwise said, upon reflection we come to consider that, at least in the direct observation case, the answer to the question in hand is "yes." If nobody were there, there would simply be no rainbow.

It is easy to see that the same holds true concerning the automatic registration variant. If, instead of setting our camera where we placed it, we had set it somewhere else, we should have photographed, not at all, seen from a different angle, the rainbow the image of which we are presently looking at but, actually, *another* rainbow, having its bases on other elements of the landscape. If many cameras had been scattered in various places, many different rainbows would have been photographed. Hence, contrary to appearances, we cannot legitimately claim that the photograph we now have in hand testifies of the fact that, when it was taken, there was a well-specified rainbow, localized as shown on the picture and which preexisted to the taking of the latter. If we are bent on expressing ourselves in terms of rainbows, all we may say is that there was there an infinite multitude of rainbows, which, however, were merely virtual. Consequently, in this case just as in the direct observation one, the matter is seen to be quite similar to what the quantum formalism suggests concerning measurements performed on (micro)objects. In either case it is

---

[10] And this, quite independently of any (nonexistent) perspective effect. To repeat: the two observers do not see the "bases" of the rainbow at the same places.

not possible to state that the observed entity—or value—actually pre-existed, in itself, to observation.

And still, we feel strong repugnance to state that the triggering of the registering device "created" the rainbow. To be sure, the device preparation implied that a given event—the taking of the picture—had to take place within it. But when this happened no real entity—no real thing—was thereby *created* elsewhere even though the word "rainbow," like any other name, refers, or seems to refer, not merely to a representation we have but to a veritable entity. We see on this example that even within the framework of purely classical physics and a reflection bearing on a well understood phenomenon,[11] our irresistible tendency at reifying—be it due to language structure or be it its cause—may well generate quite genuine difficulties.

Having become aware of the latter and reached the conclusion that the notion of creation (by us of our instruments) is inadequate when our discourse bears on rainbows, we must consider that it also is so when our discourse bears on the quantum measurement processes, which show, roughly speaking, the same features. To the above-stated question "does quantum mechanics force us to consider that the observer, or the registering device, creates the observed value?" we must therefore answer in the negative. For, as we just saw, our too thinglike traditional mode of reasoning—which here, in fact, governs the very expression of the question—entails that we may not answer "yes it creates it" (or "yes the camera creates the rainbow"). Finally, in the case of quantum objects as in the case of rainbows we may state neither that the observed "entity" pre-existed, by itself, to the operation nor that the latter created it. And the fact that, concerning rainbows, this conclusion holds true within a classical, "mechanistic" world view raising no basic conceptual problems clearly shows that, despite appearances, it is not intrinsically self-contradictory. It is therefore without qualms that we may take it up within the quantum mechanical realm.

REMARK

Still, the analogy cannot be carried too far. Classical physics suffices for studying everything concerning rainbows and, in the times when the said physics was taken to be a description of "the Real," cameras and pictures—those of rainbows included—could be taken to be inherent parts of the said "Real." This constitutes a difference with the quantum case since, assuming quantum mechanics is universal, pointers and their positions on dials may only be considered to be elements of *empirical* reality.

---

[11] The theory of the rainbow goes back to Descartes.

15-6 Removing the "Paradox of the Dinosaurs"

Within the realm of the conception this book develops it is clear that the "mind-independent reality" notion—although, as has been stressed, it is appropriate for justifying the very existence of physical laws—may never be made use of for explaining the occurrence of well-specified phenomena. From this it follows that, concerning this purpose, the conception in question has to face much the same objections as those the realists commonly address to idealism. In order to set into full light the most striking one of the latter—the one that immediately comes to mind so obvious and indisputable it looks—the realists, as already mentioned, often find it useful to call in dinosaurs. They point out that, considering the numerous discoveries that were made of dinosaur bones, unquestionably such animals really existed. Further, they existed quite independently of the human mind since, in their time, human beings did not exist. To consider dinosaurs to be mere creations of the human mind would therefore be, they stress, a sheer absurdity. Now—they ask us—isn't this, precisely, what you are doing when you claim that objects we apprehend though our senses and our scientific inference processes—dinosaurs and fossils included—are but "objects for us," that is (to within too subtle shades of meaning), just appearances?

Generally speaking, the above-developed considerations already cast doubts on the cogency of massive objections of such a kind. But those put forward near the end of section 15-4 make it possible to discard this one in a more specific manner. The point is that, between pointers (or other implements relays may contain) and dinosaur bones there is no basic conceptual difference. To complete the analogy we even might assume (although this is not necessary for the argument to go through) that during its life the considered dinosaur survived a trial similar to the one Schrödinger inflicted (by thought) on his famous cat. With respect to this dinosaur we collectively are in the situation in which, in the section 15-4 example just referred to, "I" am myself, as an observer having taken notice of the flash. We explain our observation of a fossil by saying that a dinosaur really existed, just as, in the example, I explain the flash to myself by saying that apparatus A "really was" triggered. In both cases the "explanation" refers to the past. And in both the word "really" quite normally refers to empirical reality, that is, to the only reality to which we truly have access. The only difference is that, in the dinosaur case, we have no indication as to the existence of a "being" corresponding, in the example, to the physicist who started the operation and had in mind the overall wave function of the whole set, "myself" included or, at least, the idea of this wave function. But this difference is not "philosophically significant" since, in the said example, the mind of the physicist in question is not commensurable with my own. Since, as soon as this physicist gets in touch with me his state of mind changes so as to involve a reduced wave function.

The notion of relative, alias Quine-like, ontology offers a framework into which the foregoing conception rather naturally fits. It is true that Quine focused on language his own definition of it (he stressed, to repeat, "the ontology to which one's use of a language commits him"). In the foregoing approach the language notion was not referred to. Admittedly, this difference is significant. But still, it does not blur the resemblance. Basically the latter lies in the notion of constraint. In Quine's approach as in the one here put forward, what the involved observer considers to be existing or to have existed—the dinosaur, for example—is not an "absolutely existing being" (a "reality-per-se"). But, within both, the observer in question—in fact the collectivity—is quite strictly bound, without any possibility of evading it, to an "ontology" that—even though, in the eyes of the philosopher and the physicist-philosopher (and, maybe, some mystics) it is "relative"—still exhibits all the features, space extension, efficient causality, counterfactuality and so on, that, on Earth, characterize our hard, dense, inflexible *empirical reality*.

15-7 The "False Explanation" Question

The foregoing conclusion should set at rest the minds of readers whom the previous developments may have rendered somewhat dizzy concerning the notion of explanation. Of course, we all know that there are false explanations. That occurrence, at fixed times in the year, of gifts in chimneys is not *really* due to the coming of Santa Claus. That fire does not go upwards due to a tendency it has of reaching its "natural place," that planets are not guided by angels, and so on. But we used to think that in the course of time such naive, erroneous explanations had been replaced, or were in the process of being replaced, by correct, *true* ones, in the strictest sense of the word. Like Voltaire and Victor Hugo, we tended to consider Newton's works to have been a change from obscurity to true light. Previously—in Montaigne's time and before—human beings did not *know*. After, they *knew*. And we also tended to think that, in everyday life, trivial but fundamentally true explanations are constantly given to everyday events. The dizziness above referred to is the one felt by a person who becomes suddenly aware of his or her naiveté in this respect. Who, all of a sudden, realizes that, when he/she thought she understood the falling of a book or a spoon by attributing it to the force of gravity she was mistaken since, according to general relativity, there, simply, is *no* such force. Space-time curvature is at work. Above all, it is the dizziness of the scientifically minded individuals who, through reading or thinking, found great pleasure in understanding natural phenomena and who, getting informed of the quantum formalism, discover that a number of simple, clear explanations, which they took to be final ones, are, after all, relative and of debatable validity. Experiences of such a type may generate an unpleasant feeling of "ground giving way," To repeat, the conclusion of the foregoing section may alleviate the corresponding disarray by showing that, relative as they are, still the descriptions the classical schemes yield well deserve the name "explanations." All the same, however, one quantum (and "relativist") message remains pertinent. It is that such a restored assurance may not extend so far as to impart to us the complacency of those who, at one time, thought they disposed of the final, ultimate key.

CHAPTER 16

MIND AND THINGS

§§

16-1 Empiricism, Positivism, and So On

THE PURPOSE OF THIS chapter is to examine to what extent some classical epistemological conceptions fit quantum physics and may help understanding it. Clearly, because of the important contribution empiricism made to the construction of science this ancient, protean approach ranks first among those to be considered. Of course we have especially in mind its main guiding idea, namely, the view that the whole of our knowledge comes from our senses.

In the course of time the said idea was gradually made more precise and various versions of empiricism were developed with the purpose of more acutely specifying the nature of the thus acquired knowledge. First, during the seventeenth and eighteenth centuries, all metaphysical sources of inspiration and guidance were progressively set aside. Then, in the nineteenth century, with Condillac and, later, Ernst Mach, the role of elementary sensations was set very much on the forefront. Finally, in the twentieth century the Vienna Circle positivism emphasized rejection of any Kantian synthetic a priori.

Concerning this history, and in view of the themes we are here mostly interested in, let us note once more that there is one idea—a highly natural one, to be sure!—that the empiricist "founding fathers" (the empiricists who wrote before Berkeley) seem not to have truly questioned. This is the view that, if not the secondary qualities at least the primary ones (forms, relative motions, etc.) are genuine properties of "the Real," known to us as they really *are*. Apparently the same remark holds true concerning the pre-Machian, Comte-like, positivists. With the representatives of all these schools the implicit idea seems therefore to have been that, to be sure, our knowledge does not exceed our experience but that the latter, when appropriately dealt with, somehow yields a trustworthy picture of some per se existing beings. As we noted, even today such a conception—which, after all, is nothing else than physical realism—seems to be the one of a great number of scientists. They consider that we may meaningfully speak only of what is observable, directly or indirectly. But for all that they do not question the view that by so doing science progressively lifts the veil of appearances and more and more accurately reveals how Reality really is. Note that the reason why they keep to this standpoint is easily understandable since we observed (section 13-2) that before they engage in a comprehensive examination of the facts reviewed in this book nothing really compels them to give up this commonsense view.

In fact, even the Vienna Circle epistemologists seem to have, at the beginning, but weakly rejected such a scientific realism. At any rate, to the extent that they did reject it they did so in terms that were much hazier than the ones they made use of for rejecting the Kantian synthetic a priori. This, at least, is the impression that a number of their writings strongly conveys. And the fact, if true, may perhaps be explained by the circumstance that, at that time, they considered the existence of a totally certain (in other words, fully independent of our theoretical prejudices) empirical basis to be unquestionable. It is true that, at an early stage, their mentor, Schlick (1918) had put forward the axiom according to which the meaning of a proposition *boils down* to the method of its verification. But still we may wonder whether it was not just from the moment they realized—or surmised—that any observational statement is "theoretically loaded" that they thought it fit not to stay vague concerning the meanings of such simple words as "world" and "nature," and explicitly rejected their realist acceptation (or rather, denied its meaningfulness). It is interesting to consider under this angle the difference between Carnap's two positions (here, end of 15-2-2, and beginning of 15-2-3) concerning what supports the basic laws.

Anyhow, nowadays it is, quite definitely, in an antirealist sense that the logical positivists' fundamental assertions are understood. In a way, this brings them close to those of the Kantians and, even more, neo-Kantians. On the other hand their quasi-dogmatic rejection of any innateness and therefore of all synthetic a priori judgments sets them quite apart from the latter. But on what fixed soil can a philosophy of knowledge be grounded when the empirical basis is found shaky and synthetic a priori is rejected? Schematically, this, of course, was the rock on which the great positivist endeavor came to grief. But still, even though logical positivists were too rigidly dogmatic, some of their analyses remain valuable. In my opinion contemporary physics even indicates that there was much sense in the direction they imparted to their quest. The name "phenomenalism," less sharply defined than positivism, will serve us as a guide in an attempt at recovering the sense in question.

16-2 Phenomenalism

Like most names ending in "ism," "phenomenalism" suffers from being susceptible to several different meanings. Some take it to merely designate a Comte-like rule, instructing us to look for the "how" rather than the "why," with no reference made to either realism or antirealism. But in general the word is taken to evoke a more precise philosophical position, explicitly related to antirealism, and this is the meaning that is here imparted to it. More precisely we shall define phenomenalism to be the view that knowledge is strictly limited to the (physical and mental) *phenomena*, that is, to the set of all the objects of possible perceptions (supplemented with the set of objects proceeding from introspection). In order that this definition be appropriately understood it is necessary that the meaning the word "object" takes up in it be well specified. In this context the word in question does *not* refer to a concept the sense of which is taken to be pregiven, the specification "of possible perception" just serving to limit its domain to what may be perceived, directly or indirectly. Quite on the contrary, it must be considered that this word, "object," here merely designates a stable group of perceptions and even—some would claim—of not yet analyzed ones, in other words, of sensations or, in Hume's language, "impressions." Hence it seems that, strictly speaking, it is only in its weakest version, namely, open realism, that realism may be considered to be in keeping with phenomenalism. And this, of course, is only to the extent that such perceptions or sensations are taken to exist per se; in other words, if—and to the extent that—the notion of such an existence is taken to be meaningful.

Still, we must note that adherents to phenomenalism often stay rather vague concerning the status of physical objects. As Shimony (1993, II) pointed out, only a few philosophers, Russell and Carnap in particular, tried to explain in detail how the properties of physical systems might be considered to be but groups of ideas and experimental observations. But they met with difficulties that led them to give up their proposals. One of these stemmed from the fact that the so-called "disposition properties" (such as "magnetic," "soluble," etc.) may hardly be defined without making use of counterfactual statements (of the type "if this lump of sugar were put into water it would dissolve"). However, counterfactuality goes in par with modal logic and the Vienna Circle positivists felt very suspicious about the latter, presumably in view of its relationship with realism.

Another difficulty (Russell 1929) came from the fact that, as soon as the phenomenalist rejects pure solipsism he has to make use of an unprovable inference principle that no reference to empirical data could make rationally plausible. This is just the principle according to which, in such and such circumstances, we should believe in the existence of persons other than ourselves and the truth of what they assert. Russell pointed out in this respect that, while accepting such beliefs, we might as well also believe in the existence per se of the sound waves that are the carriers of such messages, etc., which would lead us straight to physical realism. Another way of expressing the same objection is to observe that sensations are private affairs, that (as the model described in section 5-2-5 vividly illustrates) we enjoy no direct access to those of other people, and that, as a matter of principle this constitutes an obstacle to building up scientific knowledge bearing on such material since science is public by definition.

Finally, a third difficulty lies in that the program of building up objects from sense data becomes more and more complex and hard to state exhaustively as we go down to smaller and smaller scale. As Shimony (*loc. cit.*) pointed out: "if the program of exhibiting an electron as a construct were somehow fulfilled, the description of the construct would be fantastically complex, presumably consisting of an infinite set of conditional statements such as 'if a cloud chamber is prepared in such a manner, then the resulting photograph will (with a certain probability) have the following appearance.' " Moreover, the latter assertions should themselves be defined by referring to other possible observations, and so on.

Note that the last mentioned difficulty is symptomatic of quite a serious mismatch between "orthodox" or "classical" phenomenalist discourses and contemporary physics. The point is that, whether or not the former explicitly make use of the word "object," anyhow they are implicitly founded on a postulate, which is that stable groups of possible perceptions, assumed to constitute the objects of knowledge, exist. Now, at an elementary level already the Heisenberg's indeterminacy relationships hardly fit with such a hypothesis. For, within quantum mechanics, when we consider the notion of a quantum particle we are at a loss how to think of a *stable* combination of a well-defined position and a well-defined velocity the counterfactually defined perceptions of which would constitute the particle. Note moreover that the fact quantum field theory serves as a foundation for the (so-called) elementary particle theory corroborates the existence of this difficulty. For indeed, quantum field theory wipes out any conceptual difference between the dynamical properties and the very existence of a particle or combination of particles. The existence in question is merely defined, as we know, to be the contingent state in which— among other possible ones—a "Something" called "the state vector in Fock's space" happens to lie. From this it follows that the "groups of possible impressions" meant to define objects have no intrinsic stability. Those that *are* stable owe their apparent stability to contingent facts. For example, the group of possible impressions supposed to constitute one particular electron disintegrates as soon as we consider situations in which possibilities of creation and annihilation are appreciable. And when we think of either the Dirac or the Feynman picture the situation looks even worse, for we are at a loss to imagine to what group of possible impressions the notion of a Dirac sea, or the one of a particle traveling backward in time, might correspond.

16-2-1 Phenomenalism and Quantum Physics:
Operationalism Revisited

The difficulties reviewed above have been known or suspected for a long time, and ideas aimed at removing some of them were in the air. With respect, in particular, to the Russell one, it may be considered that what has been mentioned in section 5-2 concerning intersubjective agreement constitutes a step towards its solution. For it shows indeed that if—dropping for the time being any attempt at a realist interpretation—our sensations are taken to obey the quantum observational predictive rules, they are not as "private" as one might think since the said rules entail mutual agreement. On the other hand it must be observed that the very formulation of these rules implicitly refers to instruments of observation readable by anybody, so that the difficulty in hand is more displaced than truly removed. Some thinkers conjecture that it was his becoming aware of the pertinence of Russell's remark that made Heisenberg give up his first inclination toward radical phenomenalism and consider, along the lines of Bohr, the whole experimental device to obey classical physics, which enabled him, through an implicit reference to the realist language of the latter, to justify the public nature of the data read on the instrument. Clearly, however, this is a hybrid solution. And moreover, we noted (chapter 8) that to consider the experimental device to be necessarily classical raises, in some cases, difficult problems.

What, I think, follows from all this is that if phenomenalism is to remain, presumably it will be in its modified version known as operationalism. But it should be added that, seen from this angle, its future looks promising. In view of the contents of chapters 4 and 8 above it may hardly be questioned that quantum mechanics renders this approach more attractive—or at least less obstacle scattered—than any other. For what, in it, proved to be unfailingly trustworthy is the set of its observational predictive rules. Indeed while concerning the explanation and interpretation problems it raises controversy still remains open—and not without good reasons!—concerning the observational predictions it yields agreement is unanimous.

Hence it is in fact towards a basically operationalist approach to science[1] that, as we already noted, quantum physics apparently drives us. And it is worth noting that, also from a purely philosophical viewpoint, an operationalist who, on this basis, takes up a radical stand finds himself,

---

[1] Basically operationalist but which, still, may be taken to be merely methodological. This restriction then means, in line with our above stated general position, that science is considered to be a discipline that takes up no metaphysical position whatsoever. Not even negative ones such as "Reality per se is meaningless."

in some respects, in a more comfortable position than one who keeps to the, more "moderate," conventional operationalist doctrine. The point is that the latter still aims at *describing objects* and intends to define the meaning of the words used for this purpose by means of statements stipulating the way the properties they designate should be measured. And we saw (section 7-2-2) that, as stressed by many authors, one of the difficulties he or she then has to cope with is that this leads in some cases to essential ambiguities. If it were possible to do without the words in question and express scientific statements directly in the form "if we act this way we get that result" the difficulty would vanish. To be sure, it will be retorted that such an objective is merely a view of the mind, for how could we describe the contemplated measurement without making use of words describing objects? The radical operationalist thus seems doomed to keep silent. Still, it remains true that the difficulty would vanish if we disposed of some ideal operationalist method with which all scientific statements would be reducible to a few very general observational predictive rules expressed with words trivial enough to be defined by pointing out. The empirical reality notion defined in section 4-2-3 comes somewhat close to this idea since the possibility of being looked at and pointed out is what characterizes instrument pointers, which are paradigmatic instances of pieces of empirical reality.

Of course, considering such an operationalism to be an essentially correct approach implies no theoretical views bearing on, or involving, the ways the observational predictive rules constituting its framework were discovered. Unquestionably the sensations lying at the root of whole process were first interpreted by means of images, which underwent large changes in the course of time. Moreover, it must be granted that such an elaboration of images and descriptions is most likely to go on taking place, and that its usefulness as an incentive for theoretical advances is considerable. However, no *intrinsic* significance, either ontological or (even) just phenomenological, is to be imparted to it. Indeed, what differentiates the here considered "radical" operationalism from traditional phenomenalism is the fact that, contrary to the latter, it does not take the perceived forms to be the *constitutive bricks* of anything. It considers them to be appearances; appearances that, in many cases, are linked to one another by reliable empirical laws and that, in practice, may thus be raised to the level of elements of (empirical) "reality," but only if we keep carefully in mind that, basically and essentially, they are but human representations.

Moreover, the reason why, in such a framework, we basically rely on the observational predictive rules is just that, up to now, they were successful. Hence we are quite prepared to look for better ones if at some time they happen not to fit any more with the ever-increasing set of the experimental data. It is quite often pointed out that we cannot prove an assumption by referring to its consequences since the same consequences may derive from quite different assumptions. But, here, this pertinent remark is by no means a source of difficulty. When two different sets of predictive rules yield the same observable predictions we just simply should accept both (till some mathematician proves they constitute two exchangeable and mutually convertible versions of the same view, which ordinarily happens!). True, this approach imparts a considerable importance to logic and mathematics. But, as they are used in quantum physics, mathematics does not incite us to build up ontological or even just empirical objects. It essentially serves to express the general observational predictive rules and calculate the detailed predictions following from applying the rules under specified circumstances.

Conceived of along these lines, operationalism, to repeat, escapes some of the troubles encountered by the more conventional version, aimed at building up operational definitions of new scientific words for the specific purpose of inserting them within a descriptive language. This advantage is nevertheless partially counterbalanced by two inconveniences that we already analyzed. One of them (section 15-2) is that, strictly logically, observing that a rule was successful in the past yields neither a proof nor even an indication that it will be successful in the future. The other one (section 15-3) is that when, as here, the physical laws are brought to the level of mere predictive rules the mind finds it uneasy to consider they are endowed with a genuine explanatory power and tends to view them as being but explanatory relays. It follows that the need we quite normally feel to dispose—symmetrically, one might say—on the one hand of a better conceptual basis for induction and on the other hand of a "genuine explanation" concerning the laws in question is not met by such an operationalism, which therefore should be complemented.

Once more we get thereby at the idea that the notion of cause might enjoy some validity even beyond the realm of mere phenomena. Such an idea is to be found in the writings of several thinkers who do not—by far—share the same philosophical views. And of course, just for this reason, it takes up, with each one of them, a specific coloration. Hence it seems proper to situate the thesis proposed here with regard to other ones that involve the same idea of nonphenomenological causes, but in a different context. Two of them will be considered: the realist one and the thesis of the "transcendental object viewed as purely intelligible cause of the phenomena."

In the first one, upheld, for example, by Shimony (1993, I) possible objections to the limits set by Kant on the applicability of the causality notion are pointed out. One of them is grounded on the considerable diversity of existing causal relationships (from the obvious instantiations of causality involving our own bodies to the hypothetical ones suggested by most theories and the purely statistical one that quantum mechanics reveals). In Shimony's opinion, the Kantian doctrine that (strict, determinist) causality is an inherent element of the a priori forms of understanding, inapplicable to things in themselves, is much too rigid to account for such a diversity. One of the other, more general, objections he raised against Kant's views was grounded on the fact that investigation of the phenomenal self by the methods of the cognitive scientists yielded information that seem hardly compatible with Kant's sharp distinction between the phenomenal and the transcendental selves. He pointed out that, since, in Kant's doctrine, the phenomenal self is subject to the categories of the understanding (causality included), if the roles of the two selves seem to merge this ruins Kant's thesis that the categories in question do not apply to the transcendental self. Finally, like Popper, Shimony (as we already noted) observed that, according to Kant, the reward for giving up all pretence to theoretical knowledge of things-in-themselves was the undoubted possession of a knowledge of the phenomena firmly grounded on the ordering principles of mind itself. But, he noted, the developments in mathematics and physics in the last century and a half have seriously undermined the plausibility of any claim to synthetic a priori knowledge. Consequently, the giving up in question is deprived of its main justification and we may therefore question its pertinence. All these reasons clearly make Shimony consider the Kantian rule that the causes (in the plural) of the (particular) phenomena may only be phenomena to be too strict.

As we saw, the Veiled Reality thesis agrees, in substance, with the latter conclusion, even though it does so on the basis of substantially different arguments. But Shimony went farther and, following a little-known suggestion of Newton, he aimed at setting the notion of causality at the very root of those of essence and substance, or, in other words, at making it the ground notion of a new ontology. "What, he wrote, a physical system is substantially, quite apart from our knowledge of it, is best expressed in terms of the causal laws that govern its behavior." This, again, might be taken to be an element of similarity between his approach and mine since, as Lena Soler (1997) pointed out, my Veiled Reality notion draws what justification it has from the idea that our predictive laws must be rooted in something, or, in other words, have a cause. On the other hand, it remains true that my way of understanding the role of this cause seems to qualitatively differ very much from Shimony's.

The other thesis here of interest is, as above mentioned, the one of the "transcendental object viewed as purely intelligible cause of the phenomena in general." Basically, this is Kant's idea. In the chapter of the *Critique of Pure Reason* that deals with the antinomies of dialectical reasoning (and, more precisely, in the section concerning transcendental idealism and dialectical cosmology) this author explained that the objects of experience are given, not per se but only within experience. He even added that they have no existence outside the latter. In other words, they are mere representations. But, as already mentioned, Kant did not scruple to take into consideration the notion of a *cause* of these representations. True, he stressed, this cause is, to us, totally unknown. But (as we observed) this did not dissuade him from giving it a name. He called it a "transcendental object." In section 1-2, concerning Kantian idealism and following Putnam, we noted an essential feature of the said transcendental object, namely, its uniqueness. It could be called nonseparable. Incidentally, its Kantian denomination of "purely intelligible" (contrasting with "knowable") cause of the phenomena *in general* adequately expresses this fact: "phenomena" appear in it in the plural, and "trancendental object" in the singular.

As already pointed out (chapter 10), between Kant's transcendental object and my own notions of extended causality and ground Reality there exist obvious similarities, even though (to repeat!) the arguments that lead to the first and the two latter ones are different. However, as also noted, between these conceptions there is nevertheless a significant difference in that mine does not bar out the existence, within the transcendental object, of some sorts of structures implying—in an undecipherable manner—our scientific laws.

16-2-2 Laws, Phenomena, and Cognition

Our striving to adapt phenomenalism to quantum physics[2] led us to the idea of an operationalism which, though radical in the sense we stated, still is purely methodological since it is moderated by the notion of a (properly speaking) unknowable "transcendental object." Now, like most views reported on in this book, operationalism has an obvious "mentalist" or "cognitive" aspect in that the notions of operation and observational prediction imply those of a person who "operates" and another (perhaps the same) one whose mental state is changed by the observation. Incidentally, there is nothing novel in such a "mentalism." It constitutes the natural stand of any philosopher upholding the view that attempts at ontological descriptions are questionable. Still, it remains that the epithets "mentalist" and "cognitive" are somewhat ambiguous so long as it has not been specified whether they refer exclusively to individual minds or,

---

[2] This subsection focuses on analysing the approach of one particular author (J. Petitot). On first reading it may be passed for its content does not have much bearing on what comes after.

more broadly, to the concept of "mind in general." According to some, this circumstance opens a possibility of limiting their domain of meaning, and, hopefully, in such a way as to avoid that any explicitly nonontological assertion should necessarily be labeled "mentalist" or "cognitive." Does this yield a possibility of building up a conception that would bring quantum physics somewhat closer to a science—classical physics—the nonmentalist character of which is generally acknowledged? This section is devoted to an analysis of this—admittedly somewhat special—point.

When we speak of nonontological assertions we can hardly avoid thinking of Kant. However, some authors, such as Jean Petitot, consider that Kant should have disentangled his transcendentalism from mentalism more satisfactorily than he actually did. Their idea rests on the notion of (physical) lawfulness. As we already observed (section 2-3) it consists in considering (Jean Petitot 1997; Michel Bitbol 1998) that modern science has, since Galileo's time, substituted for the Aristotelian program of building up an ontology the, more human, project of discovering the legal order of phenomena. Of course, this interpretation of classical science is not the only possible one and presumably it is not even the most commonly accepted one. But it is interesting and worth discussing. In it, it is granted that reality-per-se may exist, but it is considered that it is unobservable, that our knowledge only bears on its manifestations and that therefore phenomena should be defined, as in Kantianism, by a *receptivity*. The latter may be either the, direct, one of our senses or the, more general, one of our instruments (paraphrasing Bachelard, Petitot made use of a notion of "apparatus sensibility"). However, still according to the interpretation under study what is true is that, even though classical science is "relative" (i.e., nonontological), it succeeds in building up (or "prescribing") lawlike rules that radically put this feature (that of being relative) "within brackets." Under these conditions Petitot suggested, in substance, that we should focus on the existence of the—observed—lawfulness of phenomena and consider there is no point in looking for an objective explanation of them grounded on some unreachable underlying reality. He claimed we should define objectivity on this basis and stated that "thus prescriptively defined as being legality, objectivity is distinguishable from any version of ontology."

According to this conception, and still within the realm of Galilean classical physics, space and time are not real per se. They are "mental" (in the more general sense of the word). But they are forms the role of which is just to enable us to build up the said phenomenal legality. And correlatively they are entirely "desubjectivized." No doubt, Petitot meant by this that, for example, to relate a set of visual sensations to the notion of some luminous object existing at some definite place in front of us enables us to reasonably guess what the visual impressions of somebody standing close to us actually are, etc. To situate, by thought, objects within space and time and assume, on a later step, that they obey Newton's laws thus indeed enables us to "put into brackets" the properly subjective content of the phenomenon. If we decide to save the word "mentalist" (and its synonym, "cognitive") for designating what is properly subjective in this sense, that is, what is proper to the consciousness of each individual in particular, then, in line with Petitot's remark, we may legitimately claim that the classical description of things as lying in space is neither ontological nor mentalist.

May such views be extended to (microscopic) quantum physics? At first sight a difficulty appears, related to the "desubjectivizing" role (in Petitot's words) that, within classical physics, space and time play on this matter. We just saw this role is crucial, which does not come as a surprise since the notions of space and time lie anyhow at the very basis of the whole classical physics. On the other hand, in view of the spreading of wave functions, the phenomena of entanglement at a distance, etc., they do not lie (not, at least, to such an extent) at the basis of *quantum* physics. And indeed it is hard to see how they could be made use of in the problem here under consideration. But, according to Petitot, the difficulty is not insuperable for, he claimed, in quantum physics the role in question is taken by quite different entities, namely, probability amplitudes.

Considering what we saw in section 9-6 concerning the said amplitudes such an idea may, at first sight, look surprising. For we observed there that, contrary to space and time—which we usually take to be independent of us—probability amplitudes, basically, are related to us. To the measurements we actually perform or could perform. But in fact (and this is a philosophically significant point) in the problem in hand this difference does not matter. It does not, just because, in the approach we have been investigating in this section, objectivity is taken to be in no way related to ontology. In a language already made use of in the foregoing chapters, it is just weak objectivity, alias intersubjectivity. And probability amplitudes do yield just that. Probability amplitudes inform me concerning what I should observe were I to perform such and such measurement, but they inform me just as well about what anybody else would observe under the same conditions.[3]

Hence if, same as we did above concerning the classical case, we take up, for a while, Petitot's language convention concerning the words "mentalist" and "cognitive," we may, following him, consider quantum physics

---

[3] Incidentally, note that in Petitot's approach both types of weak objectivity mentioned in section 4-2-4 appear: in descriptions of what he calls "objects" only "strongly empirical" statements may be used, and all phenomena are not "objects."

not to be "mentalist" or "cognitive" any more than classical physics was. That much being said, I consider nevertheless that two important points have to be made, which set this conclusion in its true perspective.

The first one was already stated in section 2-3. Expressed in slightly different terms it consists in stressing (once more) that a difference exists between just stating a thesis or a rule (with the view of deriving consequence from it) and *justifying* the said thesis or rule before making use of it. As we saw above, within the classical physics framework, Petitot started by *stating* that reality-per-se is unobservable and that human knowledge can bear only on the intersubjective representations we build up of it. Roughly speaking, this is the Kantian thesis. It is worth very much consideration but, as shown in section 13-2 above, within the classical physics framework it is in no way logically binding. Its nature resembles that of a mathematical axiom in that it is mainly the starting point of a chain of reasoning. For indeed, as shown in section 2-3, while Galileo's discoveries rendered it plausible in the opinion of some, they did not really weaken the opposite one, namely, realism of the accidents. And therefore, because of its intuitive character, the latter, which may also be given the name "Galilean ontology" (since it seems to reflect the main lines of Galileo's standpoint), quite normally remained—implicitly at least—the world-view of most scientists. To put it otherwise: the idea that the sciences—and physics in particular—since they could (for a long time) entirely be stated in strongly objective terms should be taken to be strongly objective descriptions of external reality hardly met, during the whole classical period, objections other than purely philosophical ones. Hence, even though the Petitot-like version of transcendentalism is most interesting on its own, to take it into consideration should not prevent us from acknowledging that, relatively to the spectrum of all admissible standpoints concerning the nature of knowledge, the findings linked with quantum physics caused a considerable change. For now the idea that science yields a strongly objective description of Reality has really become objectionable on physical grounds. True, there still exist, as we saw, ontologically interpretable theories. But they are somewhat like branches weaned from the "great trunk," that is, from the body of all the theoretical activities, based on quantum theory, that remain in symbiosis with observation and experimental research. And the latter, as we saw, are but weakly objective, not due to an axiom we set forward or a decision we took, but in virtue of their very nature.

The second point that has to be made explicit is relative to the very meaning of the words "mentalist" and "cognitive" Petitot and other transcendentalists made use of. Above, following these authors, we restricted their use to what affects an individual consciousness, that is, to subjective, as opposed to intersubjective, elements. Indeed, this is what enabled us to agree with them in considering that the interpretation Petitot proposed is not mentalist. Now could this restriction be dropped? Could "transcendentalists" keep, for their conception, the label "nonmentalist" without limiting the meaning of the word "mentalist"—alias "cognitive"—to just the private realm? I am not clear whether or not they have such an idea in mind. Maybe they do not. Conceivably they would state that it just simply does not fit with their overall standpoint. But anyhow my point is that it is not a self-consistent view. To substantiate this claim let me, just for a moment, go back to Kant. I grant, of course, that Kant took great care to distinguish his transcendentalism from Berkeley's idealism. I agree that his "transcendental idealism" is, in his words, an "empirical realism"; that, in the *Critique of Pure Reason*, he went as far as to claim that the objects of external intuition "really exist" as they are actually perceived in space. But still it is true that the realism in question is *but* empirical. That, within it (as he himself stressed) objects of experience are never given per se but only within experience. And more precisely still that, outside it, they have no existence. Obviously such statements considerably water down the significance of the expression "really exist" that we just read. Since the very concept "experience" has a meaning only through reference to one or several subjects endowed with the said experience it is clear that the very notions of Kantian (empirical) reality and objects constituting it depend on that of knowledge, that is, on cognition.[4] And this holds good even though it is true that, with Kant, the said dependence is on a "transcendental subject" considered to be the carrier of the a priori forms of sensibility and categories of understanding and not identifiable to any individual. In my opinion Petitot's analyses lead to exactly the same result. Their interest essentially but merely lies in that they justify, more explicitly than Kant himself did, Kant's view that the "transcendental subject" is essentially impersonal. But they do not do away with intersubjectivity.

16-3 Ambiguities about Innatism

Positivism, antirealism, Platonism, Phenomenalism, . . . some may feel that the philosophy of science filled the English language (and most other languages as well) with far too many names ending in ism. But even more worrying is the variation of the meanings imparted to the said names. Take, for example, the word "Platonism" and consider anew the question—already evoked in section 2-2—whether or not Galileo was a Platonist. We noted that Koyré, for one, answered positively, on the basis of

---

[4] And correlatively is seems undeniable that the very reference to the notion of deciding (or *prescribing*) in some way involves mentality.

the fact that Galileo emphatically stressed the mathematical structure of natural laws. According to Maurice Clavelin (1968), in Koyré's views Galilean science started from the firm belief that "reason, seconded by geometry, is able to attain, all by its own, intelligence of the real."

Even if such were Galileo's views at the start, what does the word "real" refer to in this context? Should we read "Ultimate Reality" (i.e., something akin to Platonic Ideas) or just "the core of phenomena"? In view of the (obviously considerable) interest Galileo took in the investigation of phenomena, thinkers who strive to link Galileo with Platonism are forced nevertheless not to set aside the second answer. In order not to be too unfaithful to the spirit of Platonism they thus have to more or less identify the two aforesaid notions. In other words, they must assume that, thanks to observational investigation, Ultimate Reality is, theoretically, knowable as it really is. However, according to many the fable of the cave suggests the opposite conception, namely, the view that, however deep our analysis of the phenomena will ever be, it will never make us know more than just a *shadow* of the Real.

Such an ambiguity is also found in the doctrines that refer to the notion of innateness, either for supporting or for rejecting it. Concerning it, in the seventeenth and eighteenth centuries the situation was, conceptually, rather simple. There were, on the one side, the empiricists, who claimed that "all of our knowledge comes from the senses" (the *tabula rasa* doctrine) and, on the other side, followers of Descartes (as well as of Saint Augustine) who considered that "we have, without any miracle being involved, intellectual ideas that were not conveyed by our senses" (Fénelon, quoted by Brunschvicg [1951]). Descartes' "clear and distinct ideas" numbered of course among the latter, and, as we noted in chapter 1, Descartes himself seems to have considered that they enable us to describe the Real as it really is. This second standpoint was called innatism. In France this conception, which had been the dominant one during the seventeenth century, was rejected in the eighteenth, and replaced by the above-mentioned empiricism. But the latter was soon to lose much of its clarity, due to the advent of, first Berkeleyian, then Kantian, idealism. If reality-per-se does not exist or is to be, forever, inaccessible, experience may not be considered to be revealing to us a reality independent of ourselves. So that, when the empiricists claim that the whole of our knowledge comes from our senses we are somewhat at a loss to grasp what they actually mean. For what can the nature be of something that we are supposed to be getting informed of but of which we are told that it partook of us right from the start? Nay, the very *meaning* of the word "knowledge" then becomes hazy.

Should we then say that to adhere to Kantianism, or, more generally, to the idea that reality-per-se in unknowable, means returning to some innatism? The answer, obviously, would be "no" if the meaning of this word were restricted (as it sometimes was in the past) in such a drastic way that only ideas bearing on reality-per-se (then identified with God) should be called "innate." But it must be "yes" when this artificial restriction is not taken up, for, then, the Kantian notions of space, time, causality, etc. must be considered innate. Trying to escape this view just by stressing the rationality of Kantianism would be hopeless since Kant's rationalism is grounded on the idea that the categories of understanding are a priori, so that it is itself an innatism (in the more general sense of the word).

All this must have an effect on what may be called our "semi-intuitive" world-view. When we try to picture the world to ourselves we aim, of course, at going beyond the level of superficial, broad frescos and are anxious not to let phantasms delude us. In that spirit we try to keep as strictly as possible to what sense data inform us of. Ideally we should abide exclusively by them. But this, as is well known, is impossible. If we stuck to such a principle we might gaze for years on Foucault's pendulum without getting any information other than just the fact that regularly spaced marks appear on the sand. To enlarge our knowledge we need at least an outline of an interpretation. Now without coming back to the general interpretation problem, let us just recall once more that the presently operative framework theory—namely, quantum mechanics—is merely weakly objective (or, in other words, antirealist). And that this implies that the conception of knowledge that seems presently adequate not only resembles that of Kantianism, but even surpasses the latter in the direction of operationalism. From this resemblance it follows, as we just observed, that it leaves room for innatism. But it should, of course, immediately be stressed that this innatism differs in the highest degree from Descartes' because, precisely, of its operational character. Essentially what, in it, is taken to be a priori is the pair of notions "generality" and "simplicity," applied to the basic mathematical formulas expressing the observational predictive rules. Considered from this angle the Veiled Reality concept (to which we shall return in chapter 19) might be viewed as being the—unprovable but gratifying and irrefutable—assumption that the said simplicity (for us) of the rules vaguely reflects some simplicity (in itself) of the Real. Such an assumption, incidentally, seems not be totally void of any relationship with what Malebranche had in mind when he spoke of "vision in God."

16-4 Poincaré, Conventionalism, and Structural Realism

In philosophy of science Henri Poincaré's name is of course reminiscent of *conventionalism*. But it is also not infrequently associated with the doctrine of *structural realism,* sketchily defined in section 1-2 and whose consistency with conventionalism is not a priori obvious. It is interesting to examine somewhat in detail these two conceptions and the relationships between them.

Most often the thought of a philosopher of science has two facets, an epistemological one and one for which the epithet "ontological" is not inappropriate (even though, in many instances, it boils down to the assertion that ontology is nonsensical!). In it, in general, the epistemological facet is the one that is the more developed, the more explicit, and holds the attention of most commentators. Which is natural since, of course, it is in it that concrete implications for research are to be found. Concerning Poincaré, in particular, I claim that his conventionalism mostly arrested the attention just because it embodies his epistemological views, but that in his work, when we read it between the lines, an ontological standpoint is to be found. We shall investigate whether, in order to designate it, the name "structural realism" is appropriate. But let us consider conventionalism first.

16-4-1 Conventionalism

We know that his interest in the nature of the axioms of geometry is what led Poincaré to conventionalism. Kant considered these axioms to be of the nature of synthetic a priori judgements. But, after his time, the discovery of non-Euclidean geometries disproved this view since, as, among others, Paul Chambadal (1979) stressed, if the said axioms had truly been a priori, building up such geometries would have been quite impossible. On the other hand, Poincaré could not consider them to be of the nature of experimental data for, as he explained in *La science et l'hypothèse* (Poincaré 1902), geometrical space and space as a framework for experience have very different properties and may not, therefore, be identified. Moreover "we do not make experiments on ideal straight lines or ideal circles." It follows that the geometrical axioms are just simply conventions. "Our choice, among all possible conventions, is guided by experimental facts; but it remains free and is merely limited by the necessity that any contradiction should be avoided. . . . One geometry cannot be truer than another one, it can only be more convenient."

This conception Poincaré extended, as we know, from geometry to physics. Indeed, he was one of those who most forcefully stressed that experimental reports and theories linking them with one another should not be taken to be descriptions of an underlying independent reality. That they are but pictures enabling us to state in the most concise, hence most convenient, way either measurement outcomes or relationships between successive such outcomes. And it is this mode of thinking that induced him, as early as 1902, to express a disinterest close to skepticism concerning the existence of ether. "We don't mind, he wrote (*loc. cit.*, chapter 12), whether ether really exists or not: this is a metaphysicians' concern. The hypothesis is convenient."

While conventionalism greatly contributed in making Poincaré one of the three co-discoverers (together with Lorentz and Einstein) of special relativity theory, we know that it also induced him to state (*loc. cit.*, chapter 3) that Euclidean geometry was and would remain the most convenient one. And that if, at some time, new astronomical data were to throw doubts on the Euclidean properties of space the choice would be to modify the properties attributed to light rather than those attributed to space. This was a most risky prediction, that the advent of general relativity theory soon proved to be erroneous. But we may hardly consider this event to constitute an argument against conventionalism since, on that occasion, Poincaré's mistake was just a misevaluation of the future. He did not foresee that what, some ten years later, would be considered most *convenient* would just be to take up a non-Euclidean geometry. So that, finally, considered independently from its author, conventionalism finds itself reinforced rather than the reverse by this episode.

The upholders of objectivist realism often object to conventionalism that it sacrifices the notion of truth to the notion of convenience. In fact, this criticism is or is not well founded according to what extension is imparted to the very notion of truth. But on the other hand this very choice cannot be left completely arbitrary. And, in particular, it is difficult to grasp on what basis one might brush the notion "truth of observational predictive rules" aside. When it is found that, within a specified domain, such a rule always works, why shouldn't we say that it is true in that domain? In my opinion this is in no way an unwarranted extension of our language. Concerning Newton's law applied to the calculation of satellite trajectories, don't we already say "truly, it works!"? Incidentally, Poincaré himself argued along these lines. In *La valeur de la science* (Poincaré 1905) to the detractors of science (who, already then, expressed themselves quite loudly) he answered "If scientific recipes are valuable for action it is because we know they are, at least generally, successful. But to know this is to know something. Why then do you tell us that we cannot know anything?"

16-4-2 Structural Realism

To the extent that Poincaré developed something akin to an ontological standpoint, the name "structural realism" is presumably the one that suits this standpoint best.[5] But do such views actually underlie this author's thought, and, if yes, what do they actually consist of?

In fact, it is not clear whether the second question is relevant, for the answer to the first one might quite possibly be negative. Indeed, while, like Kant, Poincaré often made use of the word "real," like Kant again he systematically imparted to it a connotation making it relative to us. With him its meaning even bordered on operationalism. For example, at the place (in *La science et l'hypothèse*) where he considered possible extensions of the energy conservation principle he observed that such extensions will always be possible, so that the principle will, forever, be satisfied. Is it then—he inquired—a tautology? "No," he answered, "it expresses real relationships." But, he wondered, if it has a meaning it can possibly be false. "Where are its limits? When shall we know that they are reached?" And he answered: "When it will cease to be useful, that is, to be of any help to us for discovering new phenomena. When, and if, this happens, we shall be sure that the stated relationship is no more real for, otherwise, it would be fruitful." As we see, these last words constitute a genuine definition of the word "real" and this definition is centered on the human being and his/her abilities at observing and acting.

Similar views are to be found in the last chapter of the book *La valeur de la science*, a few quotations from which will now be given.

Concerning relationships between things, Poincaré wondered: "What does it mean to ask whether such relationships are objective?" And he answered: "It just means: 'are they the same for everybody?' "

Somewhere else we read: "External objects are real in that the sensations they make us experience appear to us united by some indestructible cement rather than by day-to-day circumstances."

Further on: "In short, the sole objective reality is the set of the relationships between things. True, these relationships . . . could not be conceived of apart from a mind that conceives them. But they are objective nevertheless since they are shared by all thinking beings." Let us note, here, the use of the word "since." It seems it makes the objectivity—hence the "reality"—of relationships depend on their being totally intersubjective. In other words, it appears that Poincaré here defined it to be what I called "weak objectivity."

These quotations, to which the one given here in section 7-2-1 is to be joined, seem to show that, finally, the word "reality" was, for Poincaré, synonymous with "empirical reality" and merely meant "reality-for-us." We may feel inclined to interpret them as implying that, after all, the

---

[5] Concerning structural realism see also John Worrall's paper "Structural Realism, the Best of Both Words" (Worrall 1989), p. 99.

notion of a mind-independent reality—a reality strictly prior to us—is meaningless. Many authors interpreted them that way.

But still, this is perhaps not quite the end of the story. For indeed, in Poincaré's work there are passages that do not really convey such a view. In *La valeur de la science*, for example, there is a sentence the beginning of which has a typically "conventionalistic" flavor and the end sounds quite the reverse. Here is the beginning:

> People will claim that science is but a classification and that a classification can not be "true." That it can merely be "convenient." But it is true that it is convenient. It is true that it is so, not just for me but for all human beings. It is true that it will remain so for our descendants

And, to be sure, up to this point nothing goes beyond the framework of a "human," or "empirical" reality. But Poincaré added:

> Finally, it is true that this cannot be merely a matter of chance.

Now, what does this last phrase actually mean? If the truth in question is not "merely a matter of chance" it must follow from something. But in view of the very nature of the problem this "something" cannot be of the nature of mere phenomena (since it generates the science of phenomena or, more precisely, the "convenience" of the latter). It can therefore only be a "Reality" prior to the phenomena.

Similarly, in *La science et l'hypothèse* (and concerning the going over from Fresnel's to Maxwell's theory) we read:

> Equations express relationships and it is because the relationships preserve their reality that the equations remain true. They inform us . . . that there is a relationship between something and something else. Now, we used to call this something *motion*, and we presently call it *electric current*. But these appellations were just images, substituted to the real objects that nature will hide from us for ever. The true relationships between these real objects constitute the only reality that is attainable by us, and the only condition is that there should be the same relationships between the said objects as between the images that we are forced to substitute for them. If these relationships are known to us, what does it matter that we should find it convenient to replace one image by another one?

And a little further, concerning the case of two rival theories, Poincaré wrote:

> It may be the case that both express true relationships and that a contradiction only exists between the images with which we covered up reality.

Unquestionably, in the first one of these two quotations Poincaré referred to the real—hidden—objects, which he carefully distinguished from the images we are forced to substitute to them. And he claimed that we are able to get at the *true* relationships existing between these *real* objects. Similarly, in the second one he mentioned a reality that we cover up with images that is, with the phenomena our various theories deal with; and that therefore must be basically prior to the phenomena in question.

Quite obviously, if we rejected the very idea of some reality underlying the phenomena we would be at a loss to impart a meaning to these two last quotations. Whereas, on the contrary, they become crystal clear as soon as we accept the said idea since, then, these sentences simply mean that there exist objects having certain relationships with one another, that we cannot know these objects but that we can know the said mutual relationships. This, I believe, is the substance of the structural realism that, maybe, constituted Poincaré's—implicit!—ontology.

However, concerning the latter some remarks are still in order.

REMARK 1

It merely bears on vocabulary. In the word "ontology" there is the root "logy," which means "knowledge." If we were strictly faithful to etymology we should consider that here this word is inappropriate since, as Poincaré himself repeatedly stressed, the "real objects" he mentioned cannot be known. Still, they exist—they "are"—and this of course is what the said word is meant to express.

REMARK 2

Structural realism is closely linked to the "extended causality" and "ground Reality" notions mentioned near the end of section 16-2-1 above. Here as well as there the point is that thought, initially centered on the notion of observable, progressively realizes that consistency implies the necessity of conceptually (and merely conceptually) going somewhat beyond the strictly observational realm. However, between structural realism and my own views a basic difference, which the word "nonseparability" adequately characterizes, is to be noted. Poincaré believed in the existence of (unknowable) objects-per-se, and, as we just saw, when he mentioned them it was in the plural. It seems he did not associate the notion of their nonknowableness with that of some underlying coherence, or deep unity. Today we better realize that Poincaré's standpoint still involves an implicit assumption, the one of separability, which present day data render implausible.

REMARK 3

Poincaré stressed that equations remain true and we saw that this continues to be the case, at least approximately and within suitably specified domains of validity. Considered together with nonseparability this fact suggests a particular line of thought. Its starting point is the fact—already noted in section 9-5 above—that within equations of classical physics the mathematical symbols representing physical quantities have a dual role. Consider for example Maxwell's equations, which describe the laws of electromagnetism and in which the symbols represent the electric and magnetic fields. On the one hand, these symbols serve for writing down the structure of the said laws, which could not be expressed otherwise than by means of such equations. And on the other hand, they are said to symbolize the values possessed by these fields at any space-time point. It is primarily this second role that Poincaré questioned, by pointing out that we have no good reasons for believing that these symbols do represent values-per-se of fields existing per se, and that, in fact, they merely stand for what would be observable in such and such circumstances. We know that the advents of quantum mechanics and quantum field theory were to fully corroborate these views. Under these conditions, might we not consider building up an ontology—in the etymological sense of an (of course fragmentary) knowledge of the Real—by giving up the second role of the symbols and keeping only the first one? Might we not give to this approach the name "structural" realism?

The idea is a natural and attractive one but it would be hardly appropriate to attribute it to Poincaré since, as it seems, it was only reluctantly and when he could not avoid it that he indulged in ontological considerations. Moreover, imparting to it some real substance is not easy and attempts in this direction run a great risk of leading to unacceptable specifications. First, it must be granted that, as a rule, when theories evolve equations are kept only approximately (in astronomy, for example, the Newton laws are all the less "good approximations" as the velocities are greater). Second (and much more important), as we already noted the basic concepts of a theory that is substituted to a former one often have nothing in common with those of the superseded theory (compare the gravitation force and the space-curvature notions). And under such conditions attributing some definite conceptual structure to the "Reality" we have in mind would amount to taking up an unwarranted specification of the Reality in question. On the other hand, it remains true that we have to do with successions of theories such that (as we saw) the more recent one roughly yields the same observational predictions as the former one with the addition of new ones, and involves formulas and equations (though not concepts) from which those of the former theory may be derived. In such a case it is difficult not to evoke some substratum. In other words, the insistence with which the radical idealists claim that mentioning a substratum is meaningless looks somewhat like relentlessness. But to be more definite concerning the nature and structure of this substratum seems impossible.

Finally, the structural realism thesis thus proves justifiable only when its content is watered down to the extent of more or less coinciding with the much less ambitious one (still to be polished, see the next chapters) I called Veiled Reality. From a more general point of view we see that between the reef of classical phenomenalism—burdened with conceptual difficulties—and that of physical realism—quasi-incompatible with some present data from the very science that gave it its name—steering is far from being easy.

CHAPTER 17

PRAGMATIC-TRANSCENDENTAL VERSUS VEILED REALITY APPROACHES

17-1 Introduction

IN 1996 A TWO-DAY session dedicated to my then new book, *Veiled Reality*, took place in Paris. The proceedings of this small colloquium (Bitbol and Laugier 1997) include texts by Michel Bitbol—later reproduced by him with but a few changes in a book of his (Bitbol 1998)—containing extensive and interesting critical analyses of my own views. More recently, Hervé Zwirn devoted several pages of his own book (Zwirn 2000) to an account and an analysis of the same.[1] The ideas these author developed take those put forward in *Veiled Reality* into consideration, endorse some of the main ones, but also suggest appreciable changes. To his own conception Bitbol gave the name the *pragmatic-transcendental approach*.

The purpose of the present chapter is to report on the analyses and criticisms of my conceptions that these two authors put forward, develop appropriate replies and, symmetrically, engage into a study of their own views, so as to determine as clearly as possible the points on which we agree and those on which we disagree. Before entering this subject let me just note however that the differences of views this investigation will bring to light should finally just reveal the extent to which the three of us agree as to—at least!—the true nature of the philosophical questions raised by physics. Indeed it may well be that Michel Bitbol and Hervé Zwirn[2] were the two readers of my books who devoted the greatest critical attention to their content and most pertinently took their stand upon some of the views put forward in them for developing their own approaches. Here I shall, in turn, rely on their contributions in order to try and make my own views still a little more explicit.

17-2 Replies to Michel Bitbol's and Hervé Zwirn's Objections

Michel Bitbol (1998), to repeat, paid much attention to my views and indeed devoted several chapters to an analysis of them. He described in

---

[1] Also to be mentioned is an extensive clear and accurate earlier account of them due to the physicist Henri Ruegg (1989).

[2] Along with Henri Ruegg, of course!

detail the bulk of my ideas and stressed the points on which they depart from those most commonly received among physicists as well as among philosophers. I was pleased to note that, precisely concerning these points, Bitbol, as well as Zwirn, most often expressed agreement. The points in question were extensively dealt with in the foregoing chapters of the present book and need not be recalled here. In this section I shall therefore merely examine Bitbol's interesting and illuminative objections—for there *are* a few in his book—to which I shall, when appropriate, associate some from Hervé Zwirn. They bear on various subjects so that this investigation will at the same time constitute for me an opportunity of specifying more precisely my views in various domains.

Now it turns out that the differences between his own views and mines that Bitbol described are of two kinds. Some have to do with the core of the matter while others more or less boil down to mere questions of form, or may even proceed just from the fact that I did not explain my views clearly enough. Those that bear on the "core of the matter" mainly have to do with the question whether or not the view that reality-per-se is structured is a tenable one and, if so, in what sense. They are considered in sections 17-2-4 to 17-2-7. Those that are essentially formal mainly concern verbal issues and questions about the extent to which such and such arguments of mine are convincing. Let us begin by considering the latter.

17-2-1 First Formal Question: Dualism and "the Veil"

In chapter 2 of his above-quoted book Michel Bitbol, after having expressed agreement with quite a number of my views (including some concerning still controversial subjects such as the "consistent histories" notion and the interpretation of decoherence) examined my central notion and the name, "Veiled Reality" I gave to it. This metaphorical expression he considered to be quite radically "dualistic." And (while granting that, with me, it is somehow rectified by the coemergence notion) he disapproved of it for this reason. This of course raises three questions, namely: first, "What is dualism?"; second, "In what sense of the word 'dualism' does Bitbol consider dualism not to be any more tenable?"; and third, "am I, or was I, a 'dualist' in *that* sense?"

Answering the first two questions is simple enough. Classically, dualism is defined as the theory according to which there exist strictly two ground entities, matter and mind ("extension" and "thought" were their names in Descartes' time), neither one being more basic than the other one. Now, already in Descartes' time this conception was criticized by philosophers who pointed out that, when it is taken up, to account for the interactions between matter and mind becomes immensely difficult. And somewhat more recently (from the eighteenth century on) it was even more sharply criticized by materialists who (as we saw in chapter 12) claimed that mind is but a form of matter. Now, clearly and avowedly Bitbol does not share the materialists' standpoint. It is therefore impossible that his rejection of dualism should be grounded on their views. Hence it seems that it must be based on the first of the two just stated objections. But this implies that it is well motivated only to the extent that the said objection is pertinent, that is, is relative to dualism as above defined. Which means that the dualism Bitbol considers unacceptable must be, quite specifically, a conception according to which, to repeat, Reality is composed of two ground entities, matter and mind, neither one of which is more basic that the other one.

Having thus answered the two first questions, we may turn to the third one. Is the view I called the "Veiled Reality conception" dualistic in this sense? And—subsidiary question—does the designation "Veiled Reality" suggest it is? Is it the case that, as Bitbol wrote, "its metaphoric structure tends to automatically reproduce the dualistic scheme of the theory of knowledge"?

To the main question, the answer is obviously "no." The conception I developed in no way considers matter on the one hand—i.e., what we deal with in everyday life and physicists investigate—and mind on the other hand to be two basic entities that, taken together, would constitute "reality as it really is." Quite on the contrary it states that matter—what we make experiments on—is but an empirical reality, that is, not in the least a basic entity but merely the set of the phenomena in a Kantian sense of the word, which implies that it is, at least partly, molded by us. And it conceives of mind as emerging (but atemporally) from a *Something* (thereby making empirical reality also emerge from the latter). This *Something*, to which I gave the name (which I thought "neutral") "Independent Reality," alias "the Real," is thus conceptually prior to the mind-matter scission. For me, therefore, it is not at all the object of the precise, discursive quantitative knowledge we normally refer to when we utter the word "knowledge." In other words, in a theory of knowledge consistent with the Veiled Reality conception Independent Reality, even though it may be considered the hidden source of our physical laws, in no way is what science investigates. What science deals with is, to repeat, empirical reality, to which the notion of a "veil" does not apply.

Concerning, now, the "subsidiary question," it has in fact two facets. One is: "Is the designation 'Veiled Reality' misleading?" and the other one is: "Did my standpoint evolve from some sort of a classical dualism to the thesis just sketchily recalled, without my having sufficiently called attention to the change?"

With respect to the "first facet" I must admit that the designation in question is indeed somewhat ambiguous. It does not specify the nature of the "Reality" that it terms "veiled." It does not warn against the naive interpretation consisting in identifying it with what science investigates, which is commonly called "Reality." And, what is more, it may well be that, for the purpose of making myself quickly—even though only approximately—understood by nonexperts, I, at times, made use of allegories involving the picture of a subject placed face to face with a "world of objects." To avoid such defects the designation in question should be explained by means of some more complex picture, which, for example, would—metaphorically—involve time, would symbolically replace "the Real" by "the origins," and would evoke a veil partly hiding the said "origins." All this shows that Bitbol's criticism of the said designation is appropriate and I am glad that he and my other interlocutors expressed it, thus giving me an opportunity to state my views more precisely. On the other hand, however, a designation should be short—particularly when it is to serve as a title!—and I feel sure it will be granted that it is difficult to unambiguously compress a philosophical conception within two words.

Concerning the second "facet," quite frankly I think I always claimed that the "hard core" of quantum physics is the set of its observational predictive rules (already in *Conceptual Foundations of Quantum Mechanics* [d'Espagnat, 1976] I called them, not "laws" or "axioms," but "rules"). Hence I consider that, contrary to what, at places, Bitbol seems to believe, I did not very much change my mind in the course of years on this subject. But it is quite possible that I expressed myself inadequately. And it may also be that the ambiguity noted above inherent in the denomination "Veiled Reality" misdirected many readers.[3]

I finally think that the latter conjecture is likely to be the right one. What makes me think so is the fact that concerning my conception Hervé Zwirn (2000) and, before him, Michel Bitbol (1996, quoted by Zwirn) wrote: "it is pertinent only within the conceptual standpoint according to which the subject stands face to face with the world." Of course, in view of the foregoing, I cannot agree with such a statement. More precisely, among the various senses it may have I discovered none that would make it correct. The main reason for this is that, in contemporary physics, the word "world" is ambiguous, due to the fact that, a priori, it may refer either to "the Real" or to empirical reality, two notions that quantum mechanics shows not to coincide. Now, what we are (or at any rate have the impression of being) "face to face" with is obviously *not* some great

---

[3] Upon reflection it occurs to me that asserting I always stuck to exactly the same standpoint would not be entirely correct. It is true that in my very first book (d'Espagnat 1965) among the theories worth further examination I mentioned at least one that fits with dualism. And in *Conceptual Foundations of Quantum Mechanics* I still expressed a bent toward a non-watered-down structural realism. So yes, I did undergo an evolution. But a gentle one, I would say. By no means a revolution.

Entity whose existence we have reasons to believe in but on which we can have no precise knowledge. Unquestionably it always is a set of objects, instruments, and so on. And the very expression "face to face" suggests that the relation between the participants is not basically of the "generation" type and that indeed the subject is primarily a somewhat passive onlooker. Hence the statement in hand could only be appropriate if my conception fitted into what Hervé Zwirn (*loc. cit.*) called the "realism of phenomena," a view according to which the subject stands face to face with phenomena that "are there, take place, and that we only have to passively observe." But, as the content of many of the foregoing chapters clearly shows, this is not at all what the Veiled Reality conception asserts. And indeed it follows from various passages in Bitbol (1998) that Bitbol himself does not understand it this way any more. It is true that, in it, the empirical reality notion is linked to the use of an objectivistic language. For indeed according to it the very emergence of the said notion comes because instead of uttering sentences such as "we all saw, and shall see, the pointer lying at such and such a place" it is shorter to say "the pointer *is* at such and such a place." But it is clear that this is but a convenient way of speaking. And that the empirical reality notion basically refers to *impressions*, not to events taking place, as such, independently of the existence of thought.

Finally, it is worth noting that, having read recent texts of mine, Bitbol now considers my views to be more elaborate than what the picture of an object placed under a veil seems to suggest. In particular, he realized that when I strive to specify the kind of information quantum mechanics yields concerning Independent Reality I am lead to adopt an essentially negative standpoint. As we saw in this book first part, we cannot build up a trustworthy representation of the said Reality that would be plural, merged in space-time, "scattered" among a multitude of "atoms," etc. Below we shall come back once more on these aspects.

REMARK

For completeness sake let it be mentioned that in spite of all this a classically dualistic conception could consistently be kept. For this it would suffice that we should stop considering ontologically interpretable theories such as the Broglie-Bohm model, to be noncredible, and that we should call "matter" the particles and the universal wave function this model postulates. In a way, it could even be claimed that in the usually given descriptions of the said model a dualistic standpoint is implicitly taken up. To see this let us remember the distinction the model implies between the "true" physical quantities—the "true" momentum for example (see section 7-2-2)—and the observed values of the said quantities. In order that the model should actually predict these observed values it seems that some additional prescriptions should be added, "by hand" so to speak, into it. Such, for example, is the law that the outcome of a measurement of a physical quantity can only be an eigenvalue of the self-adjoint operator associated with this quantity. However, in the context in hand applying such rules implies referring in an entirely nontrivial way to the notion of measurement, which involves the one of "getting aware," and hence also that of consciousness. As we see, at this stage a traditional, Cartesian-like, matter-mind dualism reappears. True, these measurements take place by means of instruments that may be imagined to make automatic registrations. But even then the fact remains that we see here reappearing, in quite a basic role, the notion of instruments defined *qua* instruments (i.e., not by their structure but by their "measuring" function).

So, as we see, it is—paradoxically, some would say—an ontologically interpretable model that, if we believed in it, would bring us back to a dualistic word view. It is worth noting, however, that it still would not be a fully "classical" dualism, faithfully reproducing the scheme of some "face to face" taking place between the subject and the world. This is just because the "world" of the model is composed of hidden variables and a universal nonseparable wave function and obviously this is not the "world of objects and events" the scientist has the impression of standing "face to face" with.

17-2-2 Second Formal Question:
The Need for "Ampliative" Arguments

The above-mentioned negative standpoint is rather similar to the Kantian thesis that the "thing-in-itself" is inaccessible. But am I right to consider, as I do, that it is better justified by my reasoning, grounded on quite informative physical data, than by Kant's one, which just consisted in a priori denouncing as an illusion our tendency at identifying the phenomena with things-in-themselves? Under the title *First Tension* Bitbol (1998) pointed out that, on this point, my reasoning is not as sharp as it could ideally have been expected to be. For, he noted, my conclusions follow, not just from quantum mechanics alone, but only from the interpretations of it that remain available once those involving hidden variables have been discarded. He granted that the latter "mode of ontological projection" (of the representation on a "Real") looks much more artificial that the one of yore (the one that underlay classical physics) due to the features of experimental inaccessibility, nonlocality and contextuality that it involves. But, he wrote, this changes nothing to the fact that the model is *in principle* acceptable, and that, to discard it, considerations must be invoked that go beyond the proper domain of physics.

During the 1996 meeting this point was the subject of exchanges of views between us that Bitbol adequately summarized in section 2-3 of his book. On the said occasion, of course, I myself stressed the fact (already explained in detail in *Veiled Reality* and elsewhere) that in order to discard the hidden variable model it is necessary to make use of reasons that reach beyond physics, if the word "physics" is taken in its most restrictive sense. On the other hand I also made it clear that in the eyes of a physicist not prejudiced in favor of traditional realism the model in question *is* quite definitely artificial. I "set it aside" as well as all the other ontologically interpretable ones—I believe in none of them—just in virtue of this observation, which—I agree—reaches beyond the strict domain of physics defined in a restrictive sense but still is a valid one *even in the field of pure physical research*. Indeed, science itself implicitly but systematically sets aside many explanatory schemes that cannot be proved erroneous (those that refer to the whims of an all-powerful Zeus, for example), merely on the basis of the fact that they are unacceptably arbitrary. And moreover most hard-boiled physicists rightly take up quite a cautious attitude when they are shown several rival theories grounded on very different ideas and between which experiment seems to be unable to decide. Finally, I pointed out that I was aiming at a world-view and that I never thought that we could—and should—build up our own one by exclusively making use of proofs grounded on scientifically established facts.

Michel Bitbol stated, in substance, that he shared these views. And he went as far as to disclose their true name. Indeed he pointed out (a remark I already made use of above) that they are of the nature of the so-called "ampliative" criteria, that is, of criteria that reach beyond strict empiricism. But he stressed that ampliative criteria are not definitively frozen and that having a preference for such or such one may lead to differing ontological conceptions. He noted, for example, that under some such choices (demand for simplicity, unity, internal consistency of the conceptual scheme) the most plausible ontology is the one at which I finally get (and which was also, up to a certain point, the one of the "later Bohm" (implicit order, holomovement, etc.). But he pointed out that other choices (continuity with the classical physics theories, or just with our natural viewpoint) lead to the ontology he termed "pluralist" that was the one of the "first Bohm" (the 1952 papers). I partly agree with this way of looking at things, with, however, the reservation that the first Bohm theory was "pluralist" only in appearance. As, elsewhere in his book, Bitbol himself remarked, "at a first level it does picture the world as a set of separated corpuscles endowed with properties of their own, but nonlocality and contextuality lead to the vanishing of all the consequences of such a conceptual separation." So that finally "under the superficial appearance of a remaining philosophical atomism the approach Bohm put forward in 1952 brings the crisis of atomism to light just as forcefully as the standard versions of quantum mechanics do."

Admittedly, strictly speaking neither the said crisis nor the—in view of its generality even more serious—disproof of locality here reported on in chapter 3 necessarily imply the "inaccessibility of the thing-in-itself" mentioned at the beginning of this subsection. Indeed this is a point that I myself have been stressing for a long time. Nevertheless, I consider that the said crisis and disproof deprive the opposite thesis—the one of a progressive unveiling of the Real finally revealing it "as it really is"—of its potentially most persuasive support. I claim therefore that they constitute in favor of the said inaccessibility (or quasi-inaccessibility) a most powerful argument. True, philosophers remain free to prefer a purely philosophical reasoning, constructed along Kant's lines. But, I would say, scientifically minded people have cogent reasons to feel quite definitely more convinced by the one described here.

17-2-3 Third Formal Question: "To Have Something to Do With"

In *Veiled Reality* I wrote: " What science says has something to do with it [Independent Reality] but this information seems to be limited to some general structures of [the latter], and in this sense it can hardly be thought exhaustive." A few pages before, concerning the same Independent Reality (which I there called "something") I had written that certain facts "give us some knowledge about that something." Under the heading: "Second tension" Bitbol expressed reservations concerning such uses of the word "knowledge" and the expression "to have something to do with." A piece of knowledge, he pointed out, is knowledge *of* something. It refers, or is relative, to something. Now, he noted, while, in the quoted sentences, the idea of being relative to something is indeed present, it takes in them an intentionally vague connotation, shown for example by the use of the adverb "about" instead of "of." And which is also revealed by the fact of having written "has something to do with" instead of "deals with," a statement that would obviously have been at variance with my claim that science merely describes *empirical* reality. In short, on the whole Bitbol detected in some of my claims a semantic vagueness that he found disturbing.

This question also we discussed. I pointed out to him that the two notions "to deal with" and "to have something to do with" are distinct ones and that, according to the context, one or the other is appropriate. To make my point I used the simple example of a tourist's guide. To be spe-

cific I mentioned the guide of Saintonge, a region where many old ro-
manesque churches are to be found. The guide mentions them, of course.
And since, on that occasion, it yields indications relative to how and when
they were built I am entitled to state that its content has something to do
with the history of architecture. But I avoid bluntly claiming that it deals
with the history in question. And in this I am right, for to claim that much
would strongly suggest that its subject matter is the history of architecture, which would be false. Similarly, if, as I believe is the case, some
physical data are traces of some Independent Reality features (traces
highly distorted by the structure of our apprehending abilities) it is appropriate to state that the formers (the traces) have just "something to do"
with the latter (Independent Reality).

This simple remark removes, I guess, the difficulty. Contrary, as it
seems, to many scientists and philosophers, I consider (and people with
literary tendencies will perhaps agree on this point more readily than others) that between the notion of discursively and precisely knowing something and that of being radically ignorant of the same an intermediate one
exists. It may—imperfectly—be described as consisting in an intrinsically
vague "grasping." Indeed it seems clear, at least to thoughtful people, that
there are things that can only be "grasped" and that some of the most
important ones are of such a nature. True, to find, in our languages, a
name for the corresponding notion is difficult. But I take the notion to be
quite a valid one nevertheless. And in the present case this completes my
answer to the criticism.

On the other hand, accepting such a notion makes it difficult to radically discard any notion of a structural organization ("prestructure" in
Bitbol's language) of the Real, which leads us to the next subject. But
before taking up the latter let us note that at the place where Bitbol stated
the aforementioned reservations he associated them with a summary of
his own views (see below) and added that if the latter were accepted,
having recourse to the "naive dualistic conception" would be rendered
unnecessary. Since we saw above that in fact I do not have recourse to the
latter I do not consider this remark to be a valid supplementary argument
for preferring his conception to my own.

17-2-4 First "Core of the Matter" Question: "Prestructure"

This question relates to the way I analyzed the intersubjective agreement problem (here, chapter 5). Of course Bitbol shares my view that
quantum mechanics deprives us of the possibility (which in classical
physics times remained open) of imparting more intrinsic reality to the
so-called primary qualities (position, form, motion) than to the second-

ary ones (color, taste, etc.). He therefore agreed that we cannot anymore explain intersubjective agreement concerning our observing a teapot on the table by the fact that a teapot *really* is on the table, totally independently of our apprehension abilities. But he claimed that I did not completely draw the consequences of all this. He recalled my two-stage reasoning, the first one consisting in "explaining" the observed agreement by referring to the quantum observational predictive laws, and the second one consisting in "explaining" these very laws by referring to an Independent Reality considered to be the cause of these laws. To the first stage he expressed no objection but he clearly showed he disapproved of the second one. Here the crux of the matter is the notion—which my reasoning indeed implies—of a *prestructured* Independent Reality. Bitbol recalled the fact that such a conception encounters very strong reservations from many philosophical schools including in particular the Kantian and neo-Kantian ones. The fact is undeniable, but, of course, it does not constitute, all by itself, an objection. A genuine objection can only result from arguments showing—or indicating—*why* the conception in question is unacceptable.

In chapter 3 of his book, under the heading *Veiled Reality and its critics,* Bitbol produced a review of such arguments and stated which ones of them he considered valid. For completeness sake he first mentioned one that he did not retain. It is the (natural and apparently widespread) idea according to which it is only under the, extrascientific and logically undecidable, metaphysical hypothesis that theories such as (for example) quantum mechanics are capable of describing the Real that any assertion concerning the structural characters of the said "Real" may possibly be justified. As I pointed out then (Bitbol and Laugier 1997) and as also shown in *Veiled Reality* and in the first part of the present book, this claim is unfounded. Moreover, Bitbol made a point of explicitly granting this. He pointed out that "the thesis of the nonlocality of Independent Reality rests, not on quantum theory (nor on some other theory) but on Bell's theorem, according to which no ontologically interpretable local theory is capable of reproducing sets of results predicted by quantum mechanics and experimentally verified" (Aspect-like experiments). However he added: "of course, this thesis remains conditioned on the minimal assumption . . . that it is legitimate to refer to the phenomenal order for sketching some features of Independent Reality."

It seems to me that, although this last statement is basically right, its formulation is a little inadequate and therefore slightly misleading. It is basically right in that, of course, as soon as we undertake to state anything whatsoever (even just a fancy conjecture) concerning Independent Reality we ipso facto postulate some relationship to exist between the latter and some set of ideas of ours (possibly including sense data among them).

Indeed this corresponds to the fact, already noted in section 1-2 and used in section 3-3-4, that any "realist" world-view is in fact a *representation* of Independent Reality, that is, an *idea,* concerning which the realist conjectures that it more or less fits Reality. On the other hand, Bitbol's statement—which does not distinguish between positive and negative propositions concerning features of "the Real"—seems to me to be a little misleading due to the presence in it of the verb "to sketch." "To sketch" is, in a way, tantamount to "to interpret," and even to "to interpret as an artist," that is, in a manner the available data do not strictly prescribe. However, when we assert nonlocality, which is a very general and basically negative feature, it seems to me that no such interpretation takes place. Perhaps, what I have here in mind may be further clarified by a simple imaginary counterexample in which use of the word "sketch" would be appropriate. It goes as follows. Having observed that, at the level of the phenomena, two free massive objects are always seen to accelerate towards one another we might assume, Newton-wise, that between them "something like a force" is at work. And this would indeed be a "first sketch" of some features of (Independent) Reality. Alternatively we might suppose, Einstein-wise, that, at the Independent Reality level, there is "something like a curvature of space-time." And this would constitute another "sketch," differing from the first one, of the features of the said Reality. On the other hand, if we merely said that the observed phenomenon "must follow from some features of Independent Reality" without in the least specifying what these features are, such an assertion could not legitimately be called a "sketch." If a person puts it forward, he or she thereby merely expresses his or her acceptance of (open) realism. I think that nonlocality—the giving up of the "locality principle"—presents us with a case that has quite definite similarities with the latter one. For indeed it positively assumes nothing. And moreover, as shown in sections 3-2-2 and 3-2-4 above, within realism it is the minimal possible renunciation that we have to resign ourselves to in view of facts, independently of any theory. A more radical one would be to give up the very notion of a real space in which the notion of distance can be defined. Even more radical is the one, forcefully suggested by quantum theory as we know, consisting in altogether giving up objectivist realism. But obviously none of these options constitutes a "sketch" of Independent Reality. Quite on the contrary, they amount to dropping the idea of ever drawing such a sketch.

Admittedly the foregoing argument has specifically to do with my ideas concerning nonlocality (and its status within Independent Reality). Now, to Michel Bitbol I must grant that my Veiled Reality conception is not limited to such purely negative statements (nonlocality, nonseparability, etc.). In it I put forward the view that—qua conjectures, and somewhat in the spirit of the conservative preconceptions entertained in most scientific (though not in philosophic) circles—some scientific data may conceivably reflect (though in a distorted and incomprehensible way) some structures of "the Real." Bitbol wondered about the status of such characterizations and this induced him to examine another criticism some thinkers made to the Veiled Reality conception. It has to do with the concept of "description." Bitbol agreed, of course, that I do not take science in general, and quantum mechanics in particular, to be descriptions in the literal sense (including counterfactuality, etc.) of Independent Reality. But he observed that many of my interlocutors impart a broader meaning to the notion of "description." According to them, for a theory to be "descriptive" of the Real it suffices that its legal structures be globally isomorphic to the structures of the latter, even if the legal structures in question are merely predictive of observations. In that sense, he wrote, "it does no seem incorrect to claim" that I do attribute to physical theories some degree of descriptive power. Now, he noted, many authors consider any description to be relative to some instrumental and intellectual perceptive context. And through this remark he seemed to suggest that I might well be laboring under a delusion. That the structures I believe to exist within Independent Reality, and which I think our predictive laws are distorted reflections of, might well, in fact, only be relative to our perceptive context, that is, might come essentially from us.

My reply is that I am afraid this reasoning does not distinguish quite clearly enough between the notions of necessary and sufficient conditions. Admittedly our observational predictive rules (whether or not we choose to put the label "description" on them) depend on our perceptive context. Who would question this? But the real question is whether they depend exclusively on the said context or also on something else, which, then, could be the "Real" I have in mind. The statement that the rules in question are relative to our perceptive context is correct but too vague for yielding any information concerning this last point. True, whoever is bent on proving that the dependence in question is illusory can do so. For that it suffices to put forward a principle according to which any description, not only is relative to a perceptive context but also is relative to it alone. However, this is nothing else than just stating what, in chapter 11, I termed a "Rousseau-like" principle. The thinker who puts it forward may of course believe in it and infer most interesting propositions from it. But another thinker may, quite as legitimately, put forward a principle at variance with it and do the same. I do not consider therefore that, through the considerations mentioned here, Bitbol validly weakened the "prestructure" hypothesis (which, to repeat, is, with me, but a plausible and admittedly unverifiable conjecture).

17-2-5 Second "Core of the Matter" Question:
On the Argument that "Something Says No"

Concerning my conception, although Bitbol criticized the metaphor—
"Veiled Reality"—with which I designated it, he did realize that, in substance, what it evokes is something like a coemergence of thought and
phenomenal reality out of a Being that is conceptually prior to both. And
he seems to also have grasped (or, at least, strongly suspected) that, in my
approach, the "veil," does not lie between the two thus cogenerated terms
but between thought and an Independent Reality identified with Being
itself. But, as already noted above, he expressed strong reservations concerning the idea that, within such a conception, facts are susceptible of
providing us with some information—be it only of a negative nature—
concerning the admissible representations of the said "Real." And in support of this claim of his he added a new line of reasoning to the one
sketched above.

He developed it in two steps. In the first one he just aimed at appropriately stating his questioning under the assumption (which, we saw, is the
correct one) that my conception rests on the emergence (or coemergence)
notion. And he pointed out that under the assumption that this is the case
the question would boil down to that of knowing to what extent a subject
can indirectly apprehend, through a study of some object, the structural
preconditions of the joint appearance of himself and the said object. I am
glad that on this point there is no disagreement between us. Bitbol's wary
use of the conditional (*would* boil down) presumably just reflected the
fact that when he wrote he still somehow doubted whether I was an
"emergentist" or just a dualist in the conventional sense. But this point
has now been cleared up as we saw, and of course I fully agree that the
question does boil down to the one he stated. Needless to say I feel inclined towards a positive answer. And in fact this standpoint of mine
seems to me to be such a natural one that, here again, I shall rest content
with just backing it up with one example. Can't a sociologist meaningfully
make use of what he knows concerning the present culture of his own
ethnic group in trying to form a rough picture of what the structures and
prejudices of the society from which he and his ethnic group emerged
most likely were?

The second step in Bitbol's reasoning consisted in discarding as much
as he could what he termed my "strong faillibilism." This is the name he
gave to my observation that even though there are no "really crucial"
experiments, even though Lakatos's "protective belts" normally do pro-
tect theories from too brutal refutations, still it remains true that some
theories are finally dropped because of their being at variance with too

large a number of facts. To this conception he preferred a "weak faillibilism," which he defined and justified starting from an observation on
which almost everybody agrees. This is the fact that sometimes two theories grounded on altogether different concepts have identical consequences and that, in such cases, even a realist has no reason to consider
that one set of concepts rather than the other one truly describes "what
is real." In substance,[4] Bitbol pointed to the fact that the bearing of this
observation is wider than it seems. He referred to the fact that the replacement of one theory by another one was very rarely due to the first one
being categorically refuted. And he suggested we should assume that new
experiments merely introduce new constraints, without formally compelling us to bluntly discard even very eccentric views. Then, he pointed out,
we should have no binding reason to consider that such and such a set of
concepts rather than such and such other one conveys, through what it
adds to the phenomenal order, something of the structure of "the Real."
In other words, "under this premise" (as he himself noted) there would
be no very strong reason to adhere to the Veiled Reality conception.

Bitbol's reservation "under this premise" appropriately reminds us that
his objection is dependent on the weak faillibilism assumption, which we
are not forced to endorse. Let us examine it together with one of a rather
similar vein, also directed against my "something that says no" argument
and formulated by Hervé Zwirn (loc. cit). Zwirn essentially pointed out
that at the place where I exclaim, "this 'something' cannot be us!" I am
implicitly assuming that, being its own measure, a purely human construction can never generate contradiction. But he claimed that, strictly speaking, this is not logically necessary. To show this he first noted that, within
a conception in which everything is a purely human construction, what
we identify with reality also is an unconscious construction of the human
mind. He called it "perceptual construction" in order to mark in what
respect it is distinct from the scientific theories meant to account for it.
He then referred to the well-known fact that to ascertain the consistency
of a somewhat complex formal system is already quite difficult.[5] "By extension"—he noted—"it then should not come as a surprise that, from
time to time, we should discover contradictions revealed by discrepancies
between our theoretical constructs and the particular construct that we
call the Real."

Concerning Bitbol's objection, Bitbol himself answered it, both implicitly, as we saw, through the words "under this premise" and explicitly by

---

[4] In his text this argument is implicit.

[5] We may of course think of Gödel's theorem and the impossibility it establishes of proving the consistency of a formal system that involves arithmetic by using procedures internal
to the said system.

pointing out that I dispose of a powerful counterargument. Of course this argument is nothing else than nonlocality. For nonlocality forces us to discard some structures, namely, those that are consistent with the locality hypothesis (here, section 3-2-1). In other words, it makes it possible to apply a strong faillibilism principle, for indeed however "eccentric" it may be, no realist local theory can nowadays be kept. In other terms, nonlocality is not just simply one of the new (i.e., newly discovered) constraints that "can be taken into account in a simple consistent and efficient way only by changing the theory" (Bitbol, *loc. cit.*). It requests that the change (the shift to nonlocality) should be endorsed even by the people who, like some upholders of the hidden variables notion, would readily have sacrificed simplicity and efficiency in order to avoid it. Unless, of course, along the line of a consequence of Zwirn's argument (see below), we choose to give to elementary sensations the name "perceptual constructions" and decide that we may question their validity.

This leads us to Zwirn's objection. It is true that Zwirn's argument is extremely simple and general and that, at a purely logical level, it, at first sight, sounds convincing. But we should wonder whether its convincing power doesn't follow in part from its very generality and the qualitative nature it derives from it. For in fact, as Zwirn himself acknowledged, his notion of a "perceptual construction" is one that is not easy to characterize with some precision, nor is the way in which it is to be applied to the problem at hand. Indeed, are we really dealing here (as Zwirn seems to suggest) with discrepancies between our theoretical constructs and the intuitive, nonelaborate perceptual construct we use to call "reality"? With, for example, the one that exists between quantum theory—which is nonseparable as we know—and our spontaneous perceptual construct of reality, which obeys the "principle of analysis," in other words: is "separable"?[6] Or have we to do with discrepancies between a theory and a consistent set of data resulting from repeated, varied and intentional experimental testing of the theory in question? Between, for example, classical mechanics (or any other "separable" theory) on the one hand and the outcomes of the (various) Aspect-like experiments on the other hand? That these two types of conceivable discrepancies do indeed differ from one another is clearly shown just by the fact that we, scientists, react to them in opposite ways. In the first case we normally decide in favor of the theory (assuming it has been found to be a general and truly reliable one) whereas in the second case we always take the experimental outcome to be conclusive and hence reject the theory. But of course the only interesting case is the second one, in which, by assumption, both the theory and

---

[6] Or the one that exists between Galilean kinematics and the fact of observing that, on Earth, a moving macroscopic body not submitted to any force comes, in the long run, to rest.

the set of experimental outcomes are most seriously grounded on normal scientific criteria. And an important point is that we then decide, without any conceivable qualms, in favor of the "perceptual construct." If, for example, experiments have finally proved nonlocality (assuming all loopholes in the argument, etc., have been removed), to go on believing in a realist local theory would just be absurd; everybody agrees on this.

But then (as we already noted in section 13-3-3, Remark 1) if, as Bitbol and Zwirn seem to think, it is true that, similarly to our theoretical constructs, our perceptual constructs, including our elementary sensations, come exclusively from us, what motivates the said universal and systematic preference? What justifies the fact that we, scientists, whenever we have submitted some new theory to serious, elaborate experimental testing, always decide to believe the information yielded by the corresponding perceptual constructs rather than the one yielded by the theory? Why should we not, from time to time, take up the opposite standpoint and go on believing in an experimentally refuted theory? The hypothesis Zwirn put forward seems to logically lead to the conclusion that this would constitute an entirely rational procedure. Still, every scientist will consider that, it would be downright absurd. Some thinkers may react differently but as for me, I think it follows from this that the suggestion Zwirn put forward cannot be kept.[7]

Incidentally, in *Veiled Reality*, at the place where, concerning the "something that says *no*" argument, I asked: "How could this *something* still be us?" I added, "It seems that the degree of intellectual contortion necessary for answering such a question in any positive way exceeds what is acceptable." It is true that the expression "intellectual contortion" carries with it a pejorative hint that we must here forget about. But once it is dropped the stated commentary adequately expresses my own present standpoint on the problem at hand.

17-2-6 *Structural Invariants*

Under this heading we shall consider problems that Bitbol discussed and in which the structural invariant notion appears. Bitbol started from the remark he himself made, as we noted, that the objection grounded on the underdetermination of theories (weak faillibilism) is inoperative concerning my Veiled Reality conception. As we know, the reason is that, concerning the possible representations of "the Real," the only assertions of mine that I claim are certain are of a negative nature (essentially nonlocality).

---

[7] Or, better said, it is logically coherent but to accept it would be tantamount to negating the very foundations of science and empirical knowledge in general.

And that, as we saw in detail above, since their validity just follows from facts they hold good in any theory fitting the data. Bitbol hence called "structural invariants" the content of such assertions and considered the very general question whether or not such structural invariants necessarily reveal something relatively to a genuinely independent reality. He noted that indeed a great many searchers do share the conviction that structural invariants that hold good concerning large classes of modes of approach are plausible candidates to the role of, partly or totally, representing independent reality. And he considered that, indeed, if a faithful representation of independent reality could be obtained, it would un avoidably be expressed by means of some structural invariants. But he stressed that, strictly speaking, the reciprocal is not true. That, from the fact that we succeeded in identifying such an invariant we cannot infer that it represents some feature of independent reality.

As for me, I think that what Bitbol asserted there is quite correct but that the point of view he took up was too general. In particular, it would be appropriate to examine more closely in what sense nonlocality may be identified with a structural invariant and whether or not it has other significant aspects.

With respect to the first point, Bitbol rightly stressed that nonseparability is to be found within all the ontologically interpretable theories that yield the same observable predictions as quantum mechanics and that in virtue of the Bell theorem it *must* necessarily be found in any of them. It is clear that under these conditions the word "invariance" applies. But it is also rather clear that it is not in this sense that the mathematicians normally take it. In fact, they closely associate invariance with transformation groups. Now, it is hard to see in what way a transformation group might be specified that would have a mathematically defined relationship with nonlocality. Consequently, we are at a loss how to apply Noether's theorem to the case in hand. And indeed Bell's theorem in no way implies the existence of a mathematical invariant of the type of the baryonic or electric charge. But then, I feel I must keep an attitude of reserve with respect to a suggestion of Bitbol's. This is the idea that nonlocality, viewed by him (in the abstract) as associated to an (unspecified) transformation group, would therefore be "at least as likely to inform us about the set of contexts implied by some given research activity . . . as about the hypothetical structures of some independent reality." It seems to me that the analogy is much too weak to yield an argument—either "pro" or "contra"—endowed with genuine convincing power.

As for the second point, which bears on whether or not nonlocality has significant general features beside the one just examined, my view is that it does indeed have one. More precisely, I consider that in fact its most salient feature is not invariance in Bitbol's sense, which reflects, at best, its formal aspect. In my opinion what is most important is not the fact that nonlocality is to be found in any acceptable realist theory, it is its specific informative content. It is the fact that we must give up any hope of ever accounting for all the observed data by means of a theory that is at the same time realist and local.

17-2-7 *"Excluded Middle"?*

Now, relative to the said informative content Bitbol raised an important point. "Within it, may we"—he asked—"infer the positive from the negative and, for example, proceed from nonlocality or nonseparability to the view that independent reality is something that is unique and not embedded in space-time?"

To this question I, for one, suggested a balanced answer (Bitbol and Laugier 1997). I claimed I considered it likely that the data in hand, and quantum theory in general, yield glimpses on some structures of "the Real," and I did stress in this connection that glimpses are nothing like pictures or descriptions. But, making use of a common sense that admittedly was, in this context, a little too "thick," I wrote the phrase: "It seems clear to me that we cannot (negatively) acknowledge nonseparability (of Independent Reality) without at the same time acknowledging the presence of some unity within it."

Bitbol spotted a weak point in this argument and he was right. The acknowledgment in question, he pointed out, presupposes that the principle of excluded middle applies to Independent Reality. If this Reality is not plural and separable, then somehow it is "one." But, he stressed, this is nothing but a principle of exhaustiveness of the intelligible, according to which the rational categories entirely cover the realm of the possible. In particular, applying it to the problem in hand—which sets at stake the antithesis between the one and the multiple—implies assuming that the field of all the possible determinations of "the Real" is covered by the categories of quantity. But who knows whether or not this is really the case? "Only a postulate of transcendent intelligibility can"—he wrote—"pass over such a lack of guarantee. Without such a postulate one would rest content with seeing in nonseparability just a questioning of a traditional thesis relative to 'the Real,' to wit, the one of its spatial analysability."

Apart from its use of the word "questioning"—which I would replace by the stronger one "refuting," which I take to be the only appropriate one[8]—I consider Bitbol's remark to be quite true and I view it as an im-

---

[8] It still being, of course, understood that, as explained in section 3-3-4 above, "refutation of a thesis on the Real" can only mean "refutation of a representation of the Real."

portant contribution to the subject. But at the same time I consider that, when all is said and done, this contribution in no way reduces the conceptual bearing of the discovery of nonlocality. For indeed, while Bitbol seems to consider the refutation of the traditional thesis he mentioned to be of limited interest (as his use of the expressions "rest content" and "questioning" indicates) in my eyes it is extremely important in two respects.

One of them is that it fills what I always considered a worrying gap in the reasoning of quite a number of philosophers. When, for example, Jaspers stated that the impossibility that Being-in-itself should be objectlike is quite obvious (Jaspers 1954), to all appearances he grounded this absolute certainty of his just on the "self-evident" fact that any object is relative since there are "objects" only for "subjects." That is, he grounded it on an acceptation of the word "object" that, admittedly, is consistent with its etymology and, as is natural, is the one philosophers normally think of. But still, it is not the (a priori just as acceptable) one that people who are not philosophers normally have in mind when they utter the word "object." For them, the word "object" refers to spatially situated things existing by themselves. And in that other sense of the word, contrary to what Jaspers' statement seems to assert, it is a priori far from obvious that Being-in-itself is not, basically, of the nature of a combination of objects. Since, in Jaspers' statement, the word "object" could quite naturally be understood in the latter sense (and perhaps was, at times, so understood by this author!) I take the fact that physics now corroborates the impossibility in question also when "object" is taken in this "everyday and non philosophical" sense to be highly significant.

The other reason why I think the refutation of the traditional "analyzability" thesis is so important is that this thesis is just ontological multitudinism. Now I take the latter, extrapolated as it nowadays is by popular science books and media to the totality of "what is," to be, in the field of basic current thinking, one of the most momentous errors of our times. An error that, admittedly, in view of the philosophical message that seemed to come from classical physics is easily explainable, but which nevertheless leads to a general impoverishment of public mentality and general culture. That this thesis should now be refuted without it being necessary to invoke Comte-Sponville's known phrase "to philosophize is to think further than one knows" therefore seems to me to constitute a most significant advance.

This is why, to repeat, the said refutation satisfies me. That much being said, the fact remains that to "think further than one knows" still seems to me to be quite a gratifying activity. For this reason I am quite pleased that Bitbol unearthed an ancient metaphysician susceptible of helping us think, without going too much astray, about the advance in question. This thinker is none other than Damascius,[9] one of the last neo-Platonists, who wrote, "the One, if it is, is not even one." This means, Bitbol explained, that Plotinus' One cannot be attributed a quantitative determination implying opposition to the plural. This is why to the name "the One" Damascius preferred the expression *pantè aporeton* which, roughly speaking, means something like "the absolutely inexpressible." And Bitbol asked me what would dissuade me from identifying Independent Reality with Damascius' *pantè aporeton* rather than with the Plotinian One, as I had occasionally suggested. On this point he met, in substance, with Hervé Zwirn who, through his analysis of present-day data, was led to consider that what is conceptualizable does not "exhaust the whole," even though it is not possible to define what this "whole" is. This made him take into consideration the notion of some unknowable that we cannot even speak of, concerning which to positively claim that it exists would still be saying too much, but that, he thinks, we cannot do without nevertheless.

As for me, I consider that the views Bitbol and Zwirn expressed there are basically right. And incidentally Bitbol himself pointed out that at several places I took positions quite similar to the *pantè aporeton* one, particularly where I recommended to keep to negative notions such as nonlocality rather than introduce positive determinations (the word "implexity" was put forward) to the effect of describing Independent Reality features.[10] However, it seems to me that the views in question admit of a complement. To make my point let me go back to Bitbol's pertinent objection. As we saw it rested on the fact that to refer, in the context in hand, to the opposition between the one and the multiple amounts to granting the validity of a principle of exhaustivity of the intelligible according to which the rational categories cover the realm of the possible. And it consisted in pointing out that only a postulate of transcendent intelligibility—which he does not endorse and which I do not endorse either—might justify the principle in question. But suppose we do reject it. That is, suppose we do not take the assertion "the rational categories entirely cover the realm of the possible" to constitute a basic a priori of thinking. May we then not conceive of a "Real" that admittedly would not be describable by means of these categories but on which other "modes of quest" would yield sorts of glimpses? As for me, I think that if we follow Bitbol the whole way we cannot consider this idea to be inconsistent and reject

---

[9] Last head of the Athenian Philosophical School, under the reign of Justinian.

[10] He still criticized one point, to wit, the fact that I wrote: "Independent Reality is not in space-time." He considered that such a use of the verb "to be" imperceptibly reintroduces a positive determination of some sort. But does it really? My feeling is that only a very subtle dialectician could settle the matter!

it for that reason. And I guess some at least of the neo-Platonists must have implicitly entertained it for, if not, why should they have so profusely written about this ineffable? Anyhow, my own view about the *pantè apo-reton* is that, contrary to what one might think, this approach is not in-compatible with a *conjecture* that poetry, music, painting etc. sometimes provide us with genuine vistas in the direction of "the Real," alias sorts of glimpses concerning it.

May such a hypothesis also be made concerning the great mathematical laws of physics? Admittedly this is a hazardous step. Clearly structural realism proper, in which it is assumed that the said laws are faithful pic-tures of the great structures of "the Real," could not be justified along such lines. But, as we saw, the Veiled Reality conception in no way carries conjecture that far. It merely involves the assumption that these laws—and the basic universal constants as well—being caused by the said struc-tures, are, in a manner that may well be undecipherable, sorts of traces of them.

The foregoing pages summarized the main lines of my answers to the interesting remarks and stimulating criticisms Bitbol and Zwirn made concerning my work. It is now time that, in turn, I should consider their views.

17-3 The Pragmatic-Transcendental Approach

Toward the end of the foregoing chapter we noted Poincaré's exclamation "it cannot be by chance," meaning that it cannot be just by mere chance that science is so convenient for describing and predicting present and future experience. Now, if it is not "by mere chance" it must be because of something and—we argued—this "something" must be some ground substratum distinct from ourselves, which led us again to the basic thesis of the present book. Moreover, we considered the view that such an idea was indeed present in Poincaré's mind, even though he hardly dared grant he had it, so convinced he was of the unsuitability of expressing ourselves on subjects about which nothing precise can be stated. Today many think-ers take up, on these questions, Poincaré's reserved standpoint but go in fact much farther. According to them, not only is "the Real" unreachable but also the very notion of a mind-independent Reality is meaningless.

But, to repeat, if we give up such a notion what standpoint should we take face to the fact that science makes it possible to predict phenomena whereas other intellectual constructs fail to do so? Are there, in this do-main, avenues of approach that would enable us to avoid having recourse to open realism? All along these pages we found out that too proximate versions of realism are incompatible with recently established physical facts. The latter practically forced us to limit the correspondence between the theories and "the Real" to just a very vague homomorphism between the latter and the great legal structures. Could we not—Bitbol asked—go to the end of such a realism-weakening process without yet being driven to pure empiricism and the difficulties that go with it? "Is there not a third way, neither realist nor empiricist, of interpreting physical theories, which would consist in considering them to be less than a partial trace of the Real but more than just mere recipes?" As we see, Bitbol is in search of some intermediate standpoint, consisting in keeping to a radical meta-physical agnosticism while still sharing with realism the view that the structures of theories are highly significant. And it is to transcendental philosophy that he turns for a solution. He noted that, according to Kant, the laws of Newtonian mechanics were neither mere predictive recipes nor true descriptions of the Thing-in-itself. That they were justified by their aptitude at expressing, within the context of an application to mate-rial bodies, "general conditions of possibility of experience," that is, the three general rules (permanence of substances, causality, community of all that can be simultaneously perceived) that, according to Kant, relate perceptions to one another, thus rendering the unity of experience possi-ble. And he tried to take this as a pattern in his own quest.

Of course Michel Bitbol is fully aware of the insuperable difficulties the literal version of Kantism meets with. In the foregoing chapter we noted that if the axioms of Euclidean geometry—the only one known in Kant's time—were really a priori as Kant thought they were, we would be at a lost to understand how non-Euclidean geometries could ever have been constructed. From Kant's views he therefore decided to keep but the hard core. And as to the nature of the latter he considered, along with Petitot, that it consists in the primacy of legality, by which he means the existence of rules that, while keeping the notion that phenomena are relative to us, enable us to "set this 'relativity' within brackets" (as explained in section 16-2-2 above).

Within the framework of this most general approach the way of reactu-alizing the transcendental method that Bitbol propounds consists in what he called "making the a priori movable." This means, he explained, that to the historical, indefensible forms of the Kantian synthetic a priori we must substitute a *functional a priori*, that is "a set of basic presuppositions linked with the practiced activity" (Bitbol 1998). The existence of such a link implies that we may well have to discard a functional a priori and substitute another one to it when our activity gets modified. And ac-cording to him this explains that we may give up, at least in part, the original forms of the Kantian a priori while still remaining within the general framework of transcendental philosophy. In fact, within this con-ception the current redefinition of our experimental activities due to the enlargement of their field beyond our day-to-day environment—in Kant's time the only explored one—even makes this change compulsory.

According to Michel Bitbol this approach theoretically yields a justification of the structure of any great physical theory. And this justification is grounded, not any more on some vague and problematic referring to the structure of an unknown Independent Reality but on "its [the theory] ability at incorporating within its formalism the norms presupposed by the experimental activities it accounts for." As for these norms, he explained their nature and their evolution in the course of the advancement of science by "their aptitude at covering larger and larger cycles of operational invariance." Which seems to mean that they are, at any period of time, fitted to the pattern of the experimental outcomes of that period. And finally, as for this very pattern, according to him it must be taken to be the outcome of a "process of co-definition of the activities and the forms on which they operate." He pointed out for example that, in accordance with the weak anthropic principle, the fact that the great physical constants have precisely such and such values may be viewed as the condition for a coemergence of biological entities able to exert an epistemic activity and the objects of the said activity.

The thesis is coherent. But are we entirely convinced? Is the reference to "the norms presupposed by the experimental activities," or in other words (in Kant's language) to "the general conditions of possibility of experience," fully explanatory? According to Bitbol what speaks for this idea is, as we noted, that in Kant's eyes the laws of Newtonian mechanics were quite in accordance with it. But still, remaining within the framework of Newton's physics I think that the Kantian approach and the recourse to the a priori that it implies account, at best, only in part for the achievements of the said physics. In fact, I consider that even if we went so far as to explain the three basic laws of Newtonian mechanics merely by referring to the a priori structures of understanding, we should find it impossible to account in the same way for the precise form of the gravitation law. And I am at a loss to see how substituting a functional a priori to a static one could change anything in this respect. Maybe Bitbol's theory should be understood as implying that in some indeterminate future, when we shall have fully taken into account all symmetries, etc., observation and experiment will contribute nothing essential to the theory. The mathematical formalism will reflect perfectly and in detail all significant experimental conditions and since, by assumption, there is no external reality (or, if there is one, it plays no role) this formalism alone will determine our knowledge. Every time, experiment will corroborate its predictions. As for me, I think that such an anticipation partakes very much of utopia. The rational element on which I ground this judgement lies in the distinction introduced above (section 6-1) between "framework theories" and "theories in the ordinary sense." Newtonian mechanics was a framework theory. Newtonian gravitational theory was a "theory in the ordinary sense" taking place within the Newtonian mechanics framework once the force law has been specified. Now, to repeat, for the sake of the present discussion I am willing to grant that the first theory may have emerged just from the "general conditions of possibility of experience," alias from "norms presupposed by the experimental activities" prevailing at that time. But of course it is quite impossible for me to take up the same stand concerning the second theory, which obviously, with all due respect to Newton's sayings, basically consisted in a physical hypothesis, even if it was a fantastically brilliant one. Similarly, quantum mechanics is a framework theory. Its basic axioms (see, e.g., d'Espagnat [1976] or [1995]) are purely formal, and the various theories open to experimental tests get developed within it through specification of particular Lagrangians, etc. Under these conditions, while I do not rule out the possibility that Bitbol's conception (or rather the here proposed interpretation of it) be correct concerning quantum mechanics proper (many interesting works were done aimed at deriving it from general views concerning symmetries, etc.) I very much doubt that it should be correct concerning such particular theories.

Still, the *conjecture* may, admittedly, be considered that, in some indefinite future, even the particular theories, nay, even the universal constants, will be found to be derivable from general invariance and symmetry principles through methods comparable to the use of Noether's theorem. But even though the said principles look simple to us, we still would have to understand why they force themselves upon our mind. And we should also examine whether or not it is possible, by considering higher level symmetries, to account for the small symmetry violations that, curiously enough, affect many of the presently known symmetries. To all this Bitbol answers by observing that my way of looking at things— my "explanatory referring" to an unknown Independent Reality—is also, as I myself granted, a conjecture. This is true. But still, the conjecture in question takes place (with appropriate changes) within the continuity of an "intuition," that of the existence of some reality outside us, that all men, all civilizations, always had. To replace it by some other one would, in my opinion, be justified only if it were proved false (which is impossible) or if the alternative conception put forward as a candidate were endowed with a much greater plausibility. For the above stated reasons it seems to me that the one Bitbol proposes does not, at least at present, fulfill this condition. Hence I am reticent to endorse the substitution.

Near the beginning of the present study (section 2-3) we noted that during Antiquity and the Middle Ages the only research method taken to be valid consisted in going from the universal to the particular. To start from a world-view fully consistent at the level of ideas and deduce from it the multifarious aspects of Nature as we see them. And we observed that the crucial contribution of Galileo and his great contemporaries just consisted in replacing this too ambitious, hence ineffective, procedure by the one that we now simply call the scientific method. Now, the latter is admittedly, as Bitbol stresses, grounded on legality. But it is at least as much based on the rule that, for finding out, amidst all the possible laws, those that are the "true" ones it is necessary to compare their conse-quences to experience, that is to facts, and reject those that lead to irreme-diable discrepancies with the latter. Now, of course I do not assert that Bitbol's conception constitutes a full-fledged comeback to the ancient and mediaeval approach. The latter raised a claim at ontological significance to which Bitbol is quite opposed. But in view of the exclusive role it im-parts to human aptitudes at synthesizing it seems to me that Bitbol's con-ception still might be understood to be a partial—say, methodological—comeback to the said mediaeval approach. Which, of course, would gen-erate some suspicion concerning it in the mind of people used to more down-to-earth views.

But still, I should immediately add that, on all this, I have to stay wary, for the notion of "perceptual construction" Zwirn introduced (here, sec-tion 17-2-5) has to be kept in mind. It indicates that even if Bitbol is right it remains theoretically possible that comparing theory and experiment will everlastingly reveal tensions. That mismatches between tentative the-ories and experimental outcomes will go on taking place, even though the latter are to be considered mere perceptual constructions. However, as I already pointed out, we then still should have to find out why, in this realm, we always systematically admit that the perceptual constructions are in the right. Why is it that every time a theoretical hypothesis, be it even void of practical implications, such as a cosmological one, turns out to be flatly at variance with observed facts we reject the hypothesis (even if it is theoretically most attractive)? If, in the last analysis, everything stems from us, why not decide, sometimes in favor of the facts and some-times in favor of the (experimentally "disproved") theory? Of course such a behavior would be judged irrational, but on what ground?

Does all this mean that I finally consider the pragmatic-transcendental approach should be rejected? No, it does not, for two reasons. One of them is that it is always interesting to explore new paths, be it only to find out how far we can go along them. The other, more specific, one is that, while I think the notion of Being[11] is to be kept, it became to me clearer and clearer that such a Being—prior to the matter-mind splitting—

---

[11] Or "the One," or *pantè aporeton*, or etc.; in such matters I take the aim at preciseness to be a lure.

is not knowable as such and is extremely "far" from our current set of concepts. If, as I think is the case, the elements of our thinking that constitute the basic framework of knowledge derive from it, today physics has shown that the set of these elements does not in the least coincide with the set of the classical categories of substance, determinism, etc. Hence the idea that, as knowledge develops, an evolution of the set of concepts we take to be the basic ones should take place seems to me to be quite a reasonable one. And I consider that introducing the "movable a priori" notion constitutes at least a figurative and expressive way of conveying the said idea.

In fact I even go much farther. For indeed, quite generally speaking, I observe that as a rule our inner aims are far from clear and that, in particular, the ultimate objectives of the researches we carry out are not. And it seems to me that, in this field, we would see things more clearly if we introduced a distinction between the objectives of our general quest for understanding and the ones of scientific research. In this book my objective is general understanding, science being, for this purpose, only one reference among others (though presumably the main one). By contrast, a scientist qua scientist must aim at discursive knowledge only. And in this domain he/she may merely take into account what is clearly conceptualizable. Now the whole history of scientific research shows that this ever more strictly affirmed condition implies as a counterpart an ever increasing limitation to relative knowledge. In a quest for understanding such as the one we are here engaged in it is quite necessary to stress this limitation, which, as we saw, leads me, for one, to associate some properly scientific advances to notions such as the one of "fabricated" or "pseudo" ontology. But in the eyes of the working scientists who, to repeat, anyhow may only have what is clearly conceptualizable—hence relative—in mind, the use of such epithets—"fabricated," "pseudo"—not only is somewhat disparaging but also is rationally not entirely satisfactory. "Why 'fabricated,' why 'pseudo'?" they would ask. I think one interest of Bitbol's pragmatic-transcendental approach lies in the fact that it emphasizes what, within this "focalization on the relative," is here to stay. And that therefore it may be considered to constitute a real advance in our apprehension of what the true objective of scientific research actually is. Setting aside what, for science (in view of the nature and strictness of its truth criteria) would be a chimerical objective (i.e., the quest for Reality-in-itself), the said approach explicates the nature and meaning of what scientists, qua scientists, can really do. So that, for example, within this new framework the term "pseudo-ontology" does not fit any more the theory to which, above, this label was attached. For adepts of Bitbol's conception the notion of Ontology in the traditional sense is meaningless and if they tend to identify Feynman's approach to an "ontology" in Quine's sense, that is to a picture that is relative to us, I think they can. For them, the said theory is therefore no longer "pseudo."

That much being said, getting back to the facts to which the idea of "making the a priori movable" refers, it remains true that in virtue of my own "turn of mind" I feel I understand them better by thinking of an Independent Reality of an infinitely remote kind—but of which we catch glimpses nevertheless—than by referring to experimental practices concerning which I still fail to quite clearly grasp what, in the absence of external reality, actually guides their progress.

17-4 A Few Notes on Zwirn's Approach

Above, while I analyzed at some length Michel Bitbol's views I mentioned Hervé Zwirn's only briefly. This was mainly due to the fact that when his book came out a large part of the manuscript of the present one was already written. But still, a possibility remains for me to express myself here on this subject, which pleases me all the more as Zwirn, as already noted, took interest in my views and made on them, as on several other approaches, remarks that I found highly pertinent.

Of course, in Zwirn's case as in Bitbol's (and other authors as well) it would be quite pointless to review here all the numerous and crucial points on which we agree. Fortunately there are also some—minor!—ones, concerning which our views but partly coincide. Let me here make use of the opportunity I have to comment on them.

The first one has to do with the definition of what I called "Veiled Reality." In his section 4-7, which deals with the role of the observer, Zwirn first noted that, contrary to what, in classical physics, seemed quite conceivable, in quantum physics the "something" whose existence it seems normal to evoke for explaining our perceptions cannot be assumed to really resemble what we perceive. He, quite correctly, explained that this was the reason why I used the words "Veiled Reality" to designate what causes our perceptions but to which we have no direct access. But he added "this Veiled Reality is, partly at least, described by the quantum formalism." This, he commented in a footnote, is why it is called "veiled" rather than totally unknowable. And in the main text he continued: "it [this Veiled Reality] is indeterminate, nonseparable and not wholly understandable." This account of my views has the advantage of constituting a simple, not grossly incorrect introduction to them. However it does not convey them faithfully and is even susceptible of generating a few misinterpretations. One of them would be to believe that such and such elements of the formalism describe adequately—as they "really are"— some elements of Veiled Reality (which I often called "the Real" in the present book). For example, it would be erroneous to think that the wave function truly describes the state of a physical system, the word "state" being taken in the trivial, hence descriptive, sense it takes in both everyday life and classical physics. And that the said system would therefore be a genuine element of "the Real." We saw in the present book that this is in no way my conception. Also, Zwirn's account tends to make believe that indeterminacy is a feature of "the Real" itself. We saw that in my conception this is not the case. According to the latter, indeterminacy is essentially a feature of *empirical* reality, in other words, of the phenomena, in the Kantian sense of the word. It refers to predictions of observations and not at all to Reality-in-itself. And finally, the reason why "the Real" is said to just be veiled instead of wholly unknowable is not that we (exactly) know some, and some only, of its structures. I do not assume that we have such knowledge. The reason in question is twofold. It is that we know for sure what representations of "the Real" we are "not allowed" to adopt (they are all those that involve "locality"). And it is that, still, we may *conjecture* that between the structures of the latter and our scientific laws some, presumably vague, relationships exist.

With respect to these few misinterpretations it is however a pleasure for me to note here that, in fact, Zwirn did not really commit them and that, in his book, the disputable sentence I just commented on should in fact be considered to be a "pedagogically necessary approximation." What makes this clear is the fact that at the place in his section 6-10 where he described the Veiled Reality conception in detail his account of it did not suggest any more the misinterpretations in question. The notion of "pedagogically necessary approximation" should come as no surprise to whoever knows how difficult it is to substantially modify received views. Indeed, to that end the most convenient way may sometimes be to proceed step by step—for example, by changing one concept without changing the other ones—and such a procedure is liable to generate temporary misinterpretations that get corrected in the end.

Concerning the interesting *Commentary* Zwirn made of the Veiled Reality conception I merely have two reservations. One of them I already mentioned. It has to do with Zwirn's statement that my standpoint is pertinent only if it is considered that the subject is face to face with the world. In section 17-2-1 I stated the reason why I consider this assertion to be inadequate. The other one concerns Zwirn's statement: "For d'Espagnat . . . the deep structures are correctly apprehended by science." This would be true only if I adhered to structural realism, and we saw this is not the case.

These few remarks are not final and their content may be debated. They change nothing to the fact that Hervé Zwirn's book is in my opinion one of the clearest and most elaborate analysis of the—very novel!—conceptual problems raised by contemporary science. Moreover the three-level solution of them that this author tentatively put forward is, I think, most interesting. Incidentally, setting aside some differences in terminology (particularly concerning the empirical reality notion) it might be wondered if the said solution could not be considered to be a more detailed version of my own one. A version in which my "Veiled Reality" notion would be taken to cover really two distinct concepts: the one of the *pantè aporeton*, that is, of the Zwirnian "unknowable," and the one that he calls "empirical reality," meaning by this a partly conceptualizable but still nonrepresentable reality. It is clear, however, that such a parallelism is here put forward merely as a conjecture, which should be carefully examined before deciding whether or not it is significant.

CHAPTER 18

OBJECTS AND CONSCIOUSNESS

§℘

18-1 Introduction

UP TO NOW, AND with rare exceptions,[1] investigations concerning reality and investigations concerning consciousness were carried through very much apart and with opposite starting points. Those concerning reality were essentially undertaken by philosophers who considered that the said notion is in no way self-evident and focused therefore on inquiring to what extent and in what sense what human knowledge describes may be called "real." Which means, these thinkers investigated problems the formulation of which did not raise any question bearing on the nature of the very carrier of knowledge, namely, consciousness. Conversely, practically all the neurologists who studied the nature of the latter did so within one particular philosophical framework, to wit, the one consisting of, if not a declared materialism, at least an implicit physical realism, which means that they took reality to be a basic notion, lying beyond any possible questioning. Incidentally, in this book the itinerary that, up to this point, has been followed was just the one taken by the aforementioned philosophers. We referred to observations, verifications, and experimental outcomes without ever wondering about the nature of the entity that observes, verifies, or experiments. The only exceptions to this were the analyses carried out in chapter 10, for we there had to take into account a difference—entailed by quantum mechanics—between the standpoints of a mind that observes a phenomenon from the outside and a mind taking part in what is observed. This was indeed an incursion into the problem of the nature of consciousness, but still a somewhat marginal one.

In a way, such a partition between the two fields of research in hand may be considered a good thing. Like any other animals we physicists and we philosophers know for sure that observation and apprehension are "real." Hence we know we are in our good senses when we refer to such notions. But, being aware that they are still, in many respects, enigmas, we avoid as much as we can referring to their "real nature." On the

---

[1] Particulary Michel Bitbol's book *Physique et philosophie de l'esprit* (2000).

other hand, however, it is clear that, ticklish as it is, the question of what this nature is cannot be dodged forever. It must even be judged that the renunciation of physical realism that contemporary physics more or less dictates forces us to consider it. Conceivably some radical idealists may rest content with the observation that, consciousness being primeval, it is natural that it should not be reducible to anything else. But even they should wonder about its relationship with *empirical* reality. And for supporters of the "open realism" postulate the questioning is, of course, all the more intense as they have to wonder as well about the relationship between mind and basic Reality. This chapter is devoted to such issues. Unfortunately we shall only be able to take a few, quite insufficiently informative, glances at them. All reliable investigators whose subject matter led them to consider such problems unanimously grant that they still remain quite enigmatic.

18-2 Truth: Definitions and Criteria

People who are anxious not to delude themselves are intent on keeping an objective stand and strive to believe and state only what is true, whether they find it pleasant or not. Unfortunately it has been known for millennia that, considered within a sufficiently general framework, the notion "truth" is a tricky one. The definition of it that most naturally comes to mind—call it Definition D—consists in identifying "true" with "adequate to what reality really is." It corresponds to what philosophers call the "similitude theory of truth," which, in turn, is in keeping with *realism concerning statements* as Dummett defined it (see section 6-6). Within the world-view specifically called "physical realism" in section 1-2—a view according to which we have, at least in principle, access to "the Real"—this definition is imperative.[2] On the other hand, it is obviously meaningless within the realm of conceptions according to which we merely have access to *human representations* of "the Real." Which explains that, long since, philosophers with an empiricist or positivist bent put forward other definitions of truth, which they grounded on verifiability.

This remark, combined with the fact that quantum mechanics proves to be *the* fundamental theory, entails that Definition D could conceivably be kept only by people who would adhere to some ontologically interpretable variant of the said mechanics, or at least would not discard all of them. Now, we know (see chapter 9) that many obstacles make it difficult to accept any one of these variants. And moreover we also know that,

---

[2] At least concerning statements belonging to a wide class of statements including factlike and lawlike ones, which, for future reference, we call "Class C."

even within the latter, applying the definition in question proves conceptually ticklish (in the Broglie-Bohm model, for example, the ontology to which the model refers involves hidden variables, whose relationships with our experience is, to say the least, nonconventional). This seriously impedes adopting Definition D. Incidentally, note however that there are a few statements whose nature is such that the mentioned difficulties do not, all by themselves, bar out applying Definition D to them. Concerning these statements it is in fact the nature of the adopted philosophical outlook that makes the difference. For example, while within a radical idealist (and even a Kantian) viewpoint Definition D can be applied to, one might say, no statement whatsoever, within the veiled reality approach this, strictly speaking, is not the case. For indeed, in the latter conception there are some very general assertions—such as "existence is prior to knowledge" or "physical constants may reflect something of the Real"— that do refer to "the Real as it really is." But it remains true that, concerning all other assertions—scientific or otherwise—since, in the said conception, they merely bear on empirical reality (on objects-for-us) Definition D is inapplicable.

From all this it follows that in contemporary physics Definition D of truth could be applied only (and, even then, with reservations) within special, scientifically nonconvincing models. Another definition must therefore be looked for. To this end, the simplest is of course to keep the form D has but transpose it within the empirical reality framework (which roughly speaking amounts to define truth by means of verifiability). In other words, D should be replaced by the definition—call it D′—according to which a statement is true if it adequately describes empirical reality, in other words if it correctly pictures what intersubjectively appears to us. Stated this way Definition D′ is still extremely general. Clearly it does not require explicit, detailed acceptance of the antirealist views as defined by Dummett, even though it is fully compatible with them. D′ quite naturally applies to statements referring to ordinary properties of objects, such as those specifying positions of pointers on dials. On the other hand, at the place where, below, we shall have to consider the nature of sensations it will become clear that its use raises difficult conceptual problems.

REMARK 1

In a way, Definition D′ may be considered to be a somewhat truncated version of the definition of truth attributed by Putnam to Kant and reproduced here in section 1-2. The one according to which a statement is true if the noumenal world *as a whole* is such that the statement in question expresses what a rational being would say, given the information available to a being with our sense organs (and, of course, disposing of our instruments).

Analysis of the conditions under which Definition D may be applied shows that D might, after all, be kept if, without necessarily adopting one, definite, ontologically interpretable version of quantum mechanics, we just declined to assume, right from the start, that none of them is acceptable. But since this standpoint is an agnostic one, people who take it up may not claim, concerning some given statement,[3] that it is true in the sense of Definition D. They may merely claim that it is possible that it should be true in that sense. For example, in section 2-8 we opted in favor of the *weak* completeness hypothesis, implying that we never shall be able to experimentally decide whether hidden variables exist or not. Within the realm of the said hypothesis it should, in particular, be impossible to derive from experiment an answer to the question whether the hidden variables lying at the basis of the Broglie-Bohm model exist or not. In the case of a (generalized) measurement procedure all this implies that, concerning the statement "the pointer lies in a well-defined interval on the dial," all that may be claimed is that it is possible that it should be true in the sense of D. The hypothesis considered does not enable us to simply claim that it *is* true in this sense.

18-3 Objects and "Orders," or "Levels," of Reality

In the first part of the book the question of the nature of objects came up several times, and in particular in chapters 4, 5, 8, and 10, where we met such notions as nonlocality, quantum universality (which decoherence theory makes highly credible), weak objectivity, etc. In view of the outcome of this whole investigation it has become clear that the word "reality" may be, and is, legitimately used in two different contexts, but not with the same sense in both. More precisely, it has been found that, short of adhering to radical idealism, we must separately consider, on the one hand "mind-independent reality" (alias, "reality-per-se," alias "the Real," alias "Being") and on the other hand empirical reality, that is, the set of the phenomena. Indeed, the developments of sections 4-2-3, 5-3, 5-4, and 15-4 forced this conclusion upon us. For they made it clear that the empirical reality concept (which they define and make precise) has the observational attributes of the set of objects we use to call real while, still, it cannot

---

[3] And, in particular, one of Class C.

be identified with "the Real" in any Ontological sense. Or, to be more precise, that it could, only if the expedient were called in of resorting to models lacking scientific reliability. It therefore appears impossible to avoid having to consider at least two "orders," or "levels," of reality.

It must be granted that, in this domain, contemporary physics subjects us to considerable conceptual stress. By dissuading us from identifying empirical reality with mind-independent reality it greatly incites us to reduce the status of the former to that of a mere appearance. And indeed when, as in section 8-2-2, the question arose of the way in which decoherence themes make us consider macroscopic reality, it was the word "appearance" (in the sense of the opposite of reality) that we were led to make use of. We referred to the "building up" of empirical reality and pointed out that we thus build up, not only the present and the future, but also the past. Now, it is true that such a notion of present, future, and past being constructed by us at first sight creates in our minds a feeling of deep uneasiness, which—particularly concerning the past—borders pure and simple rejection. On the other hand, however, it is appropriate to analyze a little the reason for the uneasiness in question. When we do, we first note the obvious point that in our mind the word "appearance" is meaningful only relative to what, conversely, it considers real. And we then observe that what, in ordinary life, the mind takes to be real is the set of what it apprehends, could apprehend, or could conceivably have apprehended. Now it is precisely this prejudice of ours that generates our instinctive reservation. For, due to it, it seems to us that the statement that empirical reality is mere appearance means that there exists some other reality *of the same nature* that, actually, is the real one. Besides, we also have the impression that the said statement means negating—or at least considerably underestimating—the whole network of intersubjective agreements and causal links that we intuitively consider to lie at the core of what is more than just appearance.

What must primarily be grasped is that such is not, actually, the situation. That, when it is taken in the just sketched sense, the word "appearance," in the context in hand, is a misleading one. And that, in fact, in such a realm—in a field of discourse so far remote from ideas of everyday life—language fails us.

To see this let us first note that, even though "the Real" is merely veiled (is not a "pure x" strictly unrelated to phenomena), it nevertheless is of a nature totally differing from that of the said phenomena, the set of which constitutes by definition empirical reality. Indeed, let us remember that it is not "in" space, and hence not in space-time either. Nay it may not even be thought of as being embedded in cosmic time. Besides the fact that, today, space-time can hardly be dissociated from matter, two reasons speak against such an embedding. One of them is the fact that cosmic time can be defined only under the (admittedly very good) approximation that space is isotropic, so that it seems to partake of the nature of a representation procedure more than of that of a framework for some (vaguely defined) "absolute." And the other one is that, strangely enough, it so happens that the models that take the notion of "the Real" most seriously are the ones in which plugging in the notion of cosmic time and the correlative one of the Big Bang has been found to be most difficult. Bohm's assertion that, in his approach "this 'Big Bang' is to be regarded as actually just a 'little ripple'" (Bohm 1980) is noticeable in this respect.[4]

Next, let us keep in mind that empirical reality does exhibit all the features that our everyday intuition imparts to the notion of reality. As already noted in chapter 1, a most important one—one that crucially distinguishes empirical reality from pure observational predictability—is counterfactuality. Now, we noted in section 8-2-2 that the empirical macroscopic reality notion resulting, in quantum physics, from the study of the behavior of the so-called "collective variables" is in no way at variance with the said counterfactuality. Another one of the features in question has to do with the, tightly linked, ideas of "explanation" and "causes." As we noted, one of the reasons that make us intuitively prefer realism to mere observational predictability is that realism does give us the impression of an understandable causal link between events, which mere observational predictive laws fail to yield. However, in section 14-2 we noted that philosophical analyses of the said causal link notion have long since shown it to be intimately associated in our mind with the one of will, which means that at the level of our intuition it shows marked anthropomorphic aspects. And what, along the same lines, is perhaps even more significant is the outcome of the analysis carried out in section 15-4. For it shows that within empirical reality there is indeed room for the notion of a definite concatenation of causes and effects (conceived of, of course, as phenomena). More precisely it shows that such a concatenation ending in an observation act should normally be apprehended by the subject (or collectivity) who makes the observation. So that the notion at hand may be kept, even within the framework of generalized measurements such as those considered in chapter 4. In other words, the analysis in question showed that even in a situation resulting from a Schrödinger-cat-like interaction, concatenations of the type "the pointer acts on a relay that acts on a flasher that acts, etc." may be considered to be genuinely causal ones.

---

[4] Besides, by introducing as they do the notion of a generalized, space-time endowed with some ten dimensions of which we only perceive the three of space and the one of time, the supersymmetry and superstring theories presently under study clearly suggest something like this.

Such an interpretation admittedly is relative to us but (as already noted) *no more so than the very notion "events."*

Finally then, should we call "mere appearances" appearances—causal ones included—that are the same for all those who are able to perceive them (including, perhaps, animals)? As we know, idealists answer this question negatively, and on this particular point it seems difficult to call them wrong. In fact our judgment on this matter depends very much on the meaning we impart to words. It goes without saying that referring to things conceived as being independent of us greatly facilitates everyday life. From this it follows that we have a natural tendency toward reifying. Concerning objects, this is an approach that, with regard to practical points, is entirely legitimate. It may quite well be accepted also in philosophy, but only provided we keep in mind that, by making use of this objectivist language, we, in fact, merely refer to our communicable experience. With this reservation, empirical reality, the reality that is ours, within which we are born, live, and die, does really qualify for being called "reality." In the sense just defined it would be not only incorrect but also inconsistent to claim it is merely an "appearance." But at the same time we must remember that in view of contemporary physics such a reifying proves unwarranted when we, naively, take it strictly literally. To repeat, we have to keep in mind the fact that it finally is but a means of stating in a convenient manner some possible observational predictions (and therefore of predicting and planning possible actions). And finally, within the framework of such a conception, while the distinction between (empirical) reality and "appearances in a trivial sense" of course remains essential, the one between (empirical) reality and "appearances that are the same, at all times, for everybody" clearly ceases to be valid.

In short, we have explained and justified the use of the word "reality" for designating empirical reality. Since, on the other hand, in view of the reasons described and commented on in foregoing chapters (particularly chapters 5 and 10) we have to keep the mind-independent reality concept, we cannot, to repeat, do otherwise than take up the idea of two "orders," or "levels," of reality.[5]

18-4 A Few Remarks Concerning Sensations

That the problem of the nature of sensations ranks among the most diffi-

---

[5] Note that here the word "level" is slightly misleading since, as already pointed out, we are not dealing with mutually comparable entities such as geological strata, for example. The word may be kept only if we give up the idea of a similarity of any kind. If we were bent on stressing how different the notions of empirical reality and mind-independent reality actually are we might do so by calling the second one "the super-real."

cult ones is universally acknowledged. Are sensations identifiable to mere trigger mechanisms, as naive behaviorism would have it and simple observations—such as the one of insect-eating flowers—suggest? Or, on the contrary, do they involve some *sui generis* element, identifiable to what we call "mind," as our own intimate experience forcefully indicates? Physiologists, neurologists, and philosophers have been investigating this subject for a long time and, in this as in other fields, the tempo of discoveries is running faster and faster. But at the same time the problem proves more and more intrinsically subtle, so that it still remains with us. Here I have (of course!) no pretensions at solving it and I do not even go so far as to claim that some promising axis of research should emerge from what follows. My only purpose is to examine some a priori conceivable proposals.

18-4-1 On the "Identity Theory"

The view here called, for brevity's sake, the "identity theory" consists in asserting that, in the last analysis, any genuine sensation—that is, any "becoming aware" and, by extension, perhaps even any thought—is finally identifiable to some material structure internal to or involving neurons.[6] This theory essentially asserts that, assuming some given person—call her Alice—never had any sensation, to inform her of what the word "sensation" means it would be necessary and sufficient to provide her with some detailed enough description (practically unrealizable of course) of which neuronal states and events constitute sensations. Or, to put it differently, it would be necessary and sufficient to let her know precisely enough what detailed features the reality of the neuron system then has. The thesis is supported by indirect convincing-sounding arguments, such as the fact that no established experimental facts show souls existing independently of bodies, and the observation that, to some extent, people's thoughts and desires may be changed at will from the outside by means of drugs. But, as D. J. Chalmers, for instance, forcefully stressed (Chalmers 1996; quoted by Bitbol 2000), it remains true that a priori the neuronal events taking place when we react to some external stimuli might well occur within a totally dark cognitive night, as is (presumably) the case with computing machines and robots. Even if all mental operations

---

[6] Within the Anglo-Saxon philosophical world such a view still apparently looks new and daring. This is hardly the case in France since the view in question was already present, between the lines at least, in the works of eighteenth-century thinkers such as d'Holbach and La Mettrie. In the said country it is taken to be a significant facet of a tradition dating back to the "enlightenment" period.

were given adequate neurophysiological explanations, we still would dispose of no satisfactory answer to the question thus raised. Between such sequences of events and conscious experience lies—as Chalmers stressed—a radical "conceptual gap."

The very existence of these two conflicting sets of arguments shows how puzzling the problem actually is. Here, without (to repeat) having any pretension to solving it I aim at making clear that some pieces of information stemming from contemporary physics are relevant to its study. Those I am thinking of proceed from what was noted in section 18-2 concerning the limits of applicability of Definition D. We observed at that place that in contemporary physics the said definition may merely be applied—and, even then, under debatable conditions—to highly special conceptions, to wit, the ontologically interpretable versions of quantum mechanics such as the Broglie-Bohm model. We noted that another one has therefore to be looked for, and we considered making use of the one we called D′, consisting in identifying the truth of a statement with its adequacy to *empirical* reality.[7] But a difficulty arises at this stage. Above, we observed that according to the identity theory, in order to inform our sensation-lacking Alice of what it means to say that a person has such and such a sensation we have to provide her with some detailed enough information about the reality at the considered time of this person's neuronal system. Now, we just saw that switching from definition D to definition D′ implies substituting the words "empirical reality" for the word "reality" in the foregoing sentence. But, then, in view of the very definition of empirical reality, the just mentioned information must ultimately be of the type: "If the human collectivity performed such and such measurements on this person's neurons, what it would observe would be *xyz*" (*xyz* being one of the possible outcomes that may be got). However, this statement involves the verb "to observe." And since "to observe" implies having a sensation Alice cannot understand it since she does not know as yet what a sensation is. This, obviously, is a vicious circle. Is it possible to escape it by substituting a set of registering devices to the human collectivity? By stating that the meaning of the assertion in hand actually is that "if the set of the said devices happened to interact with the person's neuron system under conditions in which it could a priori register this or that, what it would in fact register would be *xyz*"? No, for obviously, since these registering devices exist as such merely as elements of empirical reality, the same holds true with respect to the results they register. Far from being "facts in themselves" the positions of

---

[7] Which of course implies that if in the formulation of the identity theory we introduce the word "matter" this word may only refer, in a vague, general way, to the said *empirical* reality.

their pointers have definite values only relatively to the community of observers, that is—again—relative to the *sensations* the latter have. Finally, as we see, the choice is merely between the vicious circle in question and infinite regress.[8]

Of course the difficulty under study is not surprising. As noted in section18-2, substituting Definition D', centered on empirical reality, for Definition D, centered on independent reality, is but going over from the idea that truth is adequacy to the Real to the one that truth is to be defined in terms of verifiability. But the verifiability notion implies the notion of one or several verifiers. In other words, it is meaningful only if the notion of consciousness is available. It is therefore inconsistent to try and make use of it for defining the very notions of consciousness and getting conscious.

Hence, within, at any rate, the realm of strict standard quantum mechanics—that is, of a conception of quantum mechanics that incorporates the "strong completeness" principle (section 2-8) and is therefore antirealist—there seems to be no other choice than just to reject the identity theory. If somebody still wants to try and keep the said theory he/she must therefore, either take up one of the ontologically interpretable theories or, at least, take the completeness principle only in its weak, Stapp-like form. And even then, as already noted, it can only be claimed that the identity theory may be true but that nobody shall ever be able to show it is. In other terms, we see that within the framework of the weak completeness principle the identity theory may be kept but with the proviso that it can only as a mere belief, deprived of any firm, consistent scientific basis.

REMARK 1

Biologists and neurologists alike generally agree that, although neurons (and cells in general) are structured at a surprisingly fine level, in their vast majority their elements still are macroscopic enough for being, to a very good approximation, dependent on essentially classical physical laws. Let it be stressed that this fact in no way modifies the above reached negative conclusion concerning the validity of identity theory. This is because the conclusion in question follows merely from the view that quantum mechanics is universally valid, together with the observation that, expressed in the form of observational predictive rules, the classical physi-

---

[8] I am well aware of the fact that such a rejection runs counter to ideas that, on the basis of current neurological data, may seem to be by far the most likely to be true, since within the classical framework, they are natural extrapolations of the said data. But let us remember that—even in physics!—extrapolation is often misleading. And let us observe that the neurological data in question are in no way at variance with the physical data that justify our rejection.

cal laws have been found to be mere consequences of the said mechanics. True, at first sight the argument "since classical concepts suffice for accounting for the neurological phenomena, consciousness may be interpreted in strongly objective terms" seems clear and unquestionable. But we just saw that this reasoning is faulty. The point is that what, in classical, macroscopic physics, has been shown to follow from quantum physics is only the set of its *observational predictive* laws. It is true that, considered just by themselves, these classical, macroscopic, observation-predicting (alias weakly objective) laws could be considered to trivially follow from strongly objective laws bearing on the behavior of per se existing objects (and indeed this is the interpretation normally imparted to them, at school and in everyday life). But these strongly objective classical laws do not and could not follow from the quantum mechanical rule ("could not," since philosophical realism cannot be derived from recipes). Hence, adhering to the reasoning under examination would imply assuming that, over and beyond the universal quantum mechanical "rules," or "laws," there, independently, exist classical laws, only valid within the macroscopic realm. However, the idea of such a duality meets, as we know, with a host of difficulties, mainly due to the impossibility of defining a borderline between what is macroscopic and what is not. And the whole virtue and interest of decoherence theory lies, as we saw, in that it restores unity by showing that the (observational-predictive, alias weakly objective) classical laws follow from the (equally weakly objective) quantum laws. We must therefore give up the idea that the elements composing objects in general and neurons in particular are objects-per-se. We must take them to just be elements of empirical reality. It follows, as we saw, that the truth of a statement concerning them has to be defined by means of Definition D'. And then all the rest of the argument in this section goes through without change.

REMARK 2

Admittedly the above disproof of the identity theory relies to a large extent on the principle of completeness. If it were decided to keep this principle in neither one of its two (strong and weak) versions it would of course become possible (subject to Chalmer's objection) to look upon the said theory with a more favorable eye. This is not a proper place for investigating the details of what might be attempted along these lines. We shall therefore keep to two very general remarks.

One of them concerns the possibility of inserting the identity theory into the Broglie-Bohm model. In principle this is possible, since the model is ontologically interpretable so that the truth definition D may be made use of. In the model, the identity theory would consist in identifying the state of consciousness of somebody with the fact that the parameters describing the state of this person's neurons—roughly speaking, their wave function and the values of their hidden variables—are such and such. A problem, however, is raised by the fact that the wave function in question is entangled with the one of the rest of the Universe. For indeed this implies that, theoretically, the configuration space that has to be taken into account in the theory is far greater than the one defined by the set of the coordinates of the involved neuron. In principle it is that of the whole Universe. Whether or not adequate approximations may be considered, that would restore the notion of separately existing, differently located minds, is a point that, to my knowledge, remains open.

The second remark has to do with the fact that, of course, the Broglie-Bohm model is not the only ontologically interpretable one that might, in principle, serve as a substitute for quantum mechanics. In section 9-8 we saw, in particular, that, in the last resort so to speak, some physicists bent on keeping a realist approach chose to subtly modify the Schrödinger equation by inserting in it small nonlinear terms. And some of them—particularly Roger Penrose (1995)—developed considerations that provided them with some reasons to hope that an explanation of consciousness more or less in line with the identity theory will emerge from the thus elaborated model. They may well be right. Personally I can only say that I do not believe in models of this type. Indeed, all of them have unpleasant features (one of which was briefly sketched is section 9-8), some of them are purely ad hoc, and those that are not are still in rather preliminary stages, having to face as yet unsolved riddles. It seems to me that they do not—or at least do not yet—meet all the various criteria a scientific theory is normally expected to fulfill.

18-4-2 On "Efflorescence Theory"

The theory here referred to under this name is the one according to which sensation—or at least its specific component, namely, "becoming aware" or "getting conscious of"—is just a product of neuronal activity. Disproof of the identity theory does not, by itself, entails that efflorescence theory is wrong, for (contrary to what most philosophers of Antiquity, the Middle Ages, and later have thought!) the nature of a product may quite well qualitatively differ from that of its cause. And in fact, nowadays the "received view," the view that biological and neurological approaches—and indeed just our "scientific way of looking at things"—seems to demand, is nothing else than, precisely, the efflorescence theory. It is the view that thought—including, to repeat, its first element, "getting aware of"—is a kind of efflorescence of the body. (One might think of drawing a parallel between this view and the Aristotelian one that souls are *forms* of bodies, but the difference is that Aristotle considered matter to be mere potentiality whereas present-day materialists take it to be primeval with respect to form.)

To day, a large number of people still consider that efflorescence theory commands credence. However, on the other hand, some data are now available that tend to suggest that, far from just being a mere efflorescence from neurons, thought has structures that might be, somehow, directly connected to those of "the Real." These data were noted by several authors and do not have the character of proofs. Their nature is that of mere plausibility arguments. But they are, all the same, highly worth attention. They consist of a kind of parallelism between, on the one hand, the structures of thought and, on the other hand, the structures of quantum mechanics, viewed as (dimly) reflecting those of "the Real." As Shimony noted (1993, I), this parallelism proceeds in part from the fact that the basic quantum mechanical notions are of a purely structural nature, which implies that the domains in which they may be found to have a role are not limited to the one of systems conceived of as material.

One of the said notions is entanglement. For indeed mind is today considered by neurologists to be, at the neuronal level, an essentially collective phenomenon, involving brain states of a considerable complexity, that are not separable, not even theoretically, into individual neuron (or group of neurons) states. The similarity with quantum entanglement (section 8-1-1, Remark 1) is obvious.

Another such notion is potentiality. As we know, while, by assumption, a wave function completely specifies the "state" of a quantum system, it always leaves undefined some properties of the latter. And it is only if and when one of these properties gets measured that it takes up a definite value (the wave function being simultaneously disturbed by this measurement procedure). As Heisenberg first pointed out, such a state of affairs is, to some extent, reminiscent of Aristotle's theory of act and power. For it may be said that initially the property in question is present in the system only "potentially," or "in power," and that it becomes "actually" present, or present "in act," when the measurement is performed. Now it is also known that, long since, some thinkers, notably Whitehead, in view of some analogies between human and infrahuman behaviors, considered it legitimate to attribute experience to organisms whose structure is considerably simpler that the one of human beings and (even) higher animals. The question thereby raised was whether it is conceivable to attribute some kind of a protomentality to inorganic or (even) elementary physical systems. Of course, such an idea naturally met with the objection—considered crucial at a time when classical views dominated—that one couldn't even imagine what *meaning* might then possibly be imparted to the expression "protomentality." Now, Shimony (*loc. cit.*)

pointed out that, precisely, the potentiality concept opens a possible way of removing the objection in question. His guiding idea was that the relationship of protomentality with consciousness might well merely be one of pure potentiality, and that the circumstances in which the said potentiality might change into actuality would be those in which a nervous system is present.[9]

Such considerations are, of course, mere indications. Relatively to the problem in hand they are interesting nevertheless, in that they show that—contrary to what a reflection centered on purely classical ideas would have suggested—the efflorescence theory is far from being an established truth.[10] Moreover, it turns out that, after all, it seems possible to proceed still a little further. The point is that the fundamental fact on the basis of which, in the foregoing section, the identity theory was shown inadequate is significant here as well. The said fact, which follows from the conceptual analyses carried out in this book, is, as may be remembered, that the various parts our bodies (hence also our neurons) are constituted of are themselves elements of but *empirical* reality. Now, it is difficult to imagine how such a reality, which, as we saw, is relative to consciousness, might possibly generate the latter. We are thus led to consider that if our analysis of scientific data is adequate, the, above mentioned, "received view" that thought is produced from matter, even if it remains highly useful as a model, cannot any more be raised to the level of a basic philosophical statement.

The content of the two last sections, this one and the foregoing one, was merely negative. We pointed out the difficulties the two consciousness theories there analyzed meet with, without putting forward any alternative suggestion. In theory we should now give to this a positive counterpart. We should explain what "becoming aware of" actually is. Unfortunately, in its present state human knowledge apparently is far from making it possible to do so in any strict, convincing way, and hence it might perhaps be wiser to keep silent on the subject. Man's natural tendency to speculate is irresistible, however, and is sometimes beneficent. In the next sections of this chapter I shall therefore venture to put forward

---

[9] He granted, however, that the idea will remain highly speculative as long as it is not possible to test it by means of psychology experiments same as the existence of quantum potentially effects could be inferred from interference experiments.

[10] Nowadays, quite a number of books have developed considerations going along such lines but, unfortunately, not many of them are reliable. Among those that—notwithstanding some significant reservations—I consider worth reading let me quote the one of L. Schäfer (1997) which, notwithstanding its title, mainly deals with the status of *thought* within contemporary physics.

a few, risky, "half baked" ideas that, at some time, might conceivably help to open new perspectives in a still largely unexplored field.[11]

18-4-3 In What Respects Are States of Consciousness to Be Considered Relative?

In constituting *empirical* reality the role of consciousness—or mind—is obviously primordial since the said reality essentially is a representation. At first sight this seems to make consciousness a kind of an "absolute" reference. And indeed, up to this point, implicitly we more or less kept to this view. However, it must be granted that, along with features that are attractive due to their rationality, it shows some that are much less so, as the analyses in section 10-3 already suggested and the forthcoming ones are about to confirm. We should therefore inquire whether states of consciousness are to be considered relative, at least in some of their aspects. And in fact, examining the way our conception of *physical* states evolved from the "classical age" up to now yields, by comparison, a further incitement to have a look at the question. For indeed while, within classical physics, physical states could be taken to describe reality-per-se, quantum mechanics, as noted in chapter 4, forced us to question and finally drop such a standpoint. In fact, we saw that quantum states essentially are means of predicting observational results and that therefore they are essentially relative (to preparation procedures, planned observations and so on).[12] This remark naturally prompts us to study the aspects in which also states of consciousness appear relative.[13]

To this end, let us first take a quick, retrospective look at the way in which quantum mechanics describes physical systems. And let us observe (once more) that while, as we just noted, it basically consists of a set of observational predictive rules, still, among the said rules there are some that, considered "in isolation" so to say, might be rephrased in the form of descriptive laws. They are those that do not involve probabilities and just make it possible to study static properties of systems, such as atomic and molecular energy levels.[14] The crucial point is just the obvious fact

[11] This remark is meant to try and justify the fact that the next sections in this chapter are not written in quite the same spirit as the rest of the book. Instead of keeping to an analysis of the implications of soundly established scientific data I shall, there and only there, indulge in a few provisional suggestions, inspired of course by the material we reviewed up to this point, but that, still, are of the nature of tentative speculations.

[12] This is why we decided to designate them by setting the word "state" between quotation marks.

[13] The thesis according to which these aspects are in fact representative of their innermost nature is one of the leading ideas of Bitbol's *Physique et philosophie de l'esprit* (2000).

[14] As we know, the rationale of such calculations does not basically differ from the one that, in classical physics, serves for calculating the various properties of vibrating strings.

that, in any attempt at a global understanding of the subject, the said rules can simply not be considered "in isolation." The other ones have a basic role as well, and cannot just be forgotten about. We shall find out that, concerning states of consciousness, the situation is to some extent similar.

To see how this goes, let us consider once more the Schrödinger cat or, better still, the Wigner friend problem (without jibbing at fiercely idealizing, since our quest is but a conceptual one!). More precisely, let us imagine that a physicist, Peter, is standing near a laboratory whose doors are closed and in which an experiment of the type we conventionally called a (generalized) "measurement" in section 4-2-2 is taking place. Let us assume that Peter is informed of the initial conditions relative to the said experiment and let us also assume that, within the laboratory, another physicist, Paul, looks at the instrument pointer and immediately registers what information he reads on the dial. We may make the whole setting even more sophisticated by considering that Peter and the laboratory—Paul included—are encircled within a great wall outside which a third physicist, John (also informed of the said experimental conditions), is imagined to be standing. And so on in a Chinese boxes way, ad infinitum. Considered from a certain angle that may be labeled Paul's, Paul's state of consciousness just after he looked at the dial may be taken to be an event having taken place "in the absolute." Either he saw the pointer at place A or he saw it at some other place, period. But considered from a more general angle the situation appears different. For Peter, for instance, Paul's state of consciousness still is, at that time, not just unknown but undefined. In fact, since, in conformity with our current experience, he believes it to be potentially predictive, it has to be a kind of a quantum superposition of states "having seen the pointer at A" and "having seen it elsewhere" (the said states being, of course, entangled with pointer states). Indeed for Peter, Paul's consciousness state will become definite only after he has interacted with him and asked him (say, on the phone) what he actually saw. In the meantime, if Peter were a Laplace demon he could make—and check[15]—some experimental predictions that would be incompatible with such a definite character.[16] And besides it may well be that, even after the said phone call, for John, who is outside the wall, neither Paul's nor Peter's (predictive) states of consciousness are definite

---

[15] For this, admittedly, he would have to engage into a statistical investigation, that is, he would have to reiterate the experiment a large number of times under identical conditions. But within our "fierce idealizing" there is nothing, in this and other similarly unrealistic assumptions, that should block us.

[16] More precisely, incompatible with the (natural) idea that Paul's state of consciousness is, to repeat, not only definite but also potentially predictive. We shall come back to this point (see "Important Remark" below) which will prove to be quite significant.

entities or properties. They will be, only when, in turn, he will have asked Peter or Paul.

In other words, when we claim that Paul's observation of the pointer position creates in his own mind a definite state of consciousness we are not wrong. But we inadvertently state too much since what we say implicitly conveys the idea that the said state is a "definite entity, period" the existence and nature of which may be publicly referred to, in particular for making predictions. This is not the case since, considered from Peter's angle, this very state will become "cut out of a potentiality" (as Heisenberg might have put it) only through *his* own act of taking cognizance (when, say, he phones). And the same holds true of course concerning Peter's (predictive) state of consciousness when considered from John's angle, and so on.

Finally, as we see, what is questioned here is not the idea that the keystone of physics is predicting observations. It is the subordinate idea that observation-predictive states of consciousness play in some way the role of an absolute, or of substitutes for an absolute. What we just saw is that they, in fact, are relative to "points of view," in the sense this expression takes when, in the course of a philosophical discussion, we refer to the Sirius, or the God's Eye, point of view. It is true that, in the interval between Paul's looking at the pointer and Peter's call, Paul's state of consciousness is, for him, well defined. Nevertheless, viewed as predictive it is relative, "indexed on him," as Bitbol (2000) would say.

We noted above that, in virtue of some of their characteristic features, the quantum states by means of which we strive to describe the "physical world" must be considered relative. Here we just observed that the same is true concerning states of consciousness, at least to the extent that they are considered predictive. Admittedly, this is a point of similarity between the "physical" and the "mental," but we would be very naive if we used it as an argument in favor of the identity theory. As above defined the latter is an attempt at reducing mind to a physical stuff describable by means on nonindexed concepts. The difference is manifest.

Moreover, let us note that the thus revealed generality of the notion of "being relative" strengthens the plausibility of the philosophical hypothesis introduced above, according to which the set of all living and sentient beings constitutes a "community of Wigner's friends." For indeed, in a community of Wigner's friends as defined in section 10-3-3 both the "things" and the (predictive) consciousness states are relative. None of them, as we saw, enjoy absolute reality. True, the very notion of such "relative" consciousness states of "friends" presupposes the existence of "entities"—represented, in the apologue, by the experimentalist himself and the overall wave function he prepared—playing with respect to them the role of an absolute. But it seems that the "friends" dispose of no reliable measurement procedures by means of which they might get to know the overall wave function in question. The fact that we were led to introduce the "community of Wigner's friends" notion may therefore be considered to be an indication imparting some additional likelihood to the, at first sight somewhat puzzling, notion of an "Ultimate Reality" to which we have no cognitive access.

IMPORTANT REMARK

Above, when we stated that, theoretically, Peter could perform statistical experiments—call them here "E experiments"—disproving the hypothesis that Paul is in some *definite* state of consciousness, we, admittedly, set ourselves within a framework that was idealized in several respects. Implicitly we assumed that a statistical ensemble composed of identical copies of Paul's initial state could be made available. We disregarded the environment (or we assumed that Peter is a gifted Laplace demon, able to make on the latter all appropriate measurements), etc. Or, alternatively, as suggested in section 10-3, we identified Paul with an idealized DNA molecule, endowed with consciousness but still simple enough to be manifestly subject to quantum laws. But we also assumed something else, and this is a point that a digression in the direction of the Broglie-Bohm model should help making clear.[17] For in the said model, even within the just mentioned idealized framework the disproof in question is in no way clear and obvious. In it indeed the fact, for Paul, of being in a well-defined quantum state corresponds to the representative point of his hidden variables being in one of the regions R_1 and R_2 considered in section 10-2 (for simplicity's sake we consider only two regions). Now, in the Broglie-Bohm model this in no way prevents all the observable predictions derivable from the overall wave function from being correct, including even those concerning the above defined "E experiments." In other words, in the model, the fact that these quantum mechanical predictions are what they are is *not* at variance with the idea that each one of the Pauls composing the statistical ensemble is, after he looked at the pointer and before Peter called, in a definite state of consciousness. The reason for this is, as we know, that, in the model, the representative point is driven by the overall wave. In the analogy, considered in section 10-3, with the Young slit experiment, this corresponds to the peculiarity of the model that, in it, it is true "at once" that each particle goes through but one slit and fringes develop.[18] Hence, in the said analogy the fact fringes are observed in no

---

[17] For clarity's sake I take up again here some points already considered in section 10-3.

[18] Fringes that, as we know, the presence of some gas may make vanish, just as, by virtue of decoherence, the environment renders the here considered experimental disproof impossible in practice.

way disproves the view that each particle goes through one slit only. What it disproves is merely the idea—of a basically "predictive" nature—that, for predicting the place where the various particles will hit the screen, to the ones that went through slit i ($i = 1$ or 2) and are therefore in region R_i the corresponding wave O_i should be associated. And similarly, concerning our DNA-idealized Pauls, what, within the Broglie-Bohm framework, the E experiments disprove is the thesis that the fact that Paul is in a well-defined consciousness state may be taken into account (by Peter or by "DNA-Paul" himself) when the purpose is to *predict* observational outcomes.

Now, as we saw in section 10-3 (and a moment reflection on the modal interpretation corroborates this), the above conclusion far from being specific to the Broglie-Bohm model, should be considered general. The present remark is an a posteriori explanation of the reason why we had to make use of the epithet "predictive" at several places in the foregoing developments. More precisely it explains why, when we noted that before Peter called he could not consider Paul's state of consciousness to be "well defined though unknown to him," we had to specify that the state of consciousness we were referring to is a potentially predictive one. Equivalently, it shows that when we claim that, theoretically, Peter could disprove the view that Paul is in a well-defined consciousness state, our claim is valid only when the latter state is understood to be a definite, potentially predictive entity. This is a significant point for it seems that, in theory, nothing prevents us from conjecturing that nonpredictive consciousness states exist (in our minds and perhaps in those of other living beings as well). And clearly the reasoning attributed to Peter above does not enable him to disprove the view that *these* consciousness states are well-defined ones.

Incidentally, if the conjecture in question is accepted, then, if we turn back to our above observation that, before Peter's call, Paul's state of consciousness is, for Paul himself, quite obviously well defined, we may legitimately consider that, taken as nonpredictive, this state of consciousness is something deeper than just an "epistemological reality" (section 8-1-1). And that, in some conceptions or word views, it might therefore be considered to exist *per se*.

18-4-4 Concerning the "Coemergence" Notion

The notion of a coemergence of thought and empirical reality was put forward by a number of authors but with varied connotations. For transcendentalists it often conveys the idea that we take part in the self-qualification process of a "Something" that has no intrinsic structures.

Neither contingent ones (in the form of localized objects endowed with shapes and so on), nor even just general ones (in the form of great universal laws). Bitbol's pragmatic-transcendental approach commented on in section 17-3 as well as the one Zwirn developed in his book (2000) may be considered to be examples. According to this conception, a (temporal or atemporal, depending on the authors) coemergence takes place of, on the one hand, thought (in the broadest sense) and, on the other hand, everything that thought may apprehend. Which makes these authors claim, as we saw, that, within their system, no "face to face" takes place between thought and the world. However, as I already noted, the statement is, to me, somewhat obscure since I consider the word "world" to be ambiguous.

The Veiled Reality conception resembles this one in that, to repeat, in it also, Reality is taken to have no *contingent* structures. But the difference is that, in it, Reality is thought to be endowed with *general* structures that are far from being knowable but still might constitute the "ultimate ground," not only of our great literary, artistic and mystical inspirations but also (why not?) of our great scientific laws.[19] As we see, this conception leaves some room for mystery, but this point is to be considered later. The coemergence is assumed in it same as in the above-mentioned one, with, however, the difference that the just mentioned general structures do not "emerge" since they are assumed to exist by themselves. There is still some coemergence, but merely concerning thought (or at least states of consciousness) on the one hand and empirical reality (the set of objects, events, etc.) on the other hand.

In this conception more than in more idealistic general approaches, and due precisely to the fact that, in it, states of consciousness specifically coemerge with such concrete things as objects, events and so on, attention is drawn to the question what, beyond vague words, coemergence actually is. A parallel then immediately comes to mind between the idea that consciousness somehow "emerges"—which is one of the facets of the conception in hand—and the current, received view that consciousness is produced by physical objects (neurons in particular). And then it may seem natural to adopt the latter view and (in view of all we have seen) supplement it with the symmetric one that empirical reality emerges from consciousness. This would lead us to consider that, as I wrote some time ago, consciousness and empirical reality exist in virtue of one another (d'Espagnat 1979a), or equivalently that they generate reciprocally one another (d'Espagnat 1997a).

---

[19] As we know, the proposal is not that these laws "describe" the structures in question but just that they are sort of distorted glints of them.

Without bluntly repudiating these assertions I would like to stress that they should be taken to be evocative pictures rather than literally true statements, for understanding them that way would create a problem. The difficulty does not bear on the view that empirical reality emerges from consciousness, which, in a way, represents the novel facet of the thesis. Rather, it has to do with the—again, very much "received"—view that (empirical) reality generates consciousness. The point is that, within the framework of the general ideas upheld in this book, the view in question is not defensible. How—to repeat—could mere "appearances to consciousness" generate consciousness?

In this respect, the considerations of section 2-8, together with those of the foregoing section and the accompanying remark, help putting things in the right perspective. Indeed, it follows from, in particular, the said remark that some beings may be thought to be endowed with internal, nonindexed, consciousness states provided the latter are not assumed predictive. These states are of the nature of hidden variables (remember we left open the possibility that hidden variables should exist, provided that they should be "really hidden"). In the case of our DNA-idealized Paul such states *are*, of course, strictly nonpredictive, and it is only when the involved systems become macroscopic enough for their interaction with the environment to be appreciable that they obtain some degree of public significance. This means that they gain predictive power. More precisely (as we easily realize by thinking of intermediate cases in which "not quite macroscopic" systems are involved) they make it possible to correctly predict the outcomes of a certain class—call it A—of observations. But correlatively they would yield false predictions concerning those of another class—call it B. In other words, their predictive power is but relative to a certain class of observations. Now, it is a fact that (due to the nature of the said two classes) human beings perform class A observations much more easily than class B ones. And indeed this is true to such an extent that when the involved systems are thoroughly macroscopic, class B observations are, as a rule, practically infeasible, as we know. Moreover, it is also the case that the impressions corresponding to the outcomes of class A measurements may usually be described in a realist *language*, that is, *as if* they referred to objects existing per se. The set of such intersubjective appearances is what we called "empirical reality." It is thus meaningful to speak of a kind of "coemergence" of, on the one hand, states of consciousness that are both public and predictive but merely in a practical, theoretically limited, hence relative, sense, and, on the other hand, empirical physical reality. In line with the content of section 10-4 this coemergence is to be thought of as (atemporally) taking place out of a "mind-independent reality" that itself presumably lies beyond our intersubjective abilities at describing.

On the other hand, it must be granted that, for picturing such a coemergence, the image of a reciprocal generation of consciousness and empirical reality may well be criticized. The reason, to repeat, essentially is that the received view that states of consciousness are generated by neurons is misleading, neurons being elements of just *empirical* reality. To the extent that it evokes coemergence the image in question is of course an acceptable allegory but it is one that may definitely not be carried too far.

Let us finally note that in this whole analysis we dealt with specific states of consciousness (such as seeing a pointer on a dial) more than with "thought" in general. This tallies with the fact, stressed by many philosophers, that consciousness always is "consciousness of. . . ." But still, this should not prevent us from pondering on the very notion of thought and wondering on its status. In this respect it seems that the idea of potentiality in the sense of Whitehead-Heisenberg-Shimony (an idea that sort of reappeared in our investigation of states of consciousness) offers, when considered within the framework of the veiled reality conception, an interesting way of organizing our ideas. For indeed, in the said conception it is quite possible to consider that "the Real" involves something that is not thought proper but is susceptible of giving rise to it. Although this "something" cannot be considered a "component" of the Real (which has no "parts"), still it may be viewed as being a facet of it. Its relationship with consciousness is one of pure potentiality, but on the basis of arguments similar to the above sketched one (the going over from a DNA-idealized Paul to a real, macroscopic Paul) we may speculate that this ontological potentiality somehow generates empirical actuality.

18-5 On the Question of the Plurality of Minds

Within the approach of radical idealism the reason why our sentient and thinking ego is nowhere met in our scientific world picture is, as Schrödinger put it (Schrödinger 1959), that it is itself that word picture. But then the question of the plurality of minds is known to create a problem, which the just-mentioned author summarized by noting, "there appears to be a great multitude of these conscious egos, the world however is only one." He called this puzzle *the arithmetical paradox*. Schrödinger himself considered there are but two ways out of it. One of them—with which the conception described here in section 5-2-5 shows appreciable similarities—is what he called "Leibniz' fearful doctrine of monads." The other one is the thesis of the unicity of minds, (or consciousnesses). Their multiplicity is only apparent, in truth there is only one mind, which shines in every one of us. This conception, which, he wrote, was the one of the Upanishads, was clearly the one he preferred.

The Veiled Reality conception differs very much from radical idealism. But still, at first sight the plurality of minds problem arises in it much in the same terms as in the Schrödinger approach since our minds, which, apparently, are many, take part in the emergence of empirical reality, which "proceeds," so to speak, from any one of us. Now the choice made and explained in section 10-3-2 is here of some relevance. As we may remember, the question at stake was about the way the generalized Born rule is to be interpreted when several observers take part in one and the same measurement (or, equivalently, when they measure strictly correlated observables). Should we consider that these observers' "consciousness states" get correlated by virtue of some great structure of "the Real" reflected in the Born rule, or should we assume (in the spirit of the section 5-2-5 approach) that the correlation between them is just apparent? In section 10-3-2 we opted for the first answer and, at first sight, this may be taken to be a step in the direction of something roughly similar to Schrödinger's "one mind" conception.[20] But still, even if we choose it, between my position and Schrödinger's one it seems there must be a difference. The point is that, as the Paul-Peter-John fable (section 18-4-3) showed, predictive states of consciousness are unavoidably relative. Paul is, for Paul himself, in a well-defined state of consciousness, but must not be thought of by Peter as being in such a well-defined state. It is therefore quite difficult to imagine a sense in which it could be claimed that Paul's and Peter's minds constitute "one and the same" mind.

It appears therefore that anyhow the "one mind" thesis should be, at least to some extent, watered down. However, within the above-considered laboratory imagine there are not one but several observers and each one measures one out of a set of mutually compatible physical quantities. The (compound) probability that each one gets one specified outcome is normally to be calculated with the help of the Born rule, just as if we had to do with simultaneous measurements made by one and the same observer. We may therefore—in a Schrödinger-like way—think of all "Wigner's friends" as constituting but one collective mind. And if, within the laboratory, there are still other people making no measurement at all, we may of course also include them, by thought, into the said collective mind. True, as long as the practical impossibility of performing some inordinately complex measurements is ignored it is impossible to also include Peter's mind in it, so that Peter should be considered to belong to another "mental world," or to be part of another "collective observer." On the

---

[20] This, however, is a delicate point. Within the Broglie-Bohm model the plurality of minds notion does not seem to raise any basic problem, so that if, as explained in section 2-8, we adhere to the weak completeness hypothesis we find reasons to believe that this plurality is, after all, not at variance with the basic principles of quantum mechanics.

other hand, in such a simplified picture these two different collective observers would generate two altogether different empirical realities, and it
is worth noting that, at least for each one of them separately, Schrödinger's idea removes the arithmetical paradox.

In other words we see that even within a very much idealized approach
totally ignoring decoherence, much of the Schrödinger "one mind" hypothesis remains pertinent. Actually, of course decoherence should not be
disregarded. And, obviously, by rendering quite impossible the complex
measurements the notion of which generates the conceptual difficulty at
hand it renders "purely theoretical" what inconsistency still may be considered to exist between the quantum structure of physical laws and the
fact that minds are, or appear to be, plural. It remains true, however, that
"to render purely theoretical" is not equivalent to "to remove"; and hence
I think we should grant that some efforts still have to be made before
we may claim that Schrödinger's idea truly eliminates the arithmetical
paradox.

CHAPTER 19

THE "GROUND OF THINGS"

§§

19-1 Introduction

THE LAST CHAPTER made it clear that concerning, in particular, the nature of sensations and their relationships with (empirical) reality, the views quantum mechanics speaks for and those advocated by transcendentalism-inclined thinkers have points in common. But the questioning concerning the meaningfulness of the "independent reality" notion is quite a different subject and we saw all along this book that, concerning it, my views appreciably differ from those of the said philosophers. All the same, I heartily grant that taking the arguments of the latter into account did help me to refine the views in question. It is therefore proper that my standpoint on the "Reality" issue should be presented here in its "final stage" so to say. And that, in particular, the "deep reasons" should be stated why, even after having realized the force of the antirealists' arguments, I go on considering the notion of a Real that, though nonconceptualizable, still constitutes the ultimate "ground" to be the most adequate one. Such is one of the main objectives of this concluding chapter. Of course, on this occasion I shall be led to state anew my main ideas so that this chapter will also serve as a summary of the latter.

19-2 Mystery, Affectivity, and Meaning

The presence here of the word "mystery" should attenuate the reticence that mentioning the "ground of things" notion in the chapter title may conceivably have aroused. The said notion is, to be sure, a hazy one, and the mere fact of taking it seriously might indeed be taken by some thoughtful people to denote overconfidence in the aptitudes of human reason. But, on the other hand, we observed (and below this should become even clearer) that, in the last analysis, to radically negate the pertinence of this concept—that is, to reject the notion "the Real"—is just as questionable as to engage into thoughtless discourses concerning it. This double-faced observation already made us partly side with Kant with regard, not to his mode of reasoning, but to some of his conclusions. Specifically we were induced, like him, to distinguish between phenomena and the thing-in-itself and keep both notions. But still, we parted from this philosopher on at least one point. It became clear to us that it is in no way necessary to side with him in thinking of reality-in-itself —"the Real," or "Independent Reality," in the above-used language—as being a "pure x," in other words a mere, uninteresting "limiting concept." Now, if it is not a "pure x" it may count in our eyes. But at the same time, being veiled, it is not accessible to discursive knowledge. This being the very definition of mystery it follows that these views imply that mystery exists.

Admittedly, a problem is thereby raised. In the mind of almost all scientists and most thinkers whom the latter hold to be serious, the word "mystery" carries with it a negative connotation. Scientists quite rightly point out that in backward societies the notion this word refers to is systematically used as a substitute for an explanation, and that consequently it automatically restrains, or even blocks, any yearning for organized research. And as for the philosophers, Cassirer's judgement (here reported on in section 13-3-3) yields a fair idea of their mean opinion. For him, the mystery that the notion of an unreachable thing-in-itself implies can but result in making us feel prostrated under the weight of a vague idea of something the presence of which must forever remain indescribable. And obviously this is a criticism of the Veiled Reality notion that antirealists, among others, may well take up. It is therefore appropriate that, before going through the objective information we have concerning the problem in hand, we should try and refute these two objections.

With respect to the one from the scientists, this is relatively easy for, as could be expected, practically all of them combine their natural realist bent with a great consideration for facts. Admittedly many of them would like to be able to entertain a conception of scientific research according to which, in some indeterminate future, all veils would be removed. But it is not in the spirit of science to raise such unprovable views to the level of principles, and rare are the scientists who take this stand. Practically all of them will most readily acknowledge that the Aspect-like experiments, far from obstructing the path of research, are themselves elements of a fascinating scientific quest, and, as to what the latter leads to, will just wait and see.

As to Cassirer's objection, which is more focused on the abhorrence of the unknowable, let it be granted right away that there is something correct in it. True, in any field in which it proves possible, the quest for clarity—implying rejection of what is vague and mysterious—should be eagerly carried on. Indeed, we might almost claim that the right of invoking the notion "mystery" should be saved for the persons who did long and earnest efforts of such a kind. Or at least that they have, on this matter, something like a priority right. On the other hand, however, such a (chimerical) ukase would be unjust, for in fact there are many persons whose turn of mind is altogether different from that of the just mentioned ones and whose quest may not, honestly, be ignored. They are those who judge that the phrase "some approach to mystery" is not self-contradictory; that it refers to a notion susceptible, in some cases, to having a meaning. Such is the standpoint of divines and mystics. But, at least up to recent times, it also was that of poets (think of Baudelaire and Poe, for instance), of most composers, their interpreters, and—perhaps even more genuinely—their admirers (think of those of Bach and Mozart). Only people wearing quite uncommonly thick blinders might deny that the existence of people who took such a standpoint was prodigiously fruitful. True, it might be claimed that it was but a "beneficial delusion," akin to the one that the mystical doctrine of numbers was for Kepler. Such, by the way, is the view that open-minded materialists and physicalists usually take on this issue. In their case, that is, in view of the state of their knowledge and belief, the standpoint in question is, of course, quite understandable. When you think, as they do, that you have in your possession the conceptual key that—theoretically though, of course, not practically—opens an access to knowledge of "all that is," you cannot but identify the notion of "some approach to mystery" to a fancy flight from reality. However, in the situation we actually are at present things appear in an altogether different light since we know that in fact we are not in possession of the key—which, moreover, seems not to even exist. The "reality"—empirical reality—that fits the key we dispose of, is not a preexisting entity. Quite on the contrary, it is with good reason that antirealists and instrumentalists think of themselves as being engaged in the process of building it. Hence if *they*, here, evoked "illusion" we should have to ask them: Illusion with respect to what? In other words, we now have realized that, contrary to what was once widely believed, for reaching deep truth and Reality we may not rely on just a combination of sound reasoning and paying scrupulous attention to facts. As a consequence we lack a trustworthy criterion enabling us to set aside without ado "testimonies" coming from other sources. And it is therefore impossible to imagine on what reliable ground the view could be discarded that the messages from our affective percepts (and particularly the very subtle ones) are more significant than most positively minded people of the three last centuries believed them to be.[1]

---

[1] This may be set in parallel with the well-known "demarcation problem" (between science and metaphysics) that so much worried Karl Popper and that this author never solved in a way that could entirely satisfy him.

The expression "affective percept" is one we did not, as yet, come across, and in this book it and similar ones will only be made use of quite episodically.[2] But still, it had to appear for it characterizes one of my differences with Kantism. This dissention, I must stress, is not an absolute one. In fact, on the sublime, Kant, quite contrary to his custom, wrote a few lyrical sentences that do arouse an echo of an emotional nature in most of his readers' minds. In them, moreover, Kant marked that the sublime essentially is a tension towards what radically lies beyond sense data, and this, of course, seems to point in the direction of what, roughly, I have in mind. But it is true nevertheless that for him the "noumenal world" remained completely unreachable. In his eyes, as, in particular, Alain Besançon (1994) appropriately pointed out, even the sublime lay entirely inside the mind. According to Kant it provides us with no glimpses on Being whatsoever, not even undecipherable ones. A fortiori, this also holds true relatively to other elements of our affective mind. Not being "ordered on concepts" they are in no way carriers of knowledge. As we saw, it is within such an ordering on concepts that Kant and his followers defined the reality of objects of any kind, which therefore are merely "objects for us." Incidentally, as Ferdinand Alquié pertinently showed (1979), it is also by referring to this primacy of understanding that Kant criticized the *Cogito* and called it an inconsistent syllogism. Whereas, as Alquié noted, in fact, far from having in mind any syllogism whatsoever, Descartes discovered his own existence as an unquestionable truth, just as self-evident as our joys and pains are. Indeed Malebranche long since pointed out that, contrary to what moralists frequently assert, there are no "false pleasures." There cannot be pleasures of which I could think "I am now pleased but maybe I am wrong. Maybe I am not, presently, pleased." And unquestionably there is a similarity of nature between the kind of certainty that the (core of the) *Cogito* yields and this one. I cannot deny that there is thought. That thought *is*. Is it "by itself" or is it emanating from something else, that much the *Cogito* actually does not specify.[3] But it is. Can't we say the same concerning what is most certain in it? "There is" emotion. And, similarly "there is" desire and intention.

---

[2] For expressing some ideas English, as many other languages, has pairs of epithets—audacious and rash, patriotic and chauvinistic, wary and timorous, etc.—that have the same objective meaning but are endowed with an affective coloration that is positive as regards the first ones and negative as regards the second. Unfortunately, this is not the case concerning the idea that the epithets "affective" or "emotional" convey. In "serious" books, expressions such as "purely affective" or "purely emotional" are distinctly pejorative. To evoke in a positive way the idea they convey, other words would be needed. Alas, there are none.

[3] Contrary to what Descartes seems to have thought.

To feed his/her mind, the human being needs conjectures (Comte-Sponville's inspiring phrase "to philosophize is to think farther than one knows" may here be reminded of once more). And for building up conjectures, there is no better way than to keep as close as possible to what one is quite sure of. In view of the foregoing it is thus reasonable to conjecture that, concerning Being, affective consciousness sometimes provides us with genuine elements of information—which are not obtainable from other sources since science essentially informs us on nothing but phenomena. Where may we hope to come across such elements? I, for one, have three domains in mind: mysticism, poetry and music (without barring other arts[4]). To speak of mysticism would only be possible on the basis of an experience but very few people have. Moreover, having it would hardly be of any help since all mystics assert their actual experience is ineffable. Does meditation yield some glimpses? Maybe. Fortunately, concerning poetry and music the situation is much better. In such fields even average people may have their views. Thus, Alquié (*loc. cit.*) maintained that, viewed under this light, poetry is in no way metaphorical; that it does yield genuine information. "Poetry," he wrote, "is meant to be seizure of Being: if this it is not, it is but the most idle of games." Optimistically, I tend to speculate that he is right.[5] That, in the wall of the prison within which, as Kant's thesis suggests, our understanding locks us up, top level poetry as well as top level music cut a tiny window that looks on to—indeed!—"the Real." True, what we perceive through this window shows no contours. But in the realm of Being normality goes this way.[6]

As may be remembered, this section was aimed at taking into account two prejudicial objections to the Veiled Reality conception, one from some scientists and one from Cassirer. I take it that the foregoing does refute the said objections. We are therefore in a position to go on, without restraint, critically examining the conception in question. Since the latter is grounded on the principle of open realism we have to start with an analysis of the pertinence of this principle, that is, of the notion of "ground of things."

---

[4] There are people with whom beauty—of landscapes, faces, works of art, etc.—acts as a genuine revelation, as a frequently rediscovered road to Damascus. The present-day systematic intellectualism of the "world of the arts" upper layers confines these persons to anonymity. But on the other hand, for them this very anonymity is in the long run an advantage since it saves them from the destructive blots of show business.

[5] I sometimes advocated adopting a "principle of benevolence," inspired, to some extent, from William James. It consists in *deliberately* adhering to an ontology constructed so as to be, at the same time, compatible with our knowledge and such as to impart a meaning to the quest for the infinite that characterizes human beings.

[6] As Jean Onimus (1994) wrote "It is a priori obvious that a transcendent Reality can institute with us but communion relationships, not relationships based on knowledge."

19-3 Do Things Have a "Ground"? Pro and Con Received Arguments

Admittedly, there is no proving realism. Hence "Open Realism" (section 1-2) is but a postulate. On the other hand it is true that while, in mathematics, postulates may be chosen at lib (provided only that consistency is obeyed), in physics and epistemology their pertinence has to be checked. In the present, final stage of our inquiry it is therefore appropriate that we should review several arguments that were put forward by various authors in favour of, and against, the idea that things have a "ground."

19-3-1 An—Apparent—Argument "Pro": The Stars in Their Course

"If men disappeared the stars would go on in their course." This statement, considered self-evident, is often mentioned as an objection to radical idealism (the theory that negates the very existence of reality outside mind). It might as well be used against the Veiled Reality conception since, also in the latter, forms, positions and other contingent properties of things essentially are but projections of our own modes of apprehending. On the other hand, the idealists' reply to the said objection is well known and fully consistent, even though, intuitively, we find it is difficult to accept. It consists in pointing out that the objection in question implicitly raises the concepts of space, time and objects to the level of the externally given, whereas, within the idealist approach, to claim that stars exist and were there before human beings appeared merely means that we may conveniently describe our *present* experience by expressing ourselves in such a manner. Obviously this reply also holds good within the framework of the Veiled Reality conception.

A priori we are of course inclined to discard this idealist answer by having recourse to self-evidence. For example, we feel tempted to claim that to assume things do really exist in space and time does not merely make it possible to synthesize our experience; that it also makes it possible to explain it, which idealism fails to do. However, as soon as we try and make such arguments more precise we are led to invoke either the no-miracle argument or the one that is grounded on intersubjective agreement. And we saw in chapter 5 that while these arguments have some bearing against pure idealism, neither one suffices for discarding the Veiled Reality conception. Moreover, in this domain the experience the particle physicists gained proves instructive. For, quite independently of the opinion they may have concerning the philosophical question in hand, such physicists are forced to acknowledge the fact that, quite often, they refer to particles—even at times when they are not observed—as if they were pointlike objects. That they speak of their "trajectories," of the "collisions" they suffer with other particles, etc. And all this, even though they know that this language, which, for them, is so convenient, is in fact entirely allegorical, that "in reality" the "particles" in question have, at such times, neither positions nor trajectories nor, etc., and that indeed, strictly speaking they are not truly distinct beings.[7] Under such conditions, does pure logic yield an argument prohibiting us from considering that the same might be true concerning stars? No, it does not. In this respect the most convincing argument we can find is just the observation that for analyzing our experience of objects of human or astronomical size (i.e., for predicting what will be observed in this realm) the descriptive language of macroscopic physics is both convenient and—nearly always—sufficient. But, obviously, this purely practical (not to say "technical") remark is void of philosophical significance (and it even turns out that, in cosmology, it loses much of the pertinence it still might have since typically quantum phenomena are now known to have an important role also in that realm).

Should it finally be added that, on this issue, our "plain observations" prove nothing? I "observe" that the Sun "rises" at a certain place in the morning, that it "sets" at some other place in the evening. . . .

19-3-2 Another Apparent Argument "Pro": "Pythagorism"

As we know, all theoretical physicists willingly grant that "near realism" (alias Cartesian mechanism) is false and that Reality-per-se is not reachable by means of familiar concepts only. But many of them still consider that mathematics do yield an access to the "good" concepts, (curved spaces and so on) that is, to concepts that, finally, do make it possible to describe Reality-per-se as it really is. In section 6-3, in line with other authors, we called this doctrine "Pythagorism" by reference to the famous sentence "numbers are the essence of things" attributed to Pythagoras. As noted in section 6-3, in his mature years Einstein could be considered an adept of the thus defined Pythagorism, even though, according to some

[7] A point well worth noting is that the said language is allegorical even when, philosophically, the speaker is an adept of the Broglie-Bohm model, that is, believes in the real existence of point corpuscles endowed with momenta, trajectories and so on. For indeed these "true" momenta and trajectories differ very much from the ones that the physicists have in mind and refer to when they describe experiments. Concerning, for example, a photon in a vacuum, physicists can hardly avoid considering that somehow it travels on a straight line whereas, according to the model, because of nonlocality this, in most cases, is not true.

of his writings, his position may well have been somewhat subtler. In fact, if it were not for the fact that neologisms should preferably be avoided, instead of Pythagorism we might as well (as already noted) speak of *Einsteinism*, and this appellation would even have the advantage of preventing possible misinterpretations. The point is that, in Pythagoras' and even Plato's times, no clear distinction between mathematics and physics had yet been made. Correspondingly, Pythagoras' quoted sentence may serve as a motto for the pure mathematicians with a Platonic bent as well as for the physicists. But, as we noted, it is clear that the matters on which these two groups take positions concern two quite different fields of knowledge. The mathematicians who call themselves Platonists—or Pythagorists or "adepts of mathematical realism"—consider that advances in pure mathematics are genuine discoveries, bearing on truths that preexist their finding out and are elements of a "world of pure mathematics" of which they are the explorers. This is a view that is internal to the field of pure mathematics, has nothing to do with physics and consequently qualitatively differs from Pythagorism-Einsteinism. Let us conventionally call it *mathematical Platonism*.

Mathematical Platonism is but one of the conceptions of mathematics that are considered acceptable. The investigation of its strong and weak points strictly concerns the mathematicians themselves and is not to be undertaken here. But still, it so happens that, mathematics being the main tool of the theoretical physicists, some of the latter take an interest in the conceptual foundations of both mathematics and physics and are anxious to compare the two (see, e.g., Omnès [1994a, 2002] in this connection). Ideally it might be hoped that this should lead to the notion of some "Realm of beyond," exclusively composed of pure mathematical truths, and of which our world down here would be an intelligible reflection. Or else to a duality, mathematical world—physical Reality, with a clear correspondence between the two. We would then have a "proof" that the notion of "ground of things" has a meaning. Unfortunately, these authors' analyses are quite far, as they themselves grant, from leading them to certainties of such a type. As we noted on several occasions, the main obstacle is that, being essentially predictive of observational results, the quantum formalism may not be taken to constitute a genuine apprehension of "the Real." And this observation holds true independently of whether or not the said "Real" is of a purely mathematical nature. Pythagorism is an inspiring conception but in view of what present-day physics indicates it seems it is far from yielding an assured access to "the Real."

19-3-3 An Apparent Argument "Con":
Absence of Absolute References

In *Truth, Reason and History* (Putnam 1981) Hilary Putnam made a critical analysis of conceptions of Reality akin to those that the Platonic "allegory of the Cave" typifies. He asked himself whether or not it is rationally possible to imagine that some ultimate Reality exists and that, relatively to it, the world in which we think we live is one of mere appearances. To investigate the matter he constructed a fiction and wondered about, not, of course, its factual plausibility but its acceptability "in principle" or "in theory." The hypothesis he considered was that in fact we are (and always were) animals in a vat, chained to a post, and connected to the outside world through electrodes. By means of the latter some highly gifted Superior Being, makes us have all sorts of sensorimotor perceptions resulting in that we have impressions of travelling, looking at landscapes, discussing with people and so on, while nothing about this is "real." He noted that the truth condition for the sentence (of the vat language which is the one of the chained animals we are) "there is a tree in front of me" is that an electronic image of a tree be activated. When one of us utters this sentence he/she is in the right if and only if the electronic device produces in her brain an image of a tree, which means that her assertion refers to the said image. Similarly, the condition for the sentence "this is a vat" to be true is that an electronic image of a vat be produced. In vat language the word "vat" thus refers to an electronic image. Call it "the vat image." Then, according to Putnam, when the brains (us) who, we assumed, are really in the vat, say (in vat language) "we are brains in a vat" this word "vat" they use can only have the said meaning. It can only refer to that vat image since, as regards vats, they (we) are acquainted with strictly nothing else. But, as a result, their (our) assertion does not convey the hypothesis they (we) try to express. For the said hypothesis is not "we are animals in a vat image" but "we are animals in a real vat." In other words, if we actually *were* animals in a (real) vat our assertion would be false. Now, it is clear that if we are not animals in a vat the assertion at hand is false as well. Hence it is necessarily false. In this it has something in common with self-refuting statements such as "all general statements are false." According to our author the considered hypothesis seems to be theoretically defensible only in virtue of a "magic conception of reference." If, really, there were brains thus chained in vats and getting the above-described images (which, after all, violates no known physical law), admittedly they might speculate along the here considered lines. They might utter the sentence "we are animals chained in a vat." But—

to repeat—on their lips this would in fact be an erroneous statement since, on their lips, the word "vat" could refer but to what is really an *electronic image* of a vat.

At first sight this seems to constitute a serious objection to the "ground of things" notion since it seems it entails that our conceptual frameworks are irremediably linked to us, and hence relative. However, in my opinion such a conclusion cannot be considered established, and this for several reasons. One of them is that Putnam himself stated he considered the validity of his reasoning to be dependent on, in particular, the hypothesis that mind has no access to external things and properties other than the ones provided by the senses. Now, this negation of the innate is a most reasonable and wise thesis, which amply showed its pertinence and usefulness with regard to the advancement of scientific and technical knowledge. But still, it is not proven. And, as we saw, it is not certain that it is compelling in other, quite different fields, of quite a deep nature and quite unrelated to efficiency. Another (though somewhat similar) reason for questioning the applicability of Putnam's reasoning to our problem is that it essentially concerns the intellectual process of referring to well specified concepts, that is, of producing descriptions. Now, the notion of "ground of things," or the one of a "Real" assumed veiled—which is just a special version of the former—are not of such a nature. They are intrinsically vague in the sense that within them it is impossible to "refer" in the sense Putnam's fable implies, which is the one of reference to concepts.[8] In other words they are conceptually prior to any description. They thereby escape refutation through referential arguments.

19-3-4 Another Apparent "Con" Argument: "Enaction" Theory

Considerations similar to the foregoing ones might be set forth for defending the validity of the "ground of things" notion against objections based on the neurophenomenological argument now to be described. This argument is essentially directed against the "pregiven world" hypothesis. As stated (and criticized) in the writings of diverse authors the said hypothesis consists in considering, along the lines of (traditional or "cognitivist"[9]) realism, that most features of the world exist prior to any cogni-

---

[8] In this connection it is important to remember that, within the Veiled Reality conception, great physical laws such as the Maxwell equations are not structures of Independent Reality. At best they are highly distorted reflections of the latter. In section 19-5-2 we come back to this important difference between Veiled Reality and structural realism.

[9] In short, as one knows, the cognitivist hypothesis is that the computing machine yields the best presently available model of thought, that is, of what produces knowledge. But it still is true that the said knowledge is assumed to bear on some preexisting something. So

tive activity and that the latter is, at least in part, directed at describing the said features. Against this hypothesis the founders of "neurophenomenology," F. Varela foremost, pointed out that according to recent findings the main activity of the brain is to produce changes within itself and that, correlatively, the notion of representation (of an "external world") vanishes. They claimed that this important new piece of knowledge forces us to get rid of the idea of an independent, intrinsically existing world and makes us conceive of the world as inseparable from these self-modification processes (Varela et al. 1991).

A priori it may be wondered whether it is really possible that an ontological conclusion of such a magnitude should follow from investigations exclusively bearing on neurophysiological problems, that is, on perceiving and acting processes, in short, from researches entirely focused on the human brain and mind. It is therefore appropriate to somewhat critically examine the arguments that were put forward in its favor. In so doing we first find out that in the book from which the above statement is taken what is suggested is that cognition be considered to be neither a reconstruction (of some preexisting external world) nor a projection (of ourselves). That it is an *enaction*, that is, by definition, a process in which mind and the world simultaneously come up. If we decide that, by "world" we mean *empirical* reality, such a conception obviously is akin to the one that progressively emerged all along this book and got finally summarized in section 18-4-4. And such a convergence of views is of course satisfactory. Unfortunately (and here the troublesome ambiguity of the word "world" reappears) this convergence is, in a last analysis, only partial and may generate confusion. For indeed, while we found it necessary to introduce two distinctions, one between subjectivity and intersubjectivity and one between independent reality and empirical reality, in the quoted book these distinctions do not appear. And since they are missing we are at a loss how to interpret the authors' thesis otherwise than as signifying that mind comes up together with whatever may be called "the Real." Which of course differs very much from the Veiled Reality conception.

Moreover, a detailed examination of the reasoning these authors made reveals that, surprisingly enough, they focused it on consideration of secondary qualities (in Locke's sense) such as color. This deprives it of much of the weight it might have had. For indeed it has been known for a long time that such perceptions very much depend on our own perceptual structures and up to now this knowledge did not prevent people who were in possession of it to believe in the existence of a pregiven outside world.

---

that, in cognitivism, the realist prejudice remains. According to Varela et al. [1991] it even is quite explicit in it.

Note also that in the few cases in which the authors did take primary qualities into consideration their reasoning was suggestive more than deductive. For example, at the place where, making use of a text from Maurice Merleau-Ponty, they evoked the perception of the flight of an animal, they noted, following this author, that our organism chooses by itself, in the physical world, the (visual, auditory, or other) stimuli to which it will be sensitive. To go over from this observation (correct, to be sure, but compatible, after all, with a most naive realism) to the conclusion that mind and the world arise together indicates—I would say— quite an exceptionally strong belief in the reliability of extrapolation procedures.[10] Besides, even if the said conclusion were accepted without ado, the question, "of what world is it the matter?" would unavoidably arise. Taking up again the authors' reasoning we see that the answer can only be "the world of the phenomena"—empirical reality—since it, and not Independent Reality, is what corresponds to the stimuli.

What, I think, should be retained from this quick survey is that studying but the sciences of cognition may well be insufficient for providing us with certainties concerning the question whether or not the notion of an independently existing outside world is pertinent. And it also is that, anyhow, the objections to the notion of a "pregiven" world that might be elaborated along these lines could have a bearing only relatively to the current realist thesis according to which the world is discursively knowable. They affect neither the thesis of a strictly unknowable reality-in-itself, nor even the one of a Veiled Reality having no contingent structures (positions in space, etc.) and whose general ones are such that human understanding is unable to reconstruct them.

Remark on Buddhism

In their book the authors frequently referred to Buddhist thought, of which they rightly stressed that it too rejects the notion of a "ground of things," and that it even lays stress on the opposite notion, the one of an "absence of foundation" or "emptiness." It seems to me that the last remark just made should therefore also apply to this philosophy. For it is clear that the word *sunyata* (emptiness) is used by Buddhists for denouncing—as constituting a fundamental error—the twofold idea that things exist by themselves as they are seen and that our "selves" enjoy an individ-

---

[10] The other example of this type they give is the one of kittens that were compelled to remain inactive and as a consequence did not acquire the indispensable sensory-motor reflexes. It calls, I think, for exactly the same commentary. In my opinion the conclusion in question is, in reality, trustworthy only in virtue of the objections to naive realism that come from quantum physics, as explained in the present book.

ual absolute existence. But it remains true that, apparently, present-day Buddhists are anxious to avoid that Buddhism be identified with anything like nihilism. A hope therefore exists that the Veiled Reality conception—with a sufficiently thick veil between consciousness and "the Real"!—should be found, by them, worth some attention.

19-3-5 Third Apparent Objection: Ambiguity of the "Reality" Notion

Already in chapter 4 we noted that, like the word "objectivity," the word "reality" has two meanings, which we distinguished from one another by replacing the said word by either one of the two expressions "independent reality" (or "the Real") and "empirical reality." In his book, *Ordnung der Wirklichkeit* (Heisenberg 1989) (which remained unpublished during his life) Heisenberg, right at the start, identified the notion of reality with our apprehension of the same. In other words, it was to the sole empirical reality notion that he attributed a meaning. True, he mentioned the *idea* of the existence of a "truly real" world getting depicted on the living being's consciousness as on a mirror—a mirror sometimes blurred or distorted. However, his text unambiguously shows that he conceived of such a world only as embedded within the space-time framework. Consequently he could easily show that while, in classical physics times, the idea in question looked natural, the discoveries of twentieth-century physics incite us to give it up. On this basis then, he developed a conception of knowledge grounded on the notion of *patterns of reality*. Such patterns are human works. Moreover, according to him, they cannot be grounded on mathematics or other pieces of knowledge considered indubitable, for the latter are analytic and do not inform us on reality. It ensues that these patterns are multifarious, often coexist, and follow one another as human thinking develops. Some are grounded on beliefs, others on empirical science data. The latter presuppose that the considered state of things is sufficiently separable from us for making it possible to objectivize it, be it only within the framework of what we called here "weak objectivity." Heisenberg in no way claimed that such scientifically grounded patterns are to some extent arbitrary. But, since they are human made and not grounded, as we just saw, on totally sure knowledge he nevertheless admitted of diversity among them. In fact he adopted a notion of "regions of reality" that leaves much room for the latter. Indeed it allows for the coexistence of a formally objectivist realism concerning macroscopic things, a weakly objective description of quantum objects, a comparative autonomy of biology with respect to physical sciences and a conception of mind events making them irreducible to biological phenomena (even though tightly linked with them).

Due to the fact that in Heisenberg's conception the independent reality notion is rejected, the empirical reality notion takes up, in it, a basic role. In it, to consider empirical reality to be just an appearance is therefore quite impossible and, in view of the importance this notion has for us, this circumstance makes the conception in question an intuitively attractive one. But on the other hand, as a result of this very rejection the said conception is open to the objection that it makes existence a product of knowledge. And since this objection is crucial in my eyes I consider it appropriate to point out here that the argument Heisenberg made use of for justifying the rejection in question is not binding. The point is that, as we just saw, this argument finally rested on the view that somehow independent reality has to be embedded in space-time. Now, this condition may quite well be set aside. In the Veiled Reality conception, for instance, a crucial point is that independent reality is conceived of as *not* being so embedded. It is "real," not in the sense of being something that we can touch or "kick into," but, quite at the other end of the spectrum, in a sense the (capitalized) word *Being* is meant to convey; and to which, as already noted, the expression "super-real" might fit as well. Against such a "ground of things," Heisenberg's objection does not hold.[11]

On the whole, our examination of the arguments pro and con the notion of a "ground of things" yielded nothing like a refutation. For indeed, the whole force of the objections to this notion wears itself out at refuting—successfully by the way—the notion of a world-per-se that would be "pregiven" in the sense of "describable." The idea of a "ground of things" is too vague to be thus captured and escapes the said objections as a will-o'-the-wisp escapes any would-be catcher. But it is also true that the arguments most commonly used in favor of the notion of an absolutely existing reality are in no way more convincing. Neither the ones usually invoked by believers in "commonsense" nor, as we saw, the one that mathematically oriented physicists find so attractive seem able to withstand the hard test of facts. This is why the more fact-based reasoning developed in this book may conceivably have some interest. Some of its characteristic features will be noted below, before we take a look at such and such metaphysical perspectives that the Veiled Reality conception seems to lead to.

---

[11] Nor does the one that such a notion is empty since, to repeat, it is taken (extended causality) to be the source of our observational predictive laws.

19-4 Some Consequences of the Evolution of Physics

This is the right place for noting some significant facts that are consequences of quite a general nature of the evolution of physics and do not depend on the more precise views that will be developed next.

FIRST CONSEQUENCE: THE LIMITS OF THE "EVENT" CONCEPT

Within the last half-century all various scientific activities underwent a considerable qualitative and quantitative development. With, among others, the consequence that scientists are now aware of the details—and the great diversity—of processes of which only the general nature, common to them all, had previously be apprehended. Correlatively they had to give up the illusion that, once known, the basic laws would, all by themselves, yield complete understanding of the world we live in. They found that the path leading from the said laws to the phenomena taking place on our own scale—or on that of the Universe—swarms with genuinely scientific problems the bearing of which is much greater than it was thought. In the same vein they became aware of both the true role of chaos and our ability at theoretically understanding it. In a word, they discovered complexity.

This state of affairs led people to speculate that a kind of a conceptual scientific revolution was on its way, grounded in the "discovery" that events—that is, the singular, the accidental—far from being alien to science as previously thought, lie in fact within its core. That their role is even more essential than the one imparted up to this time to general laws. According to some, this would amount to nothing less that a downright rejection of the universality notion. Others are not so radical but announce nevertheless what, using Edgar Morin's expressive phrase, they call the great "Comeback of the Event" (Morin 1982).

What is true and what false (or, if not false, at least exaggerated) in such conceptions? Concerning rejection of the universal I already expressed myself ([d'Espagnat 1990] and here in chapter 6), with appropriate supporting arguments, against the idea. Concerning the "come back of the event" I think our judgment should be definitely more shaded. What, in my opinion, should be acknowledged right at the start is that, as we saw at several places and in particular in section 15-4, there are events in the usual sense only within the realm of *empirical* reality. That is, within what, when all is said and done, is just our own way, as human beings, of apprehending the Real. Since the said empirical reality is known not to coincide with "the Real" (remember nonseparability) such a restriction of "events" to its realm obviously reduces the bearing of the "comeback of events" motto. However it should be stressed that this reducing is not a canceling. Empirical reality extends to stars and galaxies and, to repeat, it is the framework within which our own life takes place. The fact that its history, "full of sound and fury"—at least full of events!—concerns us in quite an essential way, and the fact that we are now able to decipher large portions of the history in question and partly predict what will take place clearly should not leave us indifferent. But still, it would nowadays be naive (since demonstrably false) to raise events (and more generally notions such as those of complexity and disorder) to the level of basic building blocks and attributes of "the Real." Incidentally this was finally acknowledged, nay it even was stressed, by some of the authors who most wrote about complexity and events.[12]

SECOND CONSEQUENCE: NOMINALISM QUESTIONED

It is a fact that with cultivated, literary, avant-garde people, Roscelin's and William of Ockham's nominalism is now making quite a remarkable comeback. Its brilliant adepts view it as endowed with quite a modern pep and draw many inferences from it. They like its way of discrediting rules, inspired from its denying the very existence of the universal and exalting what is particular. They appreciate its way of extolling private initiative, grounded on its principle of primacy of the individual. In their works they give us to understand that it is indeed nominalism, viewed as a product of positive knowledge, that justifies such liberating changes. But in so doing they often miss a detail. They forget to take due notice of the fact that such justifications are valid only if nominalism itself is more than just an arbitrary, questionable view, taken up as a postulate.

Those among them who would care to set their mind at rest on this point may consider using one out of two methods. One is to turn to the very creators of the doctrine and the philosophers who most discussed it. In this spirit, the works of Boetius, Bérenger de Tours, Abélard, and other Middle-Ages philosopher-theologians should be consulted. Information will thus be gained, but not of a very conclusive kind for these were early authors, uninformed of facts that we now know of and that turn out to be relevant to the question in hand. The other method consists, precisely, in inquiring whether or not present knowledge, and particularly "hard" sciences, speak in favor of nominalism. If we adopt the latter procedure we should first observe that Middle Age nominalism was aimed at being a true, that is a broad philosophy. In it, rejection of the universal was not

---

[12] In particular by Edgar Morin who, when the outcome of the Aspect experiments was published, made a point of explaining what they implied and wrote "the Real is not exhaustively described through the ideas of order, disorder and organization."

limited to the description of human beings or living beings, nor indeed to such and such a precise domain of thinking. It was quite general. Only particular data, individuals, contingent details truly exist. Common nouns, concepts, abstract terms are but words, and exist only in our thoughts. And this has to be true at any scale.

Now, is it? In the times when classical physics was at its heights the "yes" answer seemed to be, by far, the one most likely to be true, for atoms were then considered to be separately existing entities, merely linked with one another by some distance decreasing forces. True, all the atoms that pertained to a given chemical element were considered identical, which, at a pinch, might have been made use of by an opponent to nominalism[13] for claiming that the concept of an element is more than a word.[14] But the fact remains that all these atoms (or "particles," or "corpuscles," never mind the technical word) were considered to be theoretically discernible from one another, in other words, independently existing. And this observation actually seemed to reach the core of the matter, so that, in a debate engaged, in this field, with some opponent, the nominalist easily won. Now, we know that, just precisely in this field, a crucial change took place. Particles, to the extent that this word may still be used, were found to be indiscernible. According to quantum field theory they are not in any way individuals. They are but collective modes of existence—or, more precisely, of "appearance"—of some global and universal entity. Clearly, this deprives nominalism from the argument in its favor that was the most tightly related to fundamental questions. True, some people seem to consider that another argument, grounded on complexity, might replace it. They point to the above-mentioned fact that complexity has indeed been considered to constitute an objection to the universality conception. On this issue I can but refer once more to the content of sections 6-1 and 6-2. And stress that the complexity of the phenomena—which is undeniable and has a most important role in our present-day investigation of *empirical* reality—changes nothing about the fact that, conceptually, nominalism is identical to the, there defined, "realism about entities." So that the objections to realism about entities that are stated in the said sections also apply to nominalism. In other words, since a broad and self-consistent version of nominalism can hardly avoid localizing reality in the "elementary objects," thus making every particle an indi-

---

[13] In the Middle Ages such an opponent was called a "realist." But, as is well known, this name then referred to the (Plato inspired) "realism of the essences" or "of the Ideas," a conception totally opposite to the objectivist realism that, nowadays, the word "realism" most naturally conveys. In fact, nominalism is but a form of objectivist realism as defined in section 1-2.

[14] But Ockham would have pertinently retorted that such universality merely exists within our mind.

vidual, to profess nominalism amounts to falling into an error in the field, not of metaphysics (this would at present be merely a venial sin) but, more embarrassingly, of physics.

It needs time for ideas differing from "received views" to get engrained, so to speak, in the mentality of a society the more sophisticated portions of which are still more or less under the influence of the "Enlightenment" ideas. And a writer who would violate the correlative acculturation delay might entertain no hope of reaching, in his reader's brain, the zone that professionally he craves to tilt, to wit the hazy one where reason and heart overlap. Now, at first sight recent findings do—I must grant—point in two very different directions, one of which, centered on complexity and the notion of event, would lead to nominalism while the other one, grounded on nonseparability and quantum universality, indicates rejection of that same nominalism. The first one is the one that imparts the greater meaning to the world, apparent, marvelous, shimmering, and it is easily understandable that it should be the one that most attracts writers with some liking for knowledge. But seriously analyzing all the data reveals, I think, that the second one is the one that is nearest to what is true. Hence I do believe that it is the one that will finally win the game.

Third Consequence: On the Respect We Owe to Spinoza

Of this respect Einstein set the example; an example that, coming from him, carries much weight and that, moreover, is moving. For, in Spinoza, Einstein admired everything: his thought, his ethics and his life. Including of course the quest—that he himself also undertook—of the "supreme good" defined by the philosopher to be "the union of the thinking soul with the whole of Nature."

Here the word "nature" appears. It is evocative but extremely vague and susceptible therefore of generating ambiguities and misunderstandings. Incidentally, it is worth noting that such misunderstandings are today more likely to occur than was the case in Spinoza's time. For indeed the latter, who wrote in Latin, disposed of two expressions, *natura naturans* and *natura naturata*, that referred to two quite different concepts. Without it being a question of drawing a close parallel between two conceptions, Spinoza's and the Veiled Reality one, that, anyhow, are separated by a several centuries gap, it still may be noted that the notion "*natura naturans*" somewhat resembles the one of "the Real" while, by contrast, "*natura naturata*" has at least some faint similarities with the one of phenomena. Spinoza's well-known formula "*Deus sive natura*" ("God, in other words, Nature"), in view of which some people saw Spinoza as an ancestor to materialism, appears, thanks to this parallel, in a different and fairer light. For in it the word "nature" clearly refers—as the whole Spinozan context shows—to *"natura naturans,"* that is, to an entity that it is quite impossible to identify with what a chemist, a biologist, a geologist, or even an ecologist would nowadays call "nature." Spinoza's God has nothing in common with a machine (nor has Einstein's). True, it is Substance, but Substance with a capital S. Descartes imparted to the word "substance" a meaning not altogether different from the sense we give to it now. And he accepted the plurality of the substances, a decisive choice that greatly contributed giving rise to mechanism, a doctrine that, when interpreted as an ontology, we now realize is misleading. Spinoza avoided making this mistake and this, in my opinion, constitutes one of the main reasons why a great deal of consideration is due to him. Admittedly, it remains that this Substance, this God, was considered by Spinoza to be "intelligible." To be, partly at least,[15] attainable by human reason. Clearly this constitutes a considerable difference between Spinozism and the Veiled Reality conception. Anyhow, I think that whoever would care to build up a neo-Spinozism of some sort, fitted to present-day knowledge, would first have to incorporate into it the idea that near realism (mechanism) does not fit "Nature in itself." Along these lines he/she would have to emphasize the themes of *"natura naturans"* and unity. To that end, he/she might keep the term "God," although he/she would of course have to wipe out the notion, incompatible with Spinozism, of a personal, willful God.

Fourth Consequence: A Question for Phenomenology

As Jean Ladrière, for example, quite clearly showed (Ladrière, 1989) some of the ideas composing the Husserlian phenomenological approach to reality are distinctly more in keeping with the conceptions underlying quantum mechanics than with those of classical physics. This is related to the fact that within phenomenology the construction—or reconstruction—of the physical world by thought is taken to be some sort of a continuation of the *Cogito*. Schematically, adepts of phenomenology observe that our consciousness is continuously invaded by forms, colors, etc., in short, by something—conventionally called a "phenomenon"—that *offers* itself to our apprehension. In other words, they note that everything takes place as if we had to do with a creativity of some sort, through which the phenomenon manifests its independence. And this, still according to Husserl and his followers, leads reflexive consciousness to credit the phenomena—by transitivity so to speak—with the very "quality of existence" that it credited itself with when it uttered the well-known "I think, hence

---

[15] Of it we know only two attributes, thought and extension, whereas there are an infinity of them.

I am." In a way, there is continuity from the primary self-evidence of just our presence to ourselves and the secondary one constituted by the existence of phenomena. To this Ladrière added that quite often the phenomena are linked together in a natural and repetitive way that makes us discern an element—that we call a "thing"—common to them. Admittedly the perceived thing never entirely appears at once, but this precisely is why it is more than the sum of its appearances. It is this elusive way of being present that imparts to it an independence, with respect to us and the other things, in which we recognize a reality that it possesses all by itself.

Under these conditions the difficulty, on the side of classical physics, came from the fact that the latter, prompted by its demand for intelligibility, succeeded in substituting for the description, in ordinary language, of events lived through, a description of an altogether different kind, using idealized models based on mathematical objects. So that one had to do with a description of reality expressed in terms of representations (mechanist first, then mathematical) and no more in terms of direct experience of "events lived through." And the contrast was all the greater as, for constructing the said representation, classical physics had laid it down as a principle that the rational reconstruction of that experienced world was to describe the said world "as it is in itself," independently of the way we apprehend it. Hence the said physics finally introduced a sharp dividing line between the (mathematical) representation of the world and human experience. Now, Ladrière stressed, it is just the existence of this sharp dividing line that quantum physics forces us to reconsider. And indeed, today decoherence theory does to an appreciable extent corroborate this judgment of his since, as we saw, it shows it is only through an association of the quantum formalism and some elements of our mind structure that it is possible to account for our experience. Considered under this angle present-day quantum theory may therefore be said to reinforce phenomenology.

It should however be noted that this corroboration does not wipe out some significant differences. For indeed it is clear that the explanation, through decoherence, of how objects and phenomena get constructed justifies that some reservations should be made concerning the "degree of reality," so to speak, that may be attributed to the latter. Admittedly, decoherence does account for the concatenation of phenomena of which we just saw that they generate the notion of things. But, as we noted in chapter 8, it does so by referring to us. What it shows is not that the things are reasonably independent of us and of one another. In virtue of the formal structure of quantum mechanics it grants, on the contrary, that there may quite well be physical quantities the measurements of which, if we could perform them, would reveal lack of independence with respect to us. What it proves is that, in virtue of the fact that our possibilities of action are limited, the measurements in question are practically unfeasible. Under these conditions some questioning is legitimate. May we still consider that the independence of the perceived objects should make us discern in them, not just appearances but really existing things, just as the *Cogito* incited us to do concerning consciousness? It seems that the answer is "no." It seems that the reality we are thereby led to attribute to the things, the macroscopic ones included, is not one that they possess by themselves but one that we impart to them. It is, in other words, *empirical* reality.

Admittedly, it could be retorted to this that in chapters 10 and 18— with, in particular, the "community of Wigner's friends" notion—we introduced the idea of considering consciousness itself to be, partly at least, relative. And that, metaphorically speaking, this sets consciousness on an equal footing with empirical reality (didn't we speak of "coemergence"?). Still it remains that this parity is obtained at the price of making both parties relative and that, within phenomenology, making consciousness relative might raise a problem, to the extent that this doctrine is grounded on the absolute certainty of the *Cogito*.[16]

19-5 The Veiled Reality Conception Reexamined

The Veiled Reality conception was sketched in chapter 10. But in the chapters that followed many subjects were taken up, the analysis of which led to conclusions susceptible of having repercussions on the conception in question. Below, the latter will therefore be reexamined.

19-5-1 Its Basis

Concerning the "reality" notion let us note first that unfortunately our developed societies tend to adopt a not truly self-consistent approach to it. In them the elite—"the sages," say—entertain about it two opinions both of which bear the stamp of commonsense but which, finally, seem not to really fit one another. The first one consists in considering it obvious and indisputable that facts and things exist outside us, most of them (think, for example of the astronomical data) being moreover independent of us. The other one is to note that the facts and things that we do take into account cannot be devoid of any link with our perceptions since,

---

[16] Would the notion of nonpredictive, nonindexed states of consciousness tentatively considered in section 18-4-3 change the nature of the problem? Maybe. At the present stage the question, obviously, remains open.

as already noted, it would hardly be meaningful to ponder about a world having no influence whatsoever on what we can know. In virtue of this the sages posit that we should not speak of the unreachable. And most of them are even more restrictive. They posit that, disregarding tautologies, the whole of our knowledge comes from our senses and claim therefore that all of our statements should ultimately be rooted in sense data.

Up to this point, of course, a realist may consider that everything is still all right. We easily conceive of a Universe essentially independent of our own selves while some parts of it do act on us, and we find it reasonable to only talk about the latter. But on reflection we nevertheless realize that we are faced with a serious question. If our universe of discourse is indeed limited and shaped up by our perceiving aptitudes, as we just considered must be the case, what guarantees have we that we do perceive facts and things as they really are "out there"? We saw that the arguments favoring a positive answer (the no-miracle argument and the one based on inter-subjective agreement) are by no means conclusive in this respect. And clearly they look even less so when we remember various elements we came across in the first part of this book: failure of near realism, nonsepa-rability, weak objectivity of some basic axioms, etc. Certainly, as we saw, it is nowadays impossible to claim with full confidence, concerning any physical measurement correctly performed by some well-trained team, that it discloses "what, in itself, the investigated object really is quite inde-pendently of the human aptitudes at apprehending." And this remains true even when the ontologically interpretable Broglie-Bohm model is taken into account since, even within it, measurement results are, in most cases, not identifiable with the "true" values of the measured quantities. As a result, most of the sages alluded to the above fall back on a position consisting in imparting to the word "reality" merely the sense of a sort of synthesis of all of what, in the field of (probabilistically) predictable phenomena, may be metaphorically represented by means of a descriptive language. In other words, they tacitly rub out any distinction between reality (in-itself) and what I called empirical reality.[17] This way of thinking is worth analyzing. It finally amounts to hoping that it will be possible to keep the essential part of what, philosophically speaking, constitutes the doctrine of physical realism while still implicitly identifying it with a kind of positivism. More precisely, the idea seems to be to acknowledge the absolute primacy of the notion of possible observation (or experience) while simultaneously ignoring the ultimate reference to the fact of getting aware that the primacy in question normally implies. And to try and make

[17] A "reality" whose elements and structures are defined by referring to feasible or at least conceivable operations and to the data registered by conscious beings, following these experiments.

this view consistent by implicitly evoking the vague notion of some sort of an impersonal, hence objective, "observation-per-se." Thus explicated this thesis shows some similarity with the one of weak objectivity. The resemblance is superficial however, due, precisely, to the circumstance that, inconsistently perhaps, the thesis in question implies we should take the fact that the observation is human made to be but an irrelevant detail.

Now, does this idea fit with the rest or does it not? To rationally examine the question the first point to be noted is that, in their scientific acceptation, the notions "to observe" and "to experience" essentially imply selectively turning one's attention to such and such particular. This is what distinguishes experimentation from just passively "looking around." But then, may we really speak of the experience an instrument would have? Is there, in an instrument, something akin to selective attention? I mean, is, by itself, the position of its pointer more significant (for whom? for "it"?) than the one of any one of its constituting atoms, or of the fly that happened to graze its dial? These questions are akin to those Putnam raised concerning reference and, like them, call for negative answers. In other words, when all is said and done, the very notion of some "observation-per-se," taking place in a world merely composed of inanimate objects turns out to be self-contradictory. Note, finally, that to this, quite general, remark a more particular one must be added, referring specifically to the nature of contemporary physics. This is just the observation that in the context of the latter, which so forcefully incites us to consider that experience is, in the last resort, what mould objects, the inconsistency of the view that we could do without the human notions of intention and reference that characterize experience is quite specially manifest.[18]

However, this implies that a perplexing question reappears, namely, "What if, finally, the very notion of 'the Real' had no meaning? What if things had no ground?" This takes us onto, so to speak, our "second front." Having vanquished materialism thanks to physics, shall we have to surrender to radical idealism? This is (although not quite, as we shall see!) what those among my interlocutors who incline to the neo-Kantians' point of view seem to believe. And it is true that the question is a delicate one. Generally, the ancient[19] arguments in favor of a "ground of things"

[18] Even within the framework of an ontologically interpretable physics the experimentalist has to decipher the indications his instrument yields. But here, moreover, his instrument, set up as he chose, in part creates the message.

[19] By "ancient" I mean dating back to the eighteenth century. It is worth noting that in the seventeenth century the arguments in favor of the notion of "the Real," of "Being" as was then said, were altogether different. Malebranche judged that the simple idea of "Being," prior to any restriction and limitation, is, for human beings, the one that is the most straightforwardly reachable, whereas, on the contrary, "only faith is able to convince us that there really are bodies" (*Recherche de la vérité,* quoted by Brunschvicg [1951]).

made no distinction between "the Real" and "the describable" and above (sections 19-3-1 and 19-3-2) we took note of what makes them inoperative. As we saw (chapter 17 in particular), the interlocutors just alluded to therefore ask me, in substance, "Since, following the indications from physics, you go as far as to deny that objects exist per se, why don't you go to the end of this story? Why don't you give up the very idea of an existence of 'the Real'?"

In fact a substantial part of the development of the second part of this book aimed at, directly or indirectly, providing an answer to this question. This answer may be synthesized in four "steps." First, as even my critics grant, strictly speaking nothing forces me to "go to the end of the story." Second, what section 17-3 dealt with in detail was precisely the "no-ground" idea and I explained there the reasons that make me consider it unconvincing. Third, I believe that in sections 19-3-3 to 19-3-5 above I did really refute recently expressed objections to the very notion of a "ground." And fourth, last but not least, I consider that there are strong positive reasons in favor of keeping the said notion. These reasons were stated in section 10-4-1. In short they are (i) the primacy of existence over knowledge, (ii) the argument that "something says 'no,' " (iii) the difficulty of forming the notion of an evolving a priori, and (iv) the necessity of explaining the very existence of the universal predictive laws without resorting to steadily renewed miracle sequences. The purpose of the whole second part of the book was to put these reasons through the tests of criticism and a comparison with other views and, as the three aforementioned "steps" indicate, I consider that the results of the said tests were, on the whole, positive.

Incidentally, let it be noted that, after all, even the authors who part with me on this are reluctant to take the final step of actually claiming that the notion of any "Real" must be rejected. Bitbol himself wrote: "What remains screened from criticism . . . is the abstract concept of a reality conceived of as setting limits to the determining power and factual and symbolic activity of the experimentalist. Or as co-determining source of uncontrollable constraints revealed by responses to experimental impulses" (Bitbol 1998). And Zwirn (2000) took note of the need for *something,* of which he wrote that to claim it exists would be incorrect but concerning which he granted nevertheless that "to mention it amounts to indicate an existence of some sort." Finally these authors grasped that, in such matters, there is a risk of dizziness against which it is important to be on one's guard. It consists in the temptation many shrewd minds may feel of going to the very end of some reasoning that its logical coherence makes attractive, without realizing that the framework in which it takes place is not one in which strictly deductive methods fully apply. And that therefore various conceptions are tenable, so that what, on an abstract philosophical level, looks appealing may, from another angle, prove to be more seductive than convincing.

It is worth opening here a sort of a parenthesis for noting that, as already mentioned (section 6-6), this question is akin to the one, well known in philosophy, of the relationships between meaning and reference of a concept. As we know, classical instrumentalism does not distinguish between the two. It posits that the meaning of a concept is limited to its reference, that is, the set of the factual data it is meant to group together. Incidentally, in section 7-3 we observed in this connection that within the realm of an instrumentalism fitted to contemporary physics it is appropriate to substitute prediction to reference, that is, to consider that the meaning of a concept is defined by the predictions that it makes possible. But relative to the question in hand what is to be stressed is that to raise the above-mentioned limitation to the level of a fully universal philosophical rule would be an unjustified extrapolation. As already pointed out, the existence concept cannot be subjected to it in complete generality since the notion of prediction already implies the existence of some "being" who predicts.[20]

19-5-2 Its Development and Conclusion

"Nature helps powerless reason and prevents her from talking nonsense to such a degree." This somewhat skeptical but, I think, highly wise comment Blaise Pascal expressed about Pyrrhonism (*Pensées*, Article VII), applies as well to the (not infrequent) nihilistic excesses of pure reason. This is a danger that, as we just saw, Bitbol and Zwirn finally escape, since, when all is said and done, they do not cross the "red line" of rejecting the existence concept. But then I claim that, by accepting, even if reticently— merely as a "limiting concept" or an undefinable "I don't know what"— the notion of a source of constraints not wholly reducible to us, these authors take (perhaps unwillingly) a step, be it a tiny one, toward a form of Platonism. According to them (and to me) we have to say, on the one hand that objects are not things-per-se and on the other hand that we cannot consider the human beings—the prisoners in the cave—to constitute all by themselves the totality of what is.

It remains that between them, Plato, and me there are, not unsurprisingly, considerable differences. They bear, first of all, on the accessibility

---

[20] True, Hervé Zwirn (2000) pointed out that, logically speaking, it could be the case that only thought would exist. But the existence of pure thought already implies existence. Relative to this, whether or not thought alone exists is, in a way, but a "detail." To me the answer "no" seems to be the more plausible one.

question. According to Bitbol and Zwirn this "limiting concept," this indefinable "I don't know what," is really totally inaccessible, so that, about it, we cannot even have a thought. This, of course, is not Plato's view, and let me not come back in detail to the above-stated reasons why I also do not take such a thesis to be entirely convincing. In short, in view of the fact that there are universal laws—such as the Maxwell equations—that phenomena do obey[21] and that these laws remain pertinent although their interpretations evolve in time, I consider it more plausible that "the Real"—Zwirn's *something*—is structured and that some of its structure passes into our "laws." In other words, as explained in *Veiled Reality* (section 16-4) and again here in section 10-4-1, beyond Kantian causality, which underlies empirical causality and whose importance, of course, is considerable, I believe in the existence of an "extended causality" that acts, not between phenomena but *on* phenomena *from* "the Real." Clearly, since, due to nonseparability, the said "Real" may in no sensible way be considered constituted of localized elements embedded in space-time, this causality vastly differs, not only from Kantian causality but also from Einsteinian causality. Of course it does not involve eventlike efficient causes (in Aristotle's sense) since such efficient causes bring time in. But it may involve structural causes and the latter, in this approach do not boil down to mere regularities observed within sequences of phenomena. In fact these structural "extended causes"—which vaguely bring to mind Plato's Ideas—are structures of "the Real." And we saw that, in my eyes, they constitute the ultimate explanation of the fact that physical laws—hence physics—exist.

Here let us just briefly recapitulate what has been noted throughout this work relative to the relationships taking place, in virtue of the said extended causality, between "the Real" and experience. The first point to be kept in mind is that, according to the Veiled Reality conception, "the Real" is prior to mind-matter splitting.[22] Hence, while consciousness may legitimately conjecture that, through extended causality, it indirectly gleans a few glimpses of "the Real," the said glimpses totally differ by nature from the pieces of knowledge that, within the traditional dualistic conception, mind is believed to gain on an assumed "matter-per-se." Here indeed it is not on a matter taken to be lying "in front" of it that mind gathers the said glimpses. It is on what lies at its very source, which makes quite a considerable difference. All the same, as shown in section 17-2-5 gathering such glimpses is quite conceivable.

---

[21] The words "law" and "obey," though universally used, are regrettably anthropomorphic. It goes without saying that in the present context they should be disconnected from anything resembling will and commandment.

[22] This expression of course refers to the coemergence (described in section 18-4-4) of predictive consciousness states and empirical reality.

A second point worth noting is that, in virtue of the above, the Veiled Reality conception and structural realism as usually understood are really quite different views. For indeed, in the accounts given of the latter it is usually stated that great mathematical laws such as the Maxwell equations describe the true "structures of the Real" as they really are. For a reason that has been explained in section 16-4-2 (Remark 3) I think this is a questionable hypothesis. The Veiled Reality conception is much less precise. It merely involves the conjecture that our great mathematical laws are highly distorted reflections—or traces impossible to decipher with certainty—of the great structures of "the Real." Incidentally, let a surprising fact be noted in this connection. It is that (due to what resembles a semantic lacuna) within the conception at hand there is, to the question "is the Real describable?" just simply no answer. For indeed the word "describable" normally means "totally describable." Similarly the word "indescribable" normally means "totally indescribable." And the expression "partly describable" is normally taken to mean that some parts of the investigated object are describable and the others not. Obviously none of these three modes of qualifying is here correct.

As we see, my conception finally is that of a "Real" that is structured, concerning which I do not rule out the possibility that poetry, art and mysticism might yield rare and precious glimpses, but that still is, for us, human beings, basically nonconceptualizable. An objection to this that most readily comes to mind obviously is that the notion of something that human beings cannot conceptualize has no meaning. To answer it let me resort to a reasoning Hervé Zwirn (2000) put forward. This author started from the observation that, according to all appearances, our abilities at conceptualizing exceed those of dogs, monkeys, and other animals. And he wondered whether or not it is possible to envisage a conceptualizing ability exceeding ours same as ours exceeds the ones of dogs and monkeys. He pointed out that to bluntly answer "no" would be highly presumptuous. For indeed it would amount to claiming that the limit of what can be conceptualized is not reached by dogs or by monkeys but is by human beings. Now, such a thesis would be consistent only under the condition that a meaning be imparted to the notion of "conceptualizable in an absolute sense" (that is, with no reference made to such or such type of brain). And it would consist in assuming that human beings and they alone reach the limit in question. As Zwirn noted "this position is too reminiscent of the anthropomorphic and successively refuted assumptions that Earth is at the center of the Universe, that the Solar system is unique, and that, at the end of the nineteenth century, physics was complete, to be a plausible one." We are thus left with the other alternative, which is to grant that, after all, there is no absurdity in evoking the idea of a "something" that we cannot conceptualize.[23] To this it may be added that, while the idea that we can get "glimpses" on what human beings cannot conceptualize may seem questionable to many, it gives no shock to the poets. My own conjecture is that, on this point, poets are in the right.[24]

In the light of this all it is clear that if we aimed at producing a rough first idea of the Veiled Reality conception by resorting to some familiar metaphor, choosing Plato's allegory of the Cave would not be inappropriate. In it, perceived objects are shadows. But they are "shadows of . . ." and, in the fable, this is crucial.

It remains, though, that even the best allegories are misleading in some respects. Here the conception and the fable part on at least four points. The first one is that, in the fable, "the Real" is composed of distinct objects, to wit, the things, obliquely lighted by a distant fire, that porters steadily move at the threshold of the cave (and which, in a first approximation, may be identified with the Platonic Ideas). On the contrary, in the Veiled Reality conception, in virtue of nonseparability, "the Real" cannot be separated, by thought, into distinct parts. The second difference is that, contrary to the Platonic Ideas, "the Real," as we just saw, is not even conceptualizable by us. The third (related, by the way, to the first one) is that, in the fable, the shadows would exist even if there were no prisoners for they owe nothing to the presence of the latter. On the contrary, in the Veiled Reality conception (but also in some other accounts Plato himself gave of his views) phenomena as such exist only relatively to a possible experience, the one *we* have. Finally, the fourth difference is that the fable does not propose the notion of emergence of consciousness (or its coemergence with empirical reality) out of "the Real," a notion whose central role we noted within the Veiled Reality conception. For indeed, even though the fable in question adequately symbolizes "emergence" of empirical reality (the set of the shadows) from "the Real" (the set of the Ideas), it gives us no suspicion whatsoever that also consciousness (the prisoners) emerges from the said "Real."

These differences are quite considerable. They help us realize how dangerous it would be to indulge in some new version of syncretism consisting in trying to reduce the highly novel representations brought forth by the increase of our knowledge to some elements of our common cultural patrimony. On the other hand, however, it is by no way unreasonable to

---

[23] Zwirn shrank from the idea of claiming in a positive manner that there exist nonconceptualizable things, for this would seem to imply that we are able to conceptualize the nonconceptualizable. Hence he, just like me, has a preference for a negative statement, of the type "what can be conceptualizable does not cover all," without precisely stating what this word "all" refers to.

[24] In other words I rank neither Shakespeare's *Tempest*, nor Beethoven's *Ninth Symphony* nor even the ceiling of the *Sistine* Chapel among "conceptualizations."

consider that the elements in question, polished as they are by time and the intense meditation of so many thinkers, finally are kind of sesames, almost indispensable for opening the road that leads from knowing to understanding. Indeed, anybody aspiring to a well-balanced, that is, broad and consistent, inner life would appreciate the possibility of linking the word view knowledge suggests to such or such great philosophical or religious tradition the maturing of which "gave flesh" to our relationship with the said world.

In this book, as it may well be imagined, there can be no question of engaging into so attractive but so hazardous an enterprise. In the course of years, nay, of millennia, the traditions in question gave rise to so many learned, subtle, brilliant developments and controversies that to successfully carry it through would necessitate combining the professional knowledge of a physicist with the erudition of a large variety of specialists. But still: on the possible relationships between the great lesson from contemporary physics—which, I claim, the Veiled Reality image well summarizes—and the main said traditions it would undoubtedly be worthwhile to ponder. In these circumstances let us merely try and draw up a rather light and certainly nonexhaustive inventory of the questions that seem to most naturally come up in the field.

The one that comes first to mind is delicate since it concerns religions. It has to do with relativity of time and, more precisely, with the fact that, as recalled above, even cosmic time no more seems "absolute." In view of this finding the term "immortality," so often found in the said religions, has become somewhat puzzling since it seems to implicitly postulate an absolute time conceptually anterior to the human mind. The question then is, shouldn't this term rather be understood to refer, in the picturesque style religions are forced to use, to another notion that also belongs to their realm, namely that of "eternity," in the sense of *escaping* from time?

For the same reason we should perhaps (even more daringly!) examine whether or not also the notion "creation," the "creative act," might, by means of some refocusing on the one of Being, be made independent of time; at least of the human experienced time, the time of empirical reality. We should perhaps inquire whether the, well-received, "continuous creation" notion offers, after suitable reinterpretations (emergence?), some possibilities along such lines.

We shall not, here, pursue such speculations further, but just note that they are, at the same time, grounded on (or at least inspired by) physics and essentially relevant to the "quest for meaning." It therefore seems impossible to unreservedly agree with Heisenberg (1989) at the place where, in substance, he claimed that science practically yields no information whatsoever susceptible of throwing light on such a fundamentally important problem. True, this author was fully right in calling for maximal caution in this domain. The sage, he explained, realizes that all the thoughts by means of which we strive at imparting a foundation to the meaning of life come back in a circle to their starting point. And undoubtedly there is wisdom and truth in this disillusioned remark. Still it remains that present-day science largely liberates us from such blocking conceptions as materialism, some of which a less advanced science had, in fact, inspired. This counts.

Other questions are less delicate, be it only due to the fact that, in the course of years, they were mainly tackled from a purely intellectual angle. Here we think of course of those that the great philosophical systems and world-views investigated. However, the theme being too broad we shall here limit ourselves to considering the main lines of two great western traditions, Platonism (already touched upon above) and Aristotelism.[25]

Concerning Aristotelism, one of the similarities between it and the Veiled Reality conception lies in their notion of causality. For since Aristotle was, philosophically speaking, a realist, causality as he meant it was not merely operative between phenomena (that is, objects-for-us). According to him, the causes were primarily (elements of) reality-in-itself. In a sense, this is also what my "extended causes" are. Note also that, with Aristotle, causality was not theoretically limited by the condition that a cause must take place before its effect. It is well known that, far from being identical to Descartes and Galileo's somewhat "mechanist" God (the God of the "initial flip"), Aristotle's God was rather of the nature of a final cause: that toward which everything tends. Incidentally, this conception has the attractive feature that it may be taken in an allegorical sense (in the ideas of a goal to be reached, a work to be carried out, a perfection to be aimed at, the core notion, achievement, is intrinsically linked with the one of future). And then it is interesting to note that the Veiled Reality one bears some similarity with it. For since, in the said Veiled Reality conception, "the Real" is prior to time, strictly speaking the cause it constitutes is not specifically anterior to its effects. And since we are anyhow forced to think "within time," we are just as justified in situating it, by thought, in the future (which makes it likeable) as we would be in situating it, dully, in the past.

An inquiring mind may also take interest in the above noted fact that the "potentiality" notion that was central with Aristotle (he opposed "power" to "act") and came to be considered obsolete within classical

---

[25] Obviously, the similarities between the Veiled Reality conception and the great eastern philosophical systems should be considered as well. But these approaches are, for us, not traditional. Hence, for us, they are not to the same extent as those studied here the "sesames" opening a way from "knowing" to "understanding."

physics (with Newton, so it was claimed, "all is act") recovered a status in quantum physics. Admittedly, not an indisputable one but, all the same, a possibility of having some interpretative role to play. Concerning this we must remember that, contrary to the more ancient Ionic philosophy of nature, which located its explanatory principle in matter, Aristotle considered matter to merely be the seat of vague potentialities that actualize only under the influence of forms. That, in his eyes, nature was a hierarchy of existences within which simple beings, though being themselves forms ("informed" matter) played the role of matter with respect to more complex forms. That, at the very basis, lay "*materia prima*," which, being in no way "informed," was taken to be pure potentiality. Hence, for Aristotle, complex objects were more "actual," and therefore, in a way, "more real" than the simple ones from which they sprang. Whoever is in quest for a somewhat "deep" reality should therefore, according to this idea, take a great interest in complex things, without striving too much to—incorrectly—divide them by thought into smaller objects. A similarity comes immediately to mind between this conception and the empirical reality notion as we defined it.[26] For we remember that in virtue of decoherence the macroscopic objects constituting the bulk of the said reality are endowed with a "reality" that, admittedly, is but relative (empirical) but, still, is of a higher level than the mere "epistemological reality" of wave functions, etc.

As is well known Heisenberg was the first to point to the said similarity. He did not dispose of the decoherence notion so that, seen under this angle, his approach is not as elaborate as the one that physicists can nowadays put forward. But it is to him that we owe the view that wave functions enjoy a status somewhat similar to the one of the Aristotelian matter in that they are essentially "potentia," the wave function of the Universe possibly being identifiable with Aristotle's "materia prima." And, as we noted in chapter 13, some thinkers such as Abner Shimony go on taking much interest in this approach. However, this does not prevent them from acknowledging its intrinsic vagueness as well as the difficulties there are at reformulating it in a somewhat more precise way. To this, so to say, "intrinsic" trouble another, more circumstantial, one gets added. It consists in the fact that, in the course of years, some kind of a mixture (partly justified, it must be said) has taken place, in the mind of most people, between Aristotelism and a nominalism of which we saw above that it is practically incompatible with present physical data.

On the whole, then, it seems clear that a comparative analysis of the Aristotelian line of thought and the findings of contemporary physics might lead to interesting views. The above sketched comparison between

---

[26] As well, of course, as with the "realism of signification" (section 9-7).

the allegory of the cave and the Veiled Reality conception indicates that the same should be true concerning a comparative analysis of the same findings and Platonism. In fact, such an analysis would be complementary to the one concerning Aristotelism for, while the latter incites us to apply this philosophy to a deeper understanding of the empirical reality notion, the one concerning Platonism seems to point to some similarity between "the Real" and the Platonic "Good." Not, of course, that the above noted incompatibilities between the lessons of contemporary physics and the teaching of Platonism should be ignored. But they do not touch on the core of the latter, which is that "the Real" is not in the things. For Platonism this is a most significant point of convergence with today's science. In this respect quite an important fact is that while, according to Plato, "the Real" (for him, the Ideas) does not, to repeat, lie in the "things," it does not lie in "us" either. Plato was in no way a "radical idealist." It may be doubted that he would have accepted the notion of an emergence (or coemergence) of thought and the things. But if he had he then most certainly would have specified "out of some preexisting, 'Real' (the Ideas and, quite especially, the 'Good')." This is why Platonism is sometimes given the name "Realism of the essences." On this point the rainbow metaphor (section 15-5) is enlightening. A rainbow, we noted, is in no way a thing-in-itself. According to Platonism—at least to the Platonism of the cave allegory—the same is true of things. But this obviously does not mean that the rainbow depends exclusively on us. For it to exist, some meteorological conditions have to be fulfilled. And similarly, while, according to Platonism, what we perceive depends, to a considerable extent, on us, it does not exclusively depend on us. In fact it, first of all, depends on the Ideas, in other words, on "the Real." In this, the Veiled Reality conception rediscovers, in a sense, the Platonic views, the main difference being, as previously noted, that the Platonic Ideas are eminently conceptualizable by us[27] while, like Damascius' *pantè aporeton,* the "Real" of the said conception is not.

Yet, nowadays most people, including quite a large number of scientists, still believe that, by appropriately analyzing experimentally gathered

---

[27] As their very name ("Ideas") indicates. "It is remarkable," Etienne Gilson (1942) wrote, "that to the question 'what is Being?' Plato always answers by describing such and such a way of existing. For him there is being only when there is some possibility of intelligibility." Let us note, however, that this is not yet quite the end of the story. in the *Republic* (Book VI) the "narrator," Socrates, evokes "the 'Good' that all souls crave for, of the importance of which all have an intuition, and which all of them make the purpose of their action." But he goes on with the essential words: "without being able to get at certainty and define precisely what it actually is." And he owns that, as for him, to define what "the Good" is "lies beyond [his] abilities." In view of this, unique but essential, exception to the above-mentioned rule, the difference in hand in no way amounts to an opposition.

sense data we get to know basic Reality—the very ground of things—with steadily increasing accuracy. The fact that the very advances made in physics entail a rapprochement between this science and a philosophical doctrine that incites us to question this is an element well worth reflection.

All the same, it remains true that to develop such a comparative analysis would be difficult for many reasons, one of them being that the acceptation of the word "Platonism" did not always remain the same and, still today, somewhat varies from author to author. Above, we more or less identified it with the doctrine the allegory of the cave refers to. But as we noted (section 16-3) objectively it may not be reduced just to that. In fact when, today, the word "Platonist" is made use of, as a rule it serve to qualify a person who adheres, not, really, to the cave allegory doctrine but to the thesis (linked to the former, but loosely) called "Pythagorism" above. More precisely, when the said person is a physicist he or she may be an adept, either of what we called Einsteinism or of the thesis that two equally real worlds exist, one physical and the other one mathematical,[28] or even of the "three words" thesis Popper once put forward. I already explained why I maintain an attitude of reserve concerning these views. So, let me specify here that when I make use of the word "Platonism" I do not take it with such a meaning. That much being stated, it is clear of course that the doctrine that mathematical objects are truly real is *linked* with Platonism. And in the Veiled Reality conception a small something of this idea remains since, in it, it is considered that the objects in question are reflections, or "traces" of "the Real."

This leads us to Einstein. Einstein believed the world to be intelligible. True, he granted that, in practice, human mind limitations bar out the possibility of an exhaustive knowledge of what I term "the Real." But he apparently believed that "in theory" the latter is knowable as it really is, and that it is so via the channel of the great universal laws, which it is the scientist's task to discover. Moreover he attributed a great emotional value to the said task. For example, when evoking Planck's remarkable scientific perseverance he declined to attribute it to sheer force of will. According to him (Einstein 1934) the affective state of mind that makes such accomplishments possible is akin to the one in which love or faith may set us. So, he seems to have unreservedly approved of what he termed the third level of religious experience, the one reached, according to him, by religion after it has gone through the two first ones, called by him the religion centered on fear and the religion centered on morals, respec-

---

[28] Of the Platonic world of mathematical truths Penrose (1994) wrote that it is a world distinct from the physical one and on the basis of which we have to understand the latter, which does seem to imply that both worlds are equally real.

tively.[29] It is, he granted, difficult to make people "not in the know" apprehend the nature of this third level since it corresponds to no humanlike conception of God. In his eyes it essentially consisted in recognition of the sublime and marvelous nature of the order that gets reflected in nature as well as in the world of thought.

Unquestionably, since the time when Einstein wrote this much has changed in the fields of both factual data and appraisal of motivations. As may be remembered, the latter domain was lately shaken by a generalized move of suspicion cast on both the legitimacy of scientific activity and the scientists' moral conduct. Based on ideological arguments most of which were, it must be said, sheer verbalism, this worldwide campaign pell-mell uprooted mandarin-like behaviors together with great perspectives. Some time will presumably be needed before the latter again receive, within the so-called "world of thought," the attention they got during genuinely creative periods. But such fluctuations affect *doxa*, not truth. Even though they made a great impression upon the general public they are therefore less significant than the changes that took place, due to new findings, within the data. As we saw, the latter were considerable. They entail diverse consequences. One of them, which I make a point of stressing again here, is nothing else than a radical invalidation of the present scientific "vulgate," mostly disseminated by mass media and which (due to the latter being limited to instantaneous information) is grounded on the exclusive use of basic familiar concepts. As we saw, the seriously minded materialists themselves nowadays reject the so-called "scientific materialism" that this world-view disseminates. But there is another, admittedly less dramatic, consequence of the new findings. It is that some of the latter—underdetermination of theories by experiment, nonseparability, quantum weak objectivity, etc.—have an effect on the very Einsteinian construct. They shake it, not, admittedly, in its properly scientific content, which remains, as it were, untouched, but in precisely the one of its facets that we are here most interested in, namely, the world view it conveys. Is it, in this respect, outworn?

This question, of course, is not one that an exclusively objective reasoning might settle. It is therefore susceptible of being given answers differing from one another, modeled on opinions and tendencies. As for me, although the Veiled Reality conception differs from Einstein's on the important points we saw, I consider that it preserves sufficiently many features of it for Einstein's above-mentioned positions to remain in it, in

---

[29] Nobody was more aware than he of the importance of ethics and values. But in his opinion no road existed leading from knowledge of what is to knowledge of what should be (Einstein 1939). Hence according to him morals could be grounded neither on scientific nor, a fortiori, on (assumed) theological knowledge.

their broad outlines at least, pertinent. This is because, for the affective state of mind Einstein attributed to Planck (and which he himself undoubtedly experienced) to be justified it is in no way necessary that Pythagorism be one hundred percent true, that "the Real" be totally intelligible[30] by means of mathematics. It is quite sufficient that the—great!— idea should appear valid, of the existence of a "Real" toward the hidden structures and qualities of which human mind may tend with wonder, while remaining aware that it will never attain them due to the simple fact that understanding them would exceed its ability.

Indeed, we should say more. For, upon reflection, this idea clearly is even more consonant with the "third-level religious experience" dear to Einstein than is the one of a "Real" assumed to be wholly knowable. It fits to the best with what appears to be an obscure but great truth, to wit, that human mind is first of all oriented towards expectation and quest. I mean, it is a mind whose nature is to *tend*, with confidence and perseverance, towards something that it will never reach and that therefore, same as a horizon, partakes of transcendence. True, through his life, work, and writings Einstein had already shown us that even facing a classical science stamped with physical realism, some religiosity, some sense of the divine, remains possible. But, such divine still was of a mathematically limpid nature and therefore (like Beauty in Baudelaire's poem) quasi-incommensurable with mankind. Thanks to the very fact that it forces us to overstep this realism, contemporary physics shows that it is sensible, nay, we might almost say "rational," to proceed a little further. For, as we saw, it bars reducing Being to material components and thereby makes it impossible to believe consciousness to be just a product of matter, that is, of empirical reality. Hence the idea that Being is somehow prior to the mind-matter splitting becomes defensible even in front of a scientific audience. Consequently, the idea that mind may vaguely "recall" something of Being does not look any longer irremediably absurd. So that, even though it is not "reachable" the "Being" in question appears to be an "I don't know what" to which it is conceivable that the human mind is not altogether extraneous. A Being, in other words, that may constitute for it a horizon. What I mean is that, perhaps, the archetypes of some of our feelings, great longing, love, etc., are hidden there.[31] Nothing, of course, proves this to

---

[30] Incidentally, "intelligible" does not necessarily mean the same thing as "knowable" (think of the Kantian "*noumena*"). We might decide to call "intelligible" a "Real" of which we succeeded in forming a consistent conception, even if the latter is such as to imply that the said "Real" is not knowable. With such a meaning of the word, Einstein's assertion "the Real is intelligible" might become compatible with the Veiled Reality conception.

[31] In some books of mine I ventured the idea that human mind keeps some sort of a remembrance (a timeless one of course; language, here, is deficient) of this Being prior to splitting. I even went so far as to evoke bridges (but made of "spider web") linking us

be the case. Nothing even *suggests* it is. But nothing proves, nay, nothing serious any more suggests it is not. People who, explicitly or not, ground their thinking on the concepts of classical physics are quite naturally incited by the latter to consider that we are foreign to the world (or incited to call the latter absurd, which amounts to about the same). But an analysis grounded on the data of contemporary physics lead, as we see, to a conception that is quite different and highly less pessimistic.

From all this it follows that, today, even setting into operation an (indispensable) acute critical turn of mind no more results in discrediting the spiritual impetus that moves mankind. An impetus Einstein adequately evoked (*loc. cit.*) when he concisely wrote that men want to live the "whole of what is" as something endowed with both unity and meaning.

---

with Being, as well as enigmatic "calls from Being." These images are neither scientific nor philosophical but are not aimed at being either. In my eyes they still today describe fancy but not implausible conjectures.

THE BELL THEOREM

A. Proof

$\mathbf{A}$S EXPLAINED IN the text, the "Bell theorem" is the fact, proved initially by John S. Bell, that the joint hypotheses of locality and free choice, by the experimentalist, of what experiment is to be performed imply that certain measurement outcomes must obey some definite inequalities called "Bell's inequalities." Here we describe, in substance, the last version Bell gave of the proof of this theorem (1981, 1990). It has the advantage of being extremely general. Indeed it will be observed that in it nothing is assumed concerning the nature (corpusclelike, wavelike, or other) of the involved entities.

Following Bell, let us consider two space-time regions R_1 and R_2, their backward lightcones C_1 and C_2, the space-time region N common to C_1 and C_2 and a space-time slice V totally shielding N from R_1 and R_2. (fig. A1). In N there is a device that, at times t_1, t_2,. . . , t_n, emits pair of particles (or of anything else) and displays the fact that it did so. In R_1 and R_2 there are registering devices, susceptible of displaying either the response + or the response −. Assume that at times $t_1 + T$, $t_2 + T$, . . . , $t_n + T$, these devices do display one or the other of these two responses.[1] Assume also that these devices are movable and, more precisely depend on parameters, called a and b, respectively, whose values two experimentalists located in R_1 and R_2, respectively choose at will at times $t_i + T - \delta$, with $i = 1, 2,. . . , n$ and $\delta < < T$. In, for example, the dart experiment (section 3-1), in which the orientation of the dart arriving in R_1 is projected on a definite oriented axis and its sign is then detected, the parameter a defines the direction of the said axis (the parameter b having the same role in R_2). Similarly, in an Aspect-like experiment these parameters define the orientations of the two polarization analysers. Assume further that each one of the two experimentalists has, concerning the orientation of his analyzer, a choice between two directions only, to be called a and

---

[1] In order not to make the notation clumsy by superimposing indices we here assume that the time T taken by the "entities" to travel from the source to the detection region is the same for both. But this assumption has no other role in the proof and could thus easily be dropped.

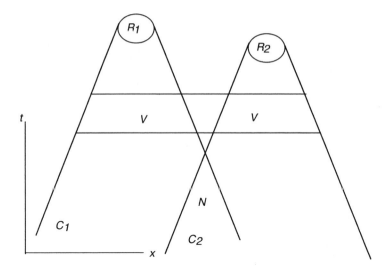

Fig. A1-1.

a' and b and b', respectively, and that they make a new choice at every time $t_i + T - \delta$. Finally, call A a response from the analyzer in R_1 and B a response from the analyzer in R_2 (A, B = + or −). A sufficient number n of such elementary operations makes it possible to test hypotheses concerning the joint conditional probability

$$P(A,B|a,b) \tag{1}$$

for the outcomes registered in R_1 and R_2 to be A and B, respectively, given orientations a and b [quite generally we denote by $P(X|Y)$ the conditional probability that X *if* Y].

Of course, we would not be surprised to find that outcomes A and B are correlated, or in other words, that $P(A,B|a,b)$ is not a product of two independent factors:

$$P(A,B|a,b) \neq P_1(A|a)P_2(B|b) \tag{2}$$

but let the symbol λ collectively designate the whole set of the parameters that completely specify all the events taking place in region V and let

$$P(A,B|a,b,\lambda) \tag{3}$$

be the joint probability of responses A and B when the values not only of a and b but also of all the λ's are specified. In virtue of the well-known general probability rule for conditional probabilities, which here takes the form

$$P_1(A|B,a,b,\lambda) \equiv \frac{P(A,B|a,b,\lambda)}{P_2(B|a,b,\lambda)} \tag{4}$$

we may write

$$P(A,B|a,b,\lambda) = P_1(A|B,a,b,\lambda)\, P_2(B|a,b,\lambda) \tag{5}$$

and the locality principle (section 3-2-1) then implies that, in the second member of this equation, certain of the conditional variables are redundant because the corresponding probabilities are in fact independent of these variables. More precisely, the said principle entails that $P_2(B|a,b,\lambda)$ is independent of a and may therefore be written $P_2(B|b,\lambda)$. And it also shows that $P_1(A|B,a,b,\lambda)$ is independent of both B and b and may therefore be written $P_1(A|,a,\lambda)$, whence, finally, we have the formula

$$P(A,B|a,b,\lambda) = P_1(A|a,\lambda)P_2(B|b,\lambda). \tag{6}$$

But of course due to the presence of the "hidden" parameters λ among their conditional variables, the probabilities $P(A,B|a,b,\lambda)$, $P_1(A|a,\lambda)$, and $P_2(B|b,\lambda)$ are not experimentally accessible. To get at the probabilities $P_1(A|a)$, $P_2(B|b)$, and $P(A,B|a,b)$ which *are* experimentally measurable—and concerning which quantum mechanics hence yields predictions—we therefore have to take into account the fact that there is a certain probability[2] $\rho(\lambda)d\lambda$ (unknown to us) for these parameters to have the values λ to within $d\lambda$. According to a well-known probability rule, to get the probabilities we are looking for we then have to separately combine $\rho(\lambda)$ with the corresponding probabilities conditioned to given values of the λ's and sum over all possible λ values. In other words, we have to write

$$P_1(A|a) = \int \rho(\lambda)\, P_1(A|a,\lambda)d\lambda \tag{7a}$$

$$P_2(B|b) = \int \rho(\lambda)\, P_2(B|b,\lambda)d\lambda \tag{7b}$$

---

[2] It is at this stage that the hypothesis that the experimentalists are completely free to chose the settings (a or a', b or b'), at the last second, as they like comes into play. For assume for a moment that this is not the case and that, quite on the contrary, the settings are fixed. We could then imagine that, due to unknown forces propagating at subluminal velocity the settings of the instruments have an influence on the hidden variables λ at the source, so that the function $\rho(\lambda)$ would in fact depend on the orientations of the settings. For example, it would not be the same in runs with orientation a as in runs with orientation a'. It is clear that then the proof given here would not work.

$$P(A,B|a,b) = \int \rho\ (\lambda)\ P(A,B|a,b,\lambda)d\lambda\ . \tag{7c}$$

Define then

$$E(a,b) = P(+,+|a,b) + P(-,-|a,b)-P(+,-|a,b)-P(-,+|a,b). \tag{8}$$

By virtue of (6), (7a), (7b), and (7c), (8) takes the form

$$E(a,b) = \int \rho\ (\lambda)\ \{P_1(+|a,\lambda) - P_1(-|a,\lambda)\}\{P_2(+|b,\lambda) - P_2(-|b,\lambda)\}d\lambda \tag{9}$$

which we may also write

$$E(a,b) = \int \rho\ (\lambda)\ G(a,\lambda)H(b,\lambda)d\lambda \tag{10}$$

with the conventions

$$G(a,\lambda) = P_1(+|a,\lambda) - P_1(-|a,\lambda) \tag{11}$$

$$H(b,\lambda) = P_2(+|b,\lambda) - P_2(-|b,\lambda). \tag{12}$$

Note that since the P's are probabilities they necessarily obey the inequalities

$$0 \leq\ = P_1 \leq 1;\ \ 0 \leq P_2 \leq 1 \tag{13}$$

from which it follows that

$$|G(a,\lambda)| \leq 1;\ \ |H(b,\lambda)| \leq 1. \tag{14a,b}$$

On the other hand (10) gives

$$E(a,b) \pm E(a,b') = \int \rho\ (\lambda)\ G(a,\lambda)[H(b,\lambda) \pm H(b',\lambda)]\ d\lambda \tag{15}$$

wherefrom, by virtue of (14a)

$$|E(a,b) \pm E(a,b')| \leq \int \rho\ (\lambda)\ |H(b,\lambda) \pm H(b',\lambda)|\ d\lambda \tag{16a}$$

and similarly

$$|E(a',b) \pm E(a',b')| \leq \int \rho\ (\lambda)\ |H(b,\lambda) \pm H(b',\lambda)|\ d\lambda. \tag{16b}$$

At this stage, let us take into account an elementary and easily proved proposition of pure arithmetic, namely, the fact that if two real numbers x and y are such that $|x| = 1$ and $|y| \leq 1$ the inequality

$$|x + y| + |x - y| \leq 2 \tag{17}$$

necessarily holds true.

From (14b) and (17) it follows that

$$|H(b,\lambda) + H(b',\lambda)| + |H(b,\lambda) - H(b',\lambda)| \leq 2. \tag{18}$$

Consider then the expression

$$|E(a,b) + E(a,b')| + |E(a',b) - E(a',b')|. \tag{19}$$

By virtue of (16a,b) this quantity is smaller than or equal to

$$\int \rho\,(\lambda)\,\{|H(b,\lambda) + H(b',\lambda)| + |H(b,\lambda) - H(b',\lambda)|\}\,d\lambda.$$

In view of (18) it is thus smaller than $2\int\rho\,(\lambda)\,d\lambda$, and since $\int\rho\,(\lambda)\,d\lambda$ is equal to 1, we finally get the inequality

$$|E(a,b) + E((a,b')| + |E(a',b) - E((a',b')| \leq 2 \tag{20}$$

which is one of Bell's inequalities or, more precisely, one of the inequalities established in 1969 by Clauser, Horne, Shimony, and Holt that generalize Bell's. Hence this inequality does follow from the combination of the two hypotheses of locality and of free choice of experiment (as we know, the latter serves to make sure that, as tacitly assumed above, the orientations a and b are independent of the λ's). As explained in chapter 3, the reason this inequality is interesting is that, surprisingly enough, it is violated by both the quantum mechanical observational predictions and the experimental outcomes (which corroborate the said predictions). So that, as soon as we grant that we are free to choose what experiment we shall do we can no longer believe in locality.

As for the just mentioned incompatibility between quantum mechanical predictions and the inequality (20) it is easily checked on an example. Indeed, in the case of photon-pair emissions by atoms during what the specialists call a cascade transition of the type $J = 0 \rightarrow J = 1 \rightarrow J = 0$, the experimentally measurable quantity $E(a,b)$ takes, according to quantum mechanics, the value cos2Θ, Θ being the angle between directions a and b. It follows that if we choose coplanar directions at angles $(a',b) = (b,a) =$

$(a,b') = \pi/8$ [whence $(a',b') = 3\pi/8$] with one another the first member in the inequality (20) takes the value $2\sqrt{2}$. The inequality is grossly violated.

From 1964, when John Bell first published them, to 1990 when he died, and even later, the Bell inequalities were made the subject of many developments, due to Bell himself and other contributors. There can be no question of reporting here on the corresponding details. It should be noted, nevertheless, that the first proof Bell gave substantially differed from the one described above. In particular, it was less general for it bore only on phenomena concerning which, when the two orientations a and b are the same, the theory predicts a strict correlation (as in the dart case mentioned in section 3-1). This first proof took place in two steps. In the first one, which was inspired by the well-known Einstein, Podolsky, and Rosen (1935) article and whose guiding idea is described below, Bell showed that *concerning the just mentioned phenomena*, the no action-at-a-distance hypothesis implies that the outcomes A and B are determined, at the source, by supplementary parameters (conventionally called "hidden"). In other words he showed that A and B must, in that case, be *functions* of a, b, and λ. And in the second step, by means of calculations similar to the ones reported above (except just that they bore on *functions* of λ) he established inequalities similar to (20).

B. A Simplified Proof

The calculations described above are simple, the mathematics they involve being downright elementary. Still, it may be the case that some readers find them tedious. For the latter's use a proof of the Bell inequality the details of which are perhaps easier to follow is described here. Like the one mentioned in the foregoing remark, it exclusively concerns the phenomena in which a strict correlation is predicted whenever orientations a and b coincide. And it also proceeds in two steps. To describe the first one, let us start, as Bell did in his first proof, from Einstein, Podolsky, and Rosen's (EPR's) most simple guiding idea. This idea just was that if we know beforehand and with certainty what the outcome will or would be of some measurement made on an object on which we cannot act, this outcome must have been predetermined (for if it were not, how could it be known in advance?). That much being granted, imagine we have to do with a set of particle pairs similar to the set of dart pairs considered in section 3-1. More precisely, assume that the only significant difference between the two is as follows, whereas, by assumption, we knew that the darts obeyed a classical, and hence deterministic, kind of physics we a priori do not know whether or not this is the case concerning our particles. On the latter let us then perform measurements exactly similar to those we first imagined being made on the darts (i.e., with both instruments oriented in the same direction; never mind the precise nature of the quantities on which these measurements bear). Assume that this has been done and that, just as in the dart case, a strict correlation is found between the results obtained on the left- and the right-hand sides. It is easy to see that, in such a case, the EPR guiding idea shows the results were predetermined. For indeed, consider some new pair emerging from the same source, let us settle down, say, "on the right" and let us make a measurement on the particle that arrives there. We get a result and, by virtue of the strict correlation we observed before—and induction!—we know what result will be obtained over there—far away—when our assistant "on the left" performs the planned measurement on the other member of the pair, since it must be identical to ours. To the extent that the distance (which may be arbitrarily large) between the places where the two measurements are made prevents us (think of locality) from having any effect on what happens "over there," the EPR guiding idea then shows that the left-hand-side measurement result must have been predetermined (as, in the dart case, except for the fact that, in the dart case, we knew beforehand that it was, for we postulated determinism). Of course the same reasoning may be made concerning the right-hand-side one.

Fortunately, it so happens that, by making use of atomic transitions or by other means it is possible to produce pairs of photons or other particles enjoying the strict correlation-at-a-distance property used in the foregoing reasoning. The latter then shows that the results of measurements (of the considered type) that may be made on particles composing such pairs are predetermined by some parameters. Incidentally, note that these parameters certainly do not have the same values on all pairs since the measurement results are not the same on all the involved pairs. Since, in most cases, all the said pairs have the same wave function it is thus clear that the parameters in question are supplementary variables, not present in the "orthodox" quantum formalism and called "hidden" for this reason.[3]

Up to this point we showed that, under the considered experimental conditions, some results of measurements that might be performed on the particles are predetermined by parameters relative to these particles. It is true that, strictly speaking, we showed this only concerning the particles that belong to a pair whose other element also undergoes a measurement.

---

[3] In other words this argument shows that the locality hypothesis is at variance with the strong completeness assumption (section 2-8). But we strive to show much more. We aim at proving that it is at variance with facts. So our reasoning must be pursued.

But since we are free to imagine the considered operation to be made on an arbitrarily large amount of pairs, it is quite natural to use here induction once more, and conclude that the elements of pairs prepared exactly in the same manner but on which no measurement has yet been made are predetermined as well.

Let us then go over to the second stage in our reasoning. Assume that the particles made use of are photons and consider three of the quantities that can be measured this way on them, namely the polarizations A, B, and C along three different directions a, b, and c. We know that measuring them can yield but two results, + or −. For the time being, consider only the photons that, in an "Aspect-type" experiment, will arrive at the detector—call it D—that is on the left. We now know that the outcomes of the measurements that could be performed on these photons are predetermined (by parameters we do not know but which objectively have such or such values). It is therefore quite meaningful to consider—by thought—within the ensemble of all these photons, all those whose parameters are such that if detector D were oriented along a the thus performed A measurement would yield result + *and* if, instead, it were oriented along b the result would also be +, *and* finally, if instead, it were oriented along c the result would also be +. Call $N(+,+,+)$ the number of these photons. Similarly, call $N(+,+,-)$ the number of photons defined just as $N(+,+,+)$ except for the fact that, on them, upon a measurement of C, the result− would be obtained. And so on. In the same spirit, Let $N(+,+,x)$ be the number of photons on which the "expected" result of a measurement of A (supposing it were made) is + and the one of a measurement of B (supposing it were made instead) is also +, no condition being specified concerning C. Obviously we have $N(+,+,x) = N(+,+,+) + N(+,+,-)$. Similarly (with notations modeled on the foregoing ones) $N(+,x,+) = N(+,+,+) + N(+,-,+)$ and $N(x,+,-) = N(+,+,-) + N(-,+,-)$. By adding term by term the two last equations and taking the first one into account we get $N(+,x,+) + N(x,+,-) = N(+,+,x) + N(+,-,+) + N(-,+,-)$. And since, obviously, none of the involved numbers is negative, we get the inequality

$$N(+,x,+) + N(x,+,-) \geq N(+,+,x). \tag{21}$$

Besides, in virtue of the above-noted strict correlation, a number such as $N(+,+,x)$ also represents the number of photon pairs in which the right-hand-side photon is such that, if made, a measurement of its polarization along a must yield result + *and* the left-hand-side photon is such that, if made, a measurement of its polarization along b must also yield +. And, with appropriate changes, the same holds true concerning the two other numbers that appear in the inequality (21). Now, the numbers thus defined are experimentally measurable. The point is that to find them, there is no need to perform two successive measurements on the same photon, an operation that would yield no usable information since the first measurement would presumably alter the parameters determining the result of the second one. It suffice to perform a measurement of A on one photon and a measurement of B on the other one in the pair. Hence, dividing both members of (21) by N, the total number of pairs in the ensemble, we get an inequality between measurable frequencies of pairs of results of polarization measurements along different directions. In the limit of large N this, in virtue of the law of large numbers, may also be considered to be an inequality between the corresponding probabilities. And in fact it is one of those that Bell derived by means of his first method [it is an easy matter to check that it is equivalent to a particular case of inequality (20)]. The orientations a, b, and c may be chosen at will and, as above, there are some choices for which the inequality (21) is violated by the quantum mechanical observational predictions. It follows that one at least of the premises used for deriving it must be false. If we are keen on preserving the ones of physical realism and induction we thus must reconcile ourselves with the idea that somehow, some influences that do not decrease when distance increases should exist, or in other words with nonlocality.

C. A Glance at the Experimental State of Things

During the last thirty or forty years quite an appreciable number of experiments meant to check the Bell theorem were performed all around the world, using various methods. With the result that the agreement between the experimental data and the quantum mechanical predictions turned out to be excellent. In other words, the experimental data corroborated the Bell's inequalities violation, that is, nonlocality. On the other hand, the involved experimentalists willingly grant that, strictly speaking, at the present stage the proof of the said violation yielded by the experimental data is not yet quite as watertight and rationally irrefutable as, ideally, they would like it to be. This is due to the fact that in this field as in any other one the measuring instruments are not ideally flawless. In particular, they do not detect all the photons. In fact it would be extremely difficult to set them in such a way that all the photons that should contribute to the statistics be actually detected. A proof of the violation is therefore effective only under the additional assumption that, in the mean, the undetected photons would behave upon detection in the same way as the detected one do (in spite of the fact that their directions of emission were different, etc.). Some devices make it possible to partly overcome such difficulties, but others then arise, the details of which need not be described here. New experiments should presumably, in the near future, fill these few gaps. But anyhow, the existence of such remaining difficulties should not hide from our view the fact that the presently available data already render fully incredible the hypothesis that, in the investigated phenomena, the Bell inequalities should be obeyed. Indeed, account being taken of the data, such a hypothesis would imply that the observable quantum mechanical predictions, which up to now were never found erroneous in any domain, not only would be false concerning these particular phenomena but also would be false in a way seemingly conceived so as to deceive us. That, for example, only the photons that, for purely contingent reasons differing from one experiment to the other one, hitherto escaped detection would violate quantum mechanics. Admittedly we must, in such fields, be quite wary. An idea that, viewed from some angle, looks incredible may, in the light of new findings, appear reasonable. This justifies that the experimental quest should be pursued. However, in the present case, to ground the view that locality can be salvaged on the existence of the above-mentioned gaps would be tantamount to making a most risky bet.

D. Historical Comments and a Short Bibliography

The ground paper in the field is the one John Bell published in 1964 in a small, newly founded journal *Physics* nowadays not easily found in libraries for it had but a few issues. Fortunately the paper in question was reproduced, together with others, in a collection of texts of the same author published in 1987 under the title *Speakable and Unspeakable in Quantum Mechanics*. Today, this little book (Bell, 1987) is an indispensable reference for anybody intending to devote himself or herself to a systematic scientific study of the question.

Concerning ideas, the next step was simultaneously taken, on the one hand by J. F. Clauser and on the other hand by M. A. Horne and the philosopher-physicist A. Shimony. The younger R. A. Holt also contributed, and the four of them chose to publish together (1989). What they did was to derive, from hypotheses very similar to Bell's original ones, inequalities, called CHSH inequalities, of which, as we know, inequality (20) is an example. It so happens that these inequalities apply to cases more general than the one (considered in part B above and hereunder called "situations SC") in which a strict correlation between A and B takes place whenever $a = b$. For this reason submitting them to an experimental test is much easier.

In the CHSH paper measurement outcomes A and B were still considered to be predetermined, or in other words, to be functions $A(a,\lambda)$ and

$B(b,\lambda)$ of the hidden parameters λ of the pair. But in situations more general than the SC ones this feature could not any more be inferred (as we did in B) from a reasoning "of the EPR type." It had to be independently postulated. In 1970, in the course of Session IL of the summer school "Enrico Fermi" in Varenna, which the Italian Physical Society had asked me to organize, Bell, while still assuming the hidden variable "theory of the Real" to be determinist, made a most suggestive remark. He pointed out that, in the said more general situations, the measuring instruments themselves might well involve hidden variables susceptible of having an influence on the outcomes. Since these instrument variables cannot be known whereas the symbols $a, b, \ldots$ represent what is known, clearly the former may not be included by thought within the set of what the latter are supposed to represent. Consequently the outcome A of an individual measurement may not be considered to be a function, in the ordinary sense, of a and λ only. Bell stressed that to take our ignorance of these instrumental hidden variables into account a probabilistic element has to be introduced. And he produced (Bell 1971) a new, more general demonstration of the CHSH inequalities, which allowed for the existence of this element. Moreover, in a footnote he pointed out that, quite independently of the small chance disturbances possibly induced in this way by the instruments, this more general demonstration should also apply to cases more general than the one in hand. Indeed, he made it clear that the thus derived inequalities provide a refutation, not only of the local hidden variable theories that are deterministic but also of those that are not. Starting from this remark, Clauser and Horne (1974) produced a general refutation of all the "realist, local" theories and Bell developed the latest version of his proof, which is, in substance, the one described in part A. The differences between the various proofs that were produced since 1974 are essentially due to the fact that their premises differ on points of detail.

The 1970s, also saw the appearances of the alternative proof reported on in part B above and of explicit analyses of its premises (Wigner 1970; d'Espagnat 1979, 1980), as well as of generalizations of the Bell inequalities (d'Espagnat 1975, 1976). They also saw the appearance of still another mode of proof, due to Henry Stapp (1977), whose interesting specificity is that, as we saw (section 3-3-3), its premises do not explicitly involve realism, the idea being to substitute counterfactuality for the latter. However, as already mentioned, it has been argued that, after all, counterfactuality, in this context, might well imply realism.

References to more recent articles on the subject are given below. There appeared a great number of them, so that an exhaustive list cannot be given, and I sincerely apologize to several authors whose important contributions could not be quoted. On the other hand, the articles whose references are given below do themselves yield a great number of references, which should enable readers to find their way in the corresponding literature.

N.D. Mermin, *Journal of Philosophy* 78, 397 (1981).

Z.Y. Ou and L. Mandel, *Physical Review Letters* 61, 50 (1988).

R.K. Clifton, *Foundations of Physics Letters* 4, 347 (1989).

D.M. Greenberger, M.A. Horne, and A. Zeilinger, in *Bell's theorem, quantum theory and conceptions of the universe,* M. Kafatos, ed., Kluwer Academic, Dordrecht, 1989.

J.D. Franson, *Physical Review Letters* 62, 2205 (1989).

R.K. Clifton, J.K. Butterfield, and M. Redhead, *British Journal of the Philosophy of Science* 41, 5 (1990).

N.D. Mermin, *American Journal of Physics* 58, 731 (1990).

J.K. Rarity and P.R. Tapster, *Physical Review Letters* 64, 2495 (1990).

D.M. Greenberger, M.A. Horne, A. Shimony and A. Zeilinger, *American Journal of Physics* 58, 1131 (1990).

A.K. Ekert, *Physical Review Letters* 67, 661 (1991).

E. Santos, *Physical Review Letters* 66, 1388 (1991).

L. Hardy, *Physical Review Letters* 68, 2981 (1992).

C.H. Bennett, G. Brassard, C. Crépeau, R. Jozsa, A. Peres, and W. Wooters, *Physical Review Letters* 70, 1895 (1993).

A. Whitaker, *Physics World* 11 (12), 29, (1998).

CONSISTENT HISTORIES, COUNTERFACTUALITY, AND BELL'S THEOREM

৪৯

IN 1989 MURRAY GELL-MANN and James B. Hartle put forward a stimulating, thought-provoking reformulation of quantum mechanics entitled *Quantum Mechanics in the Light of Quantum Cosmology* (Gell-Mann and Hartle 1989). Schematically, its defining feature is that, in it, certain sequences of events, called "consistent histories" or "history branches" are defined, to which the notion of quantum probabilities is extended.

From a conceptual (or "philosophical") point of view, an interesting question is whether or not this theory should be labeled "realist." Stated in the language we hitherto made use of the question takes the form: "is this theory weakly objective, same as conventional quantum mechanics, or is it strongly objective as classical mechanics was?" In fact, the authors of the said article did not explicitly consider the question so that the indications we may gather concerning their opinion on the subject are but indirect ones.[1] On the other hand, some such indications may be gleaned in other works of theirs. In particular, in his, most informative, popular book *The Quark and the Jaguar* (1994), Murray Gell-Mann criticized the conventional formulation of quantum mechanics for being anthropocentric and, more precisely, for ultimately resting on the notion of an outside observer. At least to nonspecialist readers, this must convey the idea that the new, proposed theory is free from this defect, which means that it stands in agreement with, at least, the leading ideas of physical realism (and in particular, at places where relativity theory is involved, with the usual, realistic interpretation of the latter). Indeed, in the mind of the same readers this impression must be strengthened by the wording of the passage in the same book meant to show that the theory removes the

---

[1] Note that in their quoted 1989 article Gell-Mann and Hartle nowhere asserted that their theory is a realist, local one, in the sense that these terms are meant in the present book.

conceptual difficulty raised by the outcome of the Aspect-like experiments. For, commenting on the said theory, Gell-Mann explained that, according to it, within the experiments in question "no action at a distance takes place" and the measurement performed on one photon "does not cause any physical effect to propagate from one photon to the other." Now, obviously it is only within the realm of physical—nay, even objectivist—realism that the notion "physical effect taking place between two physical objects" has a clear and distinct meaning. So that it is only within such a conceptual realm that the idea Gell-Mann evokes of a physical effect being propagated from one photon to the other clearly makes sense, and hence may possibly generate a difficulty (a difficulty that, to repeat, Gell-Mann claimed not to be present within his and Hartle's approach).

In view of the foregoing we are thus faced with two questions. First, is it actually true that the Gell-Mann and Hartle theory is realist (in the sense of physical realism)? And second, if "yes," is it really the case that, within it, the outcomes of Aspect-like experiments get a natural explanation involving no recourse to any novelty akin to influences at a distance?

To try and answer the first question let us first remember that, as Gell-Mann and Hartle made clear, in their theory only sequences of events to which probabilities may be attached are physically significant, and only sufficiently coarse-grained histories are susceptible of fulfilling this condition. Indeed, in his book Gell-Mann explained that, at least within the framework of the theory in question, "coarse graining typically means following only certain things at certain times and only to a certain level of detail." Now concerning the expression "following certain things only to a certain level of detail" we have to repeat what we noted in section 15-4 concerning the expression "disregarding" (both wordings being, if not strictly synonymous, at least very close to one another). That is, we must observe that the expression in hand may, in fact, be used for covering two different notions. For indeed there are situations in which "following certain things only to a certain level of detail" would just consist in not mentioning uninteresting data that we could just as well have mentioned without thereby altering the inferences we draw from the ones that we keep. And there are other situations, in which the opposite is true, in which taking the dropped data into account would indeed destroy the possibility of drawing the said inferences. Now, within the Gell-Mann and Hartle theory the procedure made use of is clearly of the second type. Indeed, the theory works only because we choose to simply ignore the fact that if we actually measured the physical quantities we decided not to be interested in we would get *definite* results. Admittedly, in view of decoherence such a procedure may seem motivated in many instances since in practically interesting cases the said physical quantities are, as a rule, but those we cannot measure in practice. But theories whose consistency depends on such a "practical" argument may not be considered strongly objective. It follows that, just as standard quantum mechanics the Gell-Mann and Hartle theory involves within its defining assumptions some that explicitly and crucially refer (collectively) to what *we* know, what *we* do and what *we* abstain from doing. In other words it is merely weakly objective. Contrary to what, at first sight might seem to be the case it does not fit into physical realism.[2]

Since, as we saw, it is essentially within the framework of physical realism that the Aspect-like experiments unquestionably convey the important information we know of, the negative answer we just found to the first question above should, in principle, exempt us from dealing with the second one. We might rest content with pointing out that since the Gell-Mann and Hartle theory lies outside the realm of physical realism it obviously has no information to convey concerning data interpretations grounded on the basic ideas of physical realism. But on the other hand Gell-Mann, in his book, expressly stressed similarities existing between what takes place in the Aspect-type experiments and macroscopic correlation effects susceptible of being observed in ordinary life. Since all of the latter are explainable in terms of physical realism—and are always explained this way; remember the dart pair example!—the very fact that Gell-Mann emphasized the said similarities may convey the impression that in his mind an interpretation (against which he made no warning) of his theory in terms of physical realism was not barred out.

Let us then imagine that there is one such interpretation. In the quoted passage Gell-Mann pointed out that, in his theory, when the circular polarization of one of the photons is measured "the circular polarization of both photons is specified with certainty," which obviously implies that, in particular, that of the other photon is specified with certainty. Now, if "is specified" just means that we know beforehand what we should read on a dial were we to measure the said polarization, the statement, of course, is true. But remember we are now interested in genuinely realist theories. In those in which the notion of a physical effect (or of the nonexistence of a physical effect) of some object on some other object may be meaningful. Now in such realist theories (like the one we automatically adopt when speaking of darts, for example) the expression "is specified," bearing on a physical quantity, in fact means more. It means, "has a value." And the expression "has a value" itself implies a kind of a counterfactual judgment. It means, among other things, that the value of the quantity in hand does not depend on whether or not we know it. And it even means that the said value depends on what we do, only when we directly or indirectly act on it. Concerning our present subject a very ele-

[2] For more details, see, e.g., *Veiled Reality*, section 12-4.

mentary question should therefore be asked, to wit: "Would the left-hand-side photon polarization be what it is if no measurement had been made on the right-hand-side photon? Or if some other measurement (that of the linear polarization along some definite direction for instance) had been made instead?" Within a local (that is, action-at-a-distance free) realist theory the answer obviously has to be "yes." In the dart experiment, for example, if the left-hand-side dart has some given orientation, as we assumed in section 3-1, it has it by itself, quite independently of what we may do or have done, or observed on the right-hand side. But in the Gell-Mann and Hartle theory this is not the case. In it, to consider the idea of (say) a linear polarization measurement being performed on the rhs photon in lieu of a circular polarization one implies switching, by thought, to another "history branch." And in this new "history branch," among the "polarization quantities" relative to the lhs photon the only one that has a definite value is this linear polarization. The initially considered circular polarization does not possess one. So that, in this theory, the answer to the question is clearly "no." Again, the conclusion is that the Gell-Mann and Hartle theory definitely does not fit local physical realism. We saw in the foregoing paragraph that it is conceptually impossible to really interpret it as just simply describing what is taking place "out there," as realists would have it. Here we realize that if we overlook this difficulty and—notwithstanding its existence—try to do so, we cannot avoid falling on something that is nothing short of a locality violation, alias something like an "action at a distance." (An action that, to be sure, we cannot use to send signals. But remember that postulating that no physical phenomenon exists except the ones that human beings can make use of would obviously set us at variance with the very spirit of physical realism.)

To repeat, Murray Gell-Mann was of course quite right in stressing that when, on some particular history branch, a given polarization is measured, hence specified with certainty, on one photon then, on the same branch, the same polarization is also specified with certainty on the other photon. But, as we saw, within any theory aimed at describing not just human experience or impressions but things as they really are (i.e., fitting physical realism), counterfactual modes of thinking, referring to "what we *should* experience *if* . . ." must be taken into account. In the two-photon case we then are forced to ask ourselves (what Gell-Mann simply did not do) what takes place when we switch, by thought, from one branch to another branch. Gell-Mann's reasoning, though entirely correct, has therefore strictly no bearing on the tentative realist interpretations of physics that constitute the subject matter Bell's theorem deals with. Hence, in particular, it does not imply that, within the said interpretations, no propagation of physical effects from one photon to the other one takes place. In other words, his analysis in *The Quark and the Jaguar* leaves the bearing and generality of Bell's theorem and the corresponding experiments, as explained, for instance, in chapter 3 of the present book, wholly untouched.

It is true that interpreting the violation of the Bell inequalities gave rise, in some articles and books, to various fanciful and regrettable extrapolations that somehow had to be denounced. And this circumstance may well explain why, at first sight, a few highly experienced physicists tended to underrate the significance of the violation in question. But on closer approximation such mistrust turns out to be unfounded. Indeed, willy-nilly we have to grant that, as I tried to show in this book, the said violation forces us to renounce all hope of ever building up any universal, realist, local theory. Which means, it forces us to consider that some of our most deeply engrained ways of thinking—of which we make use not merely in everyday life but also in science—far from reflecting self-evident structures of mind-independent reality are just elements of highly convenient models.

CORRELATION-AT-A-DISTANCE IN
THE BROGLIE-BOHM MODEL

§ §

IN SECTION 9-3-2 WE took notice of a surprising time asymmetry between the two mechanisms inducing, in the Broglie-Bohm model, the outcomes of the two distant measurements involved in an Aspect-type experiment. We considered an experimental setting inducing strict correlation (such as those in which the two polarization analyzers are oriented along the same direction). With the notations made use of in Bell (1987, chapter 15) such a strict correlation corresponds to

$$a_{m,n} = \delta_{m,n} \tag{1}$$

in which case equations (19) of the just quoted text take the form

$$\frac{dx_1}{dt} = g_1 \frac{N_1}{D_1} \tag{2}$$

$$\frac{dx_2}{dt} = g_1 \frac{N_2}{D_2} \tag{3}$$

with

$$N_1 = N_2 = - |\phi(x_1 + h_1)|^2 | \phi(x_2 + h_2)|^2$$
$$+ |\phi(x_1 - h_1)|^2 | \phi(x_2 - h_2)|^2 \tag{4}$$

$$D_1 = D_2 = |\phi(x_1 + h_1)|^2 |\phi(x_2 + h_2)|^2$$
$$+ |\phi(x_1 - h_1)|^2 |\phi(x_2 - h_2)|^2. \tag{5}$$

The functions $g_i(t)$ $(i = 1,2)$ describe the interaction between particle i and the corresponding instrument, and each one of them takes nonzero values only at times quite close to the time t_i when particle i gets measured. Hence the quantity h_i, defined as

$$h_i(t) = \int_{-\infty}^{t} g_i(t)dt \tag{6}$$

remains equal to zero up to a time close to t_i and then takes up a fixed, nonzero value. Let then x_1 and x_2 (which, in this scheme, play the role of pointer variables) be initially close to zero. Equations (2) and (3) show that they remain close to zero until time t_1, as concerns x_1, and until t_2, as concerns x_2.

Assume then that

$$t_1 \ll t_2,$$

since h_2 is zero until t_2, for $t < t_2$ the quantities $|\phi(x_2+h_2)|^2$ and $|\phi(x_2 - h_2)|^2$ are equal both in the numerator and in the denominator and may therefore be factorized. And consequently they vanish from the equations. Hence, within the time interval $0<t<t_2$, equation (2) becomes identical to equation (17) in Bell (1987, chapter 15) which deals with the one-particle case. As Bell explained there, the said equation, combined with the Schrödinger one and with the fact that, by assumption, the wave function $\phi(r)$ is a highly peaked wave packet centered at the origin, entails that, after t_1, x_1 can only take values h_1 or $-h_1$. Since equation (2) is a first-order differential equation, the value then effectively taken by x_1 is determined by its initial value, as we intuitively expected it to be.

Let us now investigate what takes place at time t_2 and after. At these times, g_1 is zero, so that x_1 keeps the value it has. To find out what happens to x_2 we must make use of equation (3). We then observe that, because ϕ is strongly peaked, it is impossible that both factors $|\phi(x_1 + h_1)|^2$ and $|\phi(x_1 - h_1)|^2$ should simultaneously be nonzero. If $x_1 = h_1$ the one that is equal to zero is the first one. Equation (3) then considerably simplifies and takes the form

$$\frac{dx_2}{dt} = g_2,$$

which shows that at the time t_2 when the measurement on the second particle takes place x_2 takes up a positive value. If, on the contrary, $x_1 = -h_1$, equation (3) becomes

$$\frac{dx_2}{dt} = -g_2,$$

which shows that x_2 gets a negative value. We see therefore that, in this case, the final value of x_2 depends not on its initial value but on that of x_1, that is, after all, on the initial value of the said x_1.

REFERENCES

Agazzi, E. (1987). *Philosophie, science, métaphysique,* Editions Universitaires, Fribourg, Switzerland.

Albert, D.Z. (1992). *Quantum mechanics and experience*, Harvard University Press, Cambridge, Mass.

Albert, D.Z., and Loewer, B. (1988). *Synthese*, 12.

Alquié, F. (1950). *La nostalgie de l'être*, P.U.F., Paris.

Alquié, F. (1979). *La conscience affective*, Vrin, Paris.

Arntzenius, F. (1990). Kochen's interpretation of quantum mechanics. In *Proceedings of the 1990 Biennal Meeting of the Philosophy of Science Association*, Vol. 1, Fine, Forbes, and Wessels, eds., East Lansing, Michigan.

Aspect, A., Dalibard J., and Roger, G. (1982a). *Physical Review Letters* 49, 1804.

Aspect, A., Grangier, P., and Roger, G. (1982b). *Physical Review Letters* 49, 91.

Aspect, A. (1983). "Trois tests expérimentaux des inégalités de Bell par mesure de corrélation de polarisation de photons." Ph.D thesis, Université de Paris-Sud, Centre d'Orsay, Paris.

Bachelard, G. (1949). *Le rationalisme appliqué*, P.U.F., Paris.

Ballentine, L.E. (1970). *Review of Modern Physics* 42, 358.

Bassi, A., and Ghirardi, G.C. (2000). *Physics Letters A* 275, 373.

Belinfante, F. (1973). *A survey of hidden variable theories*, Pergamon Press, Oxford.

Bell, J.S. (1964). *Physics,* 1, 195; reproduced in Bell (1987b).

Bell, J.S. (1966). *Review of Modern Physics* 38, 447.

Bell, J.S. (1971). "Introduction to the hidden variable question." In d'Espagnat (1971). reproduced in Bell (1987b).

Bell, J.S. (1975). *Helvetica Physica Acta* 48, 93.

Bell, J.S. (1981). "Bertlmann's socks and the nature of reality." *Journal de Physique, Colloque* C2, 42 (supplément 3), reproduced in Bell (1987b).

Bell, J.S. (1987a). "Are there quantum jumps?" In *Schrödinger, Centenary of a polymath*, Cambridge University Press, Cambridge, U.K.; reproduced in Bell (1987b).

Bell, J.S. (1987b). *Speakable and unspeakable in quantum mechanics*, Cambridge University Press, Cambridge, U.K.

Bell, J.S. (1990). La nouvelle cuisine, chap. 6 in *Between science and technology,* A. Sarlemijn and P. Kroes, eds., Elsevier Science North-Holland, Amsterdam.

Bennett, C.A., and Brassard, G. (1989). SIGACT News 15, 78.

Bennett, C.A., Brassard, G., Crépeau, C., Jozsa, C., Peres, A., and Wooters, W.K (1993). *Physical Review Letters* 70, 1895.

Besançon, A. (1994). *L'image interdite*, Fayard, Paris.

Bitbol, M. (1996). *Mécanique quantique*, Flammarion, Paris.

Bitbol, M. (1998). *L'aveuglante proximité du réel*, Flammarion, Paris.

Bitbol, M. (2000). *Physique et philosophie de l'Esprit*, Flammarion, Paris.

Bitbol, M., and Laugier, S., eds. (1997). *Physique et réalité*, Editions Frontière, Paris.

Bohm, D. (1952). *Phyical Review* 85, 165; 85, 180.

Bohm, D. (1980). *Wholeness and the implicate order*, Routledge and Kegan Paul, London.

Bohm, D., and Hiley, B.J. (1993). *The undivided universe*, Routledge, London.

Bohr, N. (1935). *Physical Review* 48, 696.

Bohr, N. (1958). Quantum physics and philosophy—causality and complementarity. Contribution to *Philosophy in the mid-century*, R. Klibansky, ed., La Nuova Italia Editrice, Florence; reproduced in *Essays 1958 1962 on atomic physics and human knowledge*, A. Bohr, ed., (Richard Clay and Co., Bungay, Suffolk, 1963).

Bouveresse, J. (1997). Le réalisme en physique. In Bitbol and Laugier (1997).

Bricmont, J. (1994). Contre la philosophie de la mécanique quantique. In *Colloquium: Faut-il promouvoir les échanges entre les sciences et la philosophie?*, Louvain-la-Neuve, Belgium, 1994.

Brown, H.R., Dewdney, C., and Horton, G. (1995). *Foundations of Physics* 25, 329.

Brune, M., Hagley, E., Dreyer, J., Maître, X., Mali, A., Wunderlich, C., Raimond, J.M., and Haroche, S. (1996). *Physical Review Letters* 77, 4887.

Brunschvicg, L. (1951). *Spinoza et ses contemporains*, P.U.F., Paris.

Bunge, M. (1990) Des bons et mauvais usages de la philosophie. *L'Enseignement Philosophique*, Paris 40, 97.

Carnap, R. (1928). *Der logische Aufbau der Welt*, Meiner, Berlin.

Carnap, R. (1950). Empiricism, semantics and ontology. *Revue Internationale de Philosophie* 4, 20; reproduced in Carnap (1958).

Carnap, R. (1958). *Meaning and necessity*, Phoenix Books.

Carnap, R. (1966). *Philosophical foundations of physics*, Basic Books, New York.

Cassirer, E. (1910). *Substanzbegriff und Funktionsbegriff*, Bruno Cassirer, Berlin; translated as *Substance and function*, Open Court, Chicago, 1923.

Cassirer, E. (1923–29). *Philosophie der Symbolischen Formen*, Bruno Cassirer, Berlin; translated as *The philosophy of symbolic forms*, Yale University Press, New Haven, 1955.

Cassirer, E. (1936). *Determinismus und indeterminismus in der modernen physik*, Göteborgs Högskolas Årsskrift 42, Göteborg, Sweden; translated as *Determinism and indeterminism in modern physics*, Yale University Press, New Haven, Conn. 1956.

Caves, C.M. (1993). In *Physical origins of time asymmetry*, J.J. Halliwell, J. Pérez-Mercader, and W.H. Zurek, eds., Cambridge University Press, Cambridge, U.K.

Chalmers, D.J. (1996). *The conscious mind: In search of a fundamental theory*, Oxford University Press, New York.

Chambadal, P. (1979). *Savoir, devoir, pouvoir*, Copernic, Paris.

Clauser, J.F., Horne, M.A., Shimony, A., and Holt, R.A. (1969). *Physical Review Letters* 23, 880.

Clauser, J.F., and Horne, M.A. (1974). *Physical Revew D* 10, 526.

Clauser, J.F., and Shimony, A. (1978). *Reports on Progress in Physics* 41, 1881.

Clavelin, M. (1968). *La philosophie naturelle de Galilée*, Armand Colin, Paris (pocket ed., Albin Michel, Paris, 1996).

Comte-Sponville, A. (1998). *Une éducation philosophique*, P.U.F., Paris.

Comte-Sponville, A. (1999). *L'être-temps*, P.U.F., Paris.

Comte-Sponville, A., and Ferry, L. (1998). *La sagesse des modernes*, R. Lafont, Paris.

de Broglie, L. (1928). La nouvelle dynamique des quanta. In *Raports et discussions du cinquième conseil de physique Solvay*, H. A. Lorentz, ed., Gauthier-Villars, Paris.

de Muynck, W.M. (2002). *Foundations of quantum mechanics: An empiricist approach*, Kluwer Academic, Dordrecht.

d'Espagnat, B. (1965). *Conceptions de la physique contemporaine*, Hermann, Paris.

d'Espagnat, B. (1966). In *Preludes in theoretical physics: In honour of V. F. Weisskopf*, A. De Shalit, H. Feshbach, and L.Van Hove, eds., North-Holland, Amsterdam.

d'Espagnat, B. ed. (1971). *Foundations of quantum mechanics, Proceedings of the International school of Physics "Enrico Fermi," Varenna (Italy)*. Course 49, Academic Press, New York.

d'Espagnat, B. (1975). *Physical Review D* 11, 1424.

d'Espagnat, B. (1976a). *Conceptual foundations of quantum mechanics*, 2nd ed., Addison-Wesley, Reading, Mass.; 4th ed., Perseus Books, Reading Mass., 1999.

d'Espagnat, B. (1976b). *Physical Review D* 18, 349.

d'Espagnat, B. (1979a). *A la recherche du réel,* Gauthier-Villars, Paris; translated as *In search of reality*, Springer, New York, 1983.

d'Espagnat, B. (1979b). *The quantum theory and reality. Scientific American*, 241 (3), 158.

d'Espagnat, B. (1989). *Reality and the physicist*, Cambridge University Press, Cambridge, U.K.

d'Espagnat, B. (1990). *Penser la science*, Dunod, Paris.

d'Espagnat, B. (1995). *Veiled reality: An analysis of present-day quantum mechanical concepts*, Addison-Wesley, Reading, Mass.; 2nd ed., Perseus Books, Reading, Mass., 2003.

d'Espagnat, B. (1997a). Essai d'une conclusion personnelle. In Bitbol and Laugier (1997).

d'Espagnat, B. (1997b). Aiming at describing empirical reality. In *Potentiality, entanglement and passion-at-a-distance*, R. S. Cohen et al., eds., Kluwer Academic, Dordrecht.

d'Espagnat, B. (2001). *Physics Letters A* 282, 133.

d'Espagnat, B. (2005). *Consciousness and the Wigner's friend problem. Foundations of Physics* 35, 1943.

Dickinson, M., and Clifton, R. (1998). Lorentz invariance in modal interpretations. In *The modal interpretation of quantum mechanics*, Dennis Dieks and Peter E. Vermaas, eds., Kluwer, Dordrecht.

Dummett, M. (1978). *Truth and other enigmas*, Duckworth, London.

Eberhard, P.H. (1977). *Nuovo Cimento* 38B, 75.

Eberhard, P.H. (1978). *Nuovo Cimento* 46 B, 392.

Einstein, A. (1934). *Mein Weltbild*, Querido, Amsterdam.

Einstein, A. (1939). Science and religion. Address at the Princeton Theological Seminar; reprinted in *Out of my later years,* Greenwood Press, Westport, Conn., 1950.

Einstein, A. (1953a). Einleitende Bemerkungen über Grundbegriffe. In *Louis de Broglie, physicien et penseur,* Albin Michel, Paris.

Einstein, A. (1953b). Elementäre Ueberlegungen zur Interpretation der Grundlage des Quanten-Mechanik. In *Scientific papers presented to Max Born on his retirement from the Tait chair of natural philosophy in the University of Edinburgh,* Oliver & Boyd, Edinburgh / Hafner, New York.

Einstein, A., Podolsky, B., and Rosen, N. (1935). *Physical Review* 47, 777.

Ekert, A.K. (1991). *Physical Review Letters* 67, 661.

Everett, H. (1957). *Reviews of Modern Physics* 29, 454

Ferry, L. (1987). Introduction to a French translation of *Critique of pure reason,* by E. Kant, Flammarion, Paris.

Feynman, R.P. (1949). *Physical Review* 76, 749; 76, 769.

Freedman, S.J., and Clauser, J.F. (1972). *Physical Review Letters* 28, 938.

Fry, E.S., and Thompson, R.C. (1976). *Physical Review Letters* 37, 465.

Gell-Mann, M. (1994). *The quark and the jaguar,* W. H. Freeman and Co., San Francisco

Gell-Mann, M., and Hartle, J.B. (1989). Quantum mechanics in the light of quantum cosmology. In *Proceedings of the Santa Fe Institute Workshop on Complexity, Entropy and the Physics of Information, 1989.*

Ghirardi, G.C.,Rimini, A., and Weber, T. (1980). *Lettere al Nuovo Cimento* 27, 293.

Ghirardi, G.C., Rimini, A., and Weber, T. (1986). Unified dynamics for microscopic and macroscopic systems. *Physical Review D* 34, 470.

Gilson, E. (1942). *Le thomisme,* Vrin, Paris.

Giulini, D., Joos, E., Kiefer, C., Kupsch, J., Stamatescu, I.O., and Zeh, H.D. (1996). *Decoherence and the appearance of a classical world in quantum theory,* Springer, Heidelberg.

Griffiths, R. (1984). *Journal of Statistical Physics* 36, 219.

Hacking, I. (1983). *Representing and intervening,* Cambridge University Press, Cambridge, U.K.

Heisenberg, W. (1989). *Ordnung der Wirklichkeit,* R. Piper, Munich.

Hepp, K. (1972). *Helvetica Physica Acta* 45, 237.

Herbert, N. (1985). *Quantum reality: Beyond the new physics,* Anchor Press/ Doubleday, New York.

Home, D., and Whitaker, M.A.B. (1992). *Physics Reports* 210, 224.

Husserl, E. (1969). *Méditations cartésiennes,* Vrin, Paris.

Jaspers, K. (1954). *Way to wisdom: An introduction to philosophy,* Yale University Press, New Haven, Conn.

Jauch, J.M. (1973). *Are quanta real?,* Indiana University Press, Bloomington, Ind.

Joos, E. (1987). *Physical Review D* 36 3285.

Joos, E., and Zeh, H.D. (1985). *Zeitschrift für Physik B* 59, 223.

Karolihazy, F. (1966). Gravitation and quantum mechanics of macroscopic systems. *Nuovo Cimento A* 42, 390.

Kochen, S., and Specker, E.P. (1967).*Journal of Mathematical Mechanics* 17, 59.

Kuhn, T. (1962). *The structure of scientific revolutions,* University of Chicago Press, Chicago.

Ladrière, J. (1989). Physical reality, a phenomenological approach. *Dialectica* (Bienne, Suisse) 43, 125–139.

Laplace, P.J. de (1814). *Essai philosophique sur les probabilités,* Paris.

Largeault, J. (1980). *Enigmes et controverses,* Aubier-Montaigne, Paris.

Laudan, L. (1977) *Progress and its problems,* University of California Press, Berkeley, Calif.

Laudan, L. (1996). *Beyond positivism and relativism,* Westview Press, Boulder, Colo.

Lévy-Leblond, J.M. (1997). Pour soulever le voile de Maya. In Bitbol and Laugier (1997).

Lurçat, F. (1990). *Niels Bohr,* Editions Criterion, Paris.

Lurçat, F. (1999). *Le chaos,* P.U.F. Paris.

Magnin, T. (1998). *Entre science et religion,* Editions du Rocher, Paris.

Mott, N.F. (1929). *Proceedings of the Royal Society of London,* A 126, 79.

Mohrhoff, U. (2000). *American Journal of Physics* 68, 728.

Mohrhoff, U. (2001). arXiv:quant-ph/0107005 v. 2.

Morin, E. (1982). *Science avec conscience,* Fayard, Paris.

Omnès, R. (1988). *Journal of Statistical Physics* 53, 893; 53, 933; 53, 957.

Omnès, R. (1994a). *Philosophie de la science contemporaine,* Gallimard, Paris.

Omnès, R. (1994b). *The interpretation of quantum mechanics,* Princeton University Press, Princeton, N.J.

Omnès, R. (2002). *Alors l'un devint deux,* Flammarion, Paris.

Onimus, J. (1994). *Béance du divin,* P.U.F., Paris.

Paty, M. (1988). Einstein et Spinoza. In *Spinoza, science et religion, actes du colloque de Cerizy-la-Salle, 1982,* Vrin, Paris.

Paty, M. (1993). *Einstein philosophe,* P.U.F., Paris.

Penrose, R. (1994). *Shadows of the mind. A search for the missing science of consciousness,* Oxford University Press, Oxford.

Pestre, D. (1997). *Histoire des sciences, histoire des techniques,* EHESS, Paris.

Petitot, J. (1997). Objectivité faible et philosophie transcendentale. In Bitbol and Laugier (1997).

Poincaré, H. (1902). *La science et l'hypothèse,* Flammarion, Paris.

Poincaré, H. (1905). *La valeur de la science,* Flammarion, Paris.

Primas, H. (1981). *Chemistry, quantum mechanics and reductionism,* Springer, Heidelberg.

Primas, H. (1994). Hierarchic quantum descriptions and their associated ontologies. In *Symposium on the Foundations of Modern Physics 1994,* K.V. Laurikainen, C. Montonen and K. Sunnaborg, eds., Editions Frontière, Paris.

Putnam, H. (1981). *Reason, truth and history,* Cambridge University Press, Cambridge, U.K.

Putnam, H. (1990). *Realism with a human face,* Harvard University Press, Cambridge, Mass.

Quine, W.V. (1943). Notes on existence and necessity, *Journal of Philosophy* 40, 113.

Quine, W.V. (1953). *From a logical point of view*, Harvard University Press, Cambridge, Mass.

Reichenbach, H. (1951). *The rise of scientific philosophy*, University of California Press, Berkeley, Calif.

Russell, B. (1929). *Our knowledge of the external world*, W. W. Norton and Co., New York.

Schäfer, L. (1997). *In search of divine reality*, University of Arkansas Press, Fayetteville, Ark.

Schlick (1918). *Allgemeine Erkenntnislehre*, Springer, Berlin.

Schrödinger, E. (1959). *Mind and matter*, Cambridge University Press, Cambridge, U.K.

Searle, J. (1995). *The construction of social reality*, Allen Lane, The Penguin Press, London.

Serres, M. (1997). Preface to *Le trésor, dictionnaire des sciences*, Flammarion, Paris.

Shimony, A. (1971). Experimental test of local hidden-variable theories. In d'Espagnat B., ed. (1971).

Shimony, A. (1990). An Exposition of Bell's theorem. In *Sixty-two years of uncertainty*, A. Miller, ed., Plenum Press, New York.

Shimony, A. (1993 I and II). *Reality, causality and closing the circle* and *Role of the observer in quantum theory*. In *Search for a naturalistic world view*, Vols. I and II. Cambridge University Press, Cambridge, U.K.

Shimony, A. and Stein, H. (2001). *American Journal of Physics* 69, 848.

Sokal, A. (1996). *Social Text* 46/47, 217.

Sokal, A., and Bricmont, J. (1997). *Impostures intellectuelles*, Odile Jacob, Paris.

Soler, L. (1997). Les régularités phénoménales: Requièrent-elles une explication? In Bitbol and Laugier (1997).

Soler, L. (2000). *Introduction à l'épistémologie*, Editions Ellipses, Paris.

Stapp, H.P. (1972) The Copenhagen interpretation. *American Journal of Physics* 40, 1098.

Stapp, H.P. (1977). *Nuovo Cimento B* 40, 191.

Stapp, H.P. (1980). *Foundations of Physics* 10, 767 (1980).

Stapp, H.P. (2001). *American Journal of Physics* 69, 854.

Toraldo di Francia, G. (1981). *The investigation of the physical world*, Cambridge University Press, Cambridge, U.K.

Uzan, P. (1998) Vers une logique du temps sémantique. Ph.D. thesis, Sorbonne, Paris.

van Fraassen, B.C. (1972). A formal approach to the philosophy of science. In *Paradigms and paradoxes: The philosophical challenge of the quantum domain*, R. Colodny, ed., University of Pittsburgh Press, Pittsburgh, Pa.

van Fraassen, B.C. (1974). The Einstein-Podolsky-Rosen paradox, *Synthese* 29, 291.

van Fraassen, B.C. (1980) *The scientific image*, Clarendon Press, Oxford.

Varela, F., Thompson, E., and Rosch, E. (1991). *The embodied mind: Cognitive science and human experience*, MIT Press, Cambridge, Mass.

Vermaas, P. (1998). The Kochen-Dieks and atomic modal interpretation. In *The modal interpretation of quantum mechanics*, Dennis Dieks and Peter E. Vermaas, eds. Kluwer, Dordrecht.

Vuillemin, J. (1955). *Physique et métaphysique kantiennes*, P.U.F., Paris.

Wheeler, J.A. (1984). Bits, quanta, meaning. In *Problems in theoretical physics*, A. Giovannini, F. Mancini and M. Marinaro, eds., University of Salerno Press, Salerno, Italy.

Whitehead, A.N. (1933). *Adventure of ideas*, Macmillan, London.

Whorf, B.L. (1956). *Language, thought and reality*, MIT Press, Cambridge, Mass.

Wigner, E. (1961). Remarks on the mind-body question. In *The scientist speculates*, I.J. Good, ed., W. Heinemann, London.

Wigner, E. (1970). *American Journal of Physics* 38, 1005.

Wittgenstein, L. (1921). Tractatus logico-philosophicus, *Annalen der Naturphilosophie*; reprinted by Routledge and Kegan Paul, London, 1961.

Worrall, J. (1989). Structural realism: The best of both worlds? In *Dialectica* (Bienne, Switzerland) 43, 99–124.

Zeh, H.D. (1970). *Foundations of Physics* 1, 69.

Zurek, W.H. (1998). *Philosophical Transactions of the Royal Society of London, A* 356, 1793.

Zwirn, H. (2000). *Les limites de la connaissance*, Odile Jacob, Paris.

NAME INDEX

SUBJECT INDEX

∮∮

Page numbers in bold italic type contain more detailed explanations.